Labour and capital in

CO

D1442997

HD
8105
P45
1981

DOUGLAS COLLEGE LIBRARY

LABOUR AND CAPITAL IN CANADA 1650 ~ 1860

H.CLARE PENTLAND

Edited and with an Introduction by Paul Phillips

James Lorimer & Company, Publishers
Toronto, 1981

DOUGLAS COLLEGE LIBRARY

Copyright © 1981 by James Lorimer & Company, Publishers

All rights reserved. No part of this book may be reproduced or transmitted in any form or by any means, electronic or mechanical, including photocopying, or by any information storage or retrieval system, without permission in writing from the publisher.

ISBN 0-88862-379-8 cloth
        0-88862-378-X paper

Cover design: Laurel Angeloff

Canadian Cataloguing in Publication Data

Pentland, H. Clare.
   Labour and capital in Canada, 1650-1860

Manuscript ed., 1960, had title: Labour and
industrial capitalism in Canada.
Bibliography: p. 264
Includes index.
ISBN 0-88862-379-8 (bound). — ISBN 0-88862-378-X
(pbk.). — ISBN 0-88862-430-1 (University)

1. Labour and labouring classes — Canada — History.
2. Canada — Industries — History. 3. Canada —
Economic conditions — To 1867.* I. Title.
II. Title: Labour and industrial capitalism in
Canada.

HD8105.P45   331'.0971   C81-094276-3

This book has been published with the help of a grant from the Social Science Federation of Canada, using funds provided by the Social Sciences and Humanities Research Council of Canada.

James Lorimer & Company, Publishers
Egerton Ryerson Memorial Building
35 Britain Street
Toronto M5A 1R7, Ontario

Printed and bound in Canada

6 5 4 3 2 1   81 82 83 84 85 86

Contents

Introduction

> I have been conscious, at times, of writing against the weight of prevailing orthodoxies....
>
> I am seeking to rescue the poor stockinger, the Luddite cropper, the "obsolete" hand-loom weaver, the "utopian" artisan, and even the deluded follower of Joanna Southcott, from the enormous condescension of posterity.[1]

With minor changes, E. P. Thompson's introduction to his classic work on the English working class would make a fitting opening for H. Clare Pentland's seminal study of the emergence of the Canadian industrial working class. He too broke with the prevailing orthodoxies, not only of North American labour studies but also of Canadian economic history, to produce a work of enduring influence.

In the first place, Pentland too seeks to rescue the anonymous labourer from the neglect of historians who have devoted their energies to the exploits of the ruling elite. As he notes himself in the introduction to his first influential article published in 1948, "The Lachine Strike of 1843":[2]

> Historians have paid considerable attention to the English capital that made possible Canada's canal and railway building in the eighteen-forties and fifties, and some attention, too, to the Scottish contractors who supervised the work. But there has been almost complete neglect of the real builders of Canadian public works, the thousands of labouring men, mainly Irish, who toiled with pick and shovel at a time when the application of power machinery to construction had scarcely begun.

But while the Irish labourers remain a focal point of this present volume, *Labour and Capital in Canada, 1650-1860*, a slightly revised version of his thesis, "Labour and the Development of Industrial Capitalism in Canada",[3] he ranges much further afield. Beginning in the seventeenth and eighteenth centuries, he traces the evolution of the institutions of labour supply in Canada, encompassing early slavery; indentured and convict labour; pre-industrial, "personal labour" of the French period (which he initially called "feudal labour"[4]); the non-Irish worker and the skilled artisan as well as the labourer. Perhaps more importantly the scope was expanded to include the critical process by which the mass of immigrants and native born were transformed into a modern industrial working class with the emergence of a capitalist labour market.

Secondly, this focus on the workingman in pre-Confederation Canada pointed in a different direction from what was up to that time the overwhelming preoccupation of Canadian economic historians: the great staple trades and the political institutions that emerged from these trades.[5] Nevertheless it is misleading to imply, as some current interpreters do, that Pentland turned his back on the staple interpretation of Canadian history. Clearly, he did not, as he argues himself:

> Commercial capitalism has been ... a system that asked remarkably few questions about the method of production of the goods in which it dealt. It could readily integrate slave societies, feudal societies and incipient capitalist societies into its structure. It is not apparent that the European settlers of Canada were prohibited in any absolute sense from arranging production upon a basis of slavery, or upon a basis of small-scale capitalism: European settlers elsewhere had chosen such courses.... The basis of choice of method in Canada's case (and in the other cases) lay in the nature of the supply of, and demand for, labour.[6]

Thus, rather than ignoring staples or, even more incongruously, being credited with opposing the commercial capitalist interpretation of Canadian history associated with the staple interpretation, Pentland began his work by studying the organization of labour in the period of the staple trades. But as he points out (and as Buckley in another context has also argued[7]), the commercial capitalist interpretation of Canadian economic development becomes increasingly less convincing or applicable by the second half of the nineteenth century, particularly as it relates to the labour market.

> The tasks of the thesis are to delineate the feudal organization of labour as it appeared in various Canadian employments up till about 1850, and to trace the transformation that produced a capitalistic labour market and a well-developed capitalistic economy shortly after 1850.[8]

It is in the context of this changed focus of interest that Pentland had his greatest impact upon the course and direction of studies of Canadian political economy, although an impact that developed very slowly.

This can be ascribed to the fact that his thesis has remained unpublished for two decades although it has circulated fairly widely in manuscript form.[9] In addition, some of the important elements developed in this work were presented in article form before the complete work was presented.[10]

It was with the revival of interest in political economy, labour and Marxism in the 1970s in Canadian universities that Pentland's work and his approach began to acquire the stature and attention that would be revealed when, on his death, issues of three academic journals were dedicated to his life and work.[11] The basis of this reputation lay in the originality of his approach and of his subject in relation to the existing body of Canadian economic history. In recent years this situation has changed markedly and at least part of the credit resides with Clare Pentland.

The Man and His Work

The apparent break with the Innisian tradition centred in the University of Toronto, or with the hinterland response associated with Saskatchewan, was just one of the intellectual idiosyncracies that have been credited to Pentland. Although born and raised in the agricultural west, his initial preoccupation was with the central Canadian industrial worker. While the Laurentian School concerned itself with commercial trade, transportation and resources, Pentland concerned himself with local industry and the anonymous worker rather than the ruling elite. While the limited field of labour studies concentrated on labour organizations and the regulation of industrial conflict, Pentland was concerned with the labour process by which the working class and its pre-capitalist progenitors were propagated, shaped, moulded, disciplined, skilled, allocated and rewarded, and how the worker responded to the dehumanizing progression of the impersonal labour market. And in the repressive political atmosphere of the Cold War, Pentland incorporated the analytic approach of British Marxist historians.

Kealey has noted that "although Pentland's work stands apart starkly from the Toronto political economy tradition, there are no clear explanations for this in his intellectual biography."[12] Perhaps there is little to explain these apparent contradictions in Pentland's interests and approaches in his academic training but this is not the case in his personal background. Clare Pentland was born October 17, 1914, on a farm near Justice, Manitoba, a town some ten miles north-east of Brandon. His father was a farmer, later a trucker, his mother a school teacher. The Pentland family, however, were not recent immigrants to Canada. Clare's great-great grandfather, an Ulster-Scot hand-loom weaver, emigrated to Canada from County Down in Ireland in 1821, settling first at Amherst Island, near Kingston, Ontario, where he practised the dual vocations of farmer and weaver.[13] His son, John, continued the agrarian-artisan tradition, becoming a carpenter. In fact, he built his father a hand loom that still exists, residing in the Royal Ontario Museum. In 1843, the family moved to homestead in the Huron Tract, eight miles north of Goderich.

John's son (Clare's grandfather), Thomas, continued the westward move to the frontier, homesteading near Justice in 1881 where he combined farming with blacksmithing. This was the limit of the westward movement. The Pentland family became well established in the Elton municipality around Justice and a Pentland has been reeve of the area for a good part of its political history. It was there that Clare's father grain-farmed and began his trucking business. While Clare was still a child, his family moved to Brandon to develop the business, largely in shipping cattle to the packers. What is striking about this family background is the overlay with Pentland's intellectual interests — his concern with the Irish and Scottish immigrant to central Canada in the nineteenth century and with the artisan and industrial worker.

There are other aspects of Pentland's family background that obviously influenced his perspective. As Ulster-Scots, the family had a strong historic connection with the Orange Order (Clare's uncle was a member), a fact that no doubt contributed to Pentland's sympathetic interest in the Order in the nineteenth century. Allied with this heritage was a staunch Conservatism and, particularly in his mother a strict Weslyan Methodism, both of which Clare was to rebel against at a fairly early age. But the rebellion was within the tradition of the social gospel movement of western Canada, humanitarian, left- inclined and concerned with the welfare of the workers. This Clare revealed both in his academic work and in his own actions.

Clare grew up in Brandon, graduating from the Collegiate in 1931 and the Brandon Normal School in 1933. This was followed by three years of teaching in small country school houses at Whirlpool, a soldier settlement area near Clear Lake, and at Ericson. Here he saw the intense poverty of the depression-wracked prairies, an experience which strongly influenced his critical approach to contemporary society. He returned to university in 1936 and four years later, in 1940, graduated with a B.A. in Economics from Brandon College after receiving the University of Manitoba Gold Medal in Arts and Science, the Ephraim Buckwold Gold Medal in Economics and Silver Medals in History, Political Economy and general proficiency. While he attended university he worked as an attendant at the Brandon Mental Hospital, at least until the summer of 1939 when he tried to organize the employees into a staff association, an activity for which he was summarily fired by the provincial government. It was also at the hospital that he met a young nurse, Harriet Brook, who was later to become his wife. The following summer found him working as a brakeman on the CPR running between Brandon and Broadview, beginning, as with many westerners, a lifelong love of the railway — though not necessarily of the CPR.

The outbreak of war did not immediately interrupt Pentland's renewed educational program. From 1940 to 1942, he attended the University of Oregon where he obtained his Masters.[14] The direction of his interests is indicated by the subject of his thesis, "The Adequacy of Unemployment Benefits in Oregon". Perhaps more influential was his teacher and, subsequently long-term friend, Louis A. Wood, best known in Canada as author of *A History of Farmers' Movements in Canada*. What is interesting is the similarity in Wood's background to Pentland's, particularly the Southern Ontario protestant heritage and, though greatly supplemented by theological studies, the bridging of history and economics.[15] Wood taught labour economics and courses in socialism which may have provided Pentland with the background in European Marxist scholarship.[16] Possibly more important was Wood's activity during the depression on the Oregon Unemployment Commission and in support of Roosevelt's "New Deal". In fact, in his reform and pro-labour and farmer sentiments he also represented the social gospel tradition.[17]

Almost immediately after completing his thesis in the early summer of 1942, Pentland enlisted in the Army and while undergoing training in British Columbia married Harriet in the fall of 1942 in Vancouver. After officer training near Victoria and artillery training at Brandon and Brockville, he went overseas in February of 1944 where he was transferred to the infantry as an education officer. He returned to Canada and to university, this time at Toronto, in the spring of 1946, under the veterans assistance program and by 1948 completed all the requirements but the thesis for his Ph.D. At Toronto he felt the full influence of the Innisian and Toronto tradition[18] and although he had already chartered his direction toward labour studies, he was no doubt directed towards the historical approach. He lectured briefly at Toronto, from 1947 to 1949, before returning to his native province as Assistant Professor of Economics at the University of Manitoba in 1949. He remained at Manitoba for the remainder of his career.

But the completion of his thesis did not progress quickly. In part this may have been the result of attempting too much. Certainly, he was continually advised to narrow and confine his topic. Initially (1947) he picked as his topic "A History of Labour in Canada to 1867", but within a year this had been narrowed to "A Study of Canadian Labour (Organized and Unorganized), in the Period 1830-1860". By 1951, it was still further narrowed to "The Irish Labourers on the Canadian Canals and Railways, 1830-1860". Yet when the thesis was finally presented as "Labour and the Development of Industrial Capitalism" almost ten years later, it had again been broadened more or less to the scope of his original proposal. Even at that, Pentland never considered the thesis "complete". But completion was also slowed by his quest for perfection, in part indicated by the copious and detailed footnotes to the thesis, but which also extended to his writing and his teaching. Finally, however, and at the urging of his supervisors, the thesis was presented and defended late in 1960 and the degree conferred in 1961.

It is apparent from this brief biographical sketch that much of what initially appears as intellectual idiosyncracy originated not in academic influences but in his personal background, particularly his Irish-artisan-immigrant ancestry, his strict Methodist upbringing, and his exposure to poverty and the authoritarianism of employers. But what of the academic or intellectual influences that led him away, both in subject and methodology, from the prevailing Canadian school of political economy? Wood may have had, as already indicated, some influence. But according to his family, Harold Innis was the most influential person in Pentland's intellectual development. Yet, as already mentioned, his work did break with the staple tradition despite some recent attempts to place Pentland back in the mainstream.[19]

One might suggest two aspects in which his approach was similar to the Toronto staple school. First, of course, was the historical approach to the Canadian economy which relegated abstract and partial equilibrium theorizing to a decidedly minor role[20] and which adopted a strongly interdisciplinary approach to economic history.[21] Secondly, Pentland was fascinated with the impact of technology, and Innis, after all, was preoccupied not with staples *per se*, but rather with technology, a preoccupation which led to visits to observe in detail the physical and technical conditions in the industry he was examining.[22] But in most of the staple literature there is little or no attention paid to the question of class — with the possible exception of the merchant class, especially in the work of Creighton — and particularly with the working people, a neglect which weakens the analytic power of the staple model increasingly after the eighteenth century.[23] Pentland's work, on the other hand, concentrates on exactly this question. At the same time, he was more concerned in this thesis with the question of the emergence of a working class *in itself* (that is, the creation of a class of workers alienated from ownership of land and capital, a working class objectively defined by its relation to the means of production) rather than with the question of a working class *for itself* (the creation of a class-conscious, politicized stratum of workers), though he was to direct more attention to the latter question in some of his later work.

It is not clear how Pentland became exposed to and influenced by what Kealey calls "the English Marxist tradition of historical writing"[24] with its use of class as the fundamental tool of political-economic analysis and its concern with the role of the state in capitalistic transformation and accumulation. Obviously he became familiar quite early with Marx and the writings of other leading Marxists, particularly Maurice Dobb, and he utilized the works of Dobb, Hobsbawm, Thompson and others in this tradition quite freely in his teaching. On the other hand, Pentland never considered himself a Marxist despite the fact that some contemporary analysts have so characterized him (much to the surprise of his own family). Indeed he even avoided the label of socialist. He joined no political party although his sympathies lay with the CCF and later the NDP, but even then not necessarily with the left wing of the Party. His remained, throughout, an inconoclastic and independent mind.

Perhaps this is the real clue to his reputation as an intellectual radical. He was an independent thinker who rebelled against the constraints of any orthodoxy, Marxist, Methodist or Conservative. But given his interest in studying his own roots and the history of Canadian labour, he adopted the methodology and analytic tools most suited to the task, the study of the labour process and the making and shaping of the Canadian working class. This would lead automatically to the Marxist historical tradition which has been the only tradition in which this process has occupied a central and, indeed, critical place.

It is worth noting in this regard that Marxism can be both an analytic framework and a political commitment (what Bryan Palmer has in a similar context called an *engaged* history).[25] The former does not necessarily imply the latter. It is possible, though perhaps not that common, to use Marxian concepts of class and labour process to analyze the evolution of the capitalist economic system without requiring a commitment to the goals associated with the abolition of the capitalist wage system. Class and process are powerful tools for the critical investigation of the capitalist system even where the purpose is the limited one of humanizing and liberalizing that very system.[26]

What this suggests is that Pentland's dialectic approach to the evolution of the Canadian labour market was not rooted in an adherence to radical scholarship but rather the reverse — that his radical interpretation was the result of his choice of topic. This is indicated in the introduction to one of his articles:

> It has been the custom of economists since the Mercantilists to assume the existence of a capitalist labour market. It appears that this abstraction was adopted, not from ignorance of labour conditions...but because it simplified the economists' main task, as they conceived it, of expounding how a capitalist economy works. The abstraction had the disadvantage, however, of diver-

ting attention from how a capitalist labour market might develop.*
This question, besides its historic and sociological interest, is relevant to the study of labour markets and of "economic development".[27]

Since he defined economic development in terms of systemic changes
in the labour market, and argued that these changes result from "capital
accumulation, importation and investment"[28], the obvious dialectic process of accumulation is made clear.

If this line of argument is correct, therefore, Pentland was not in the
conventional sense a "Marxist" but rather an independent and original scholar who did not disdain any tradition, but whose own
Canadian analysis coincided rather well with the radical or Marxist
tradition in European and even American scholarship.

Pentland's analysis rests on one major theoretical premise. The
organization of production in society (and hence the behaviour
patterns imposed on members of society) follows from "the nature of
the supply of, and demand for, labour."[29] Changes in supply and
demand conditions, therefore, invite social change. Hence the investigation of economic transformation begins with the study of labour
market conditions. The determinants of supply and demand
conditions are complex and non-mechanistic, however, and encompass both quantitative and qualitative variables. Demography, technology, skill levels, economic stability, migration, social and cultural
patterns of the work force all became grist for his mill. But in the
1960s, Pentland increasingly turned his attention to historical demography and in particular the relation between labour supply and
economic change in Britain.[30] In part this was a natural extension of
his work on Canada, but it was also stimulated by contact with British
economic historian Peter Mathias.[31] In 1962-63, Pentland spent a
sabbatical in Cambridge, England. Again, in 1969-70, he spent a
sabbatical in England, this time at the University of Sussex working
on parish population studies. Unfortunately, failing health prevented
him from completing this work and further refining and developing
the ideas introduced in his 1965 paper to the Third International
Conference on Economic History in Munich.[32]

Despite his pursuit of historical demography in the 1960s, Pentland
was also able to research and write his second major and influential
unpublished manuscript, "A Study of the Changing Social, Economic,
and Political Background of the Canadian System of Industrial Relations", commissioned by the federally-appointed Task Force on Labour

---

*Pentland's footnote in the original reads "However see Karl Marx, *Capital* 1,
Chs. xxvi ff." Ed.

Relations under the chairmanship of Professor H.D. (Bus) Woods.[33]
Its limited circulation in mimeographed form is unfortunate because
his report remains one of the most original, innovative and comprehensive investigations of the history of Canadian industrial relations.
Yet in spite of its all-encompassing scope in both time and subject
matter indicated in its title, it was completed in a couple of years, a far
cry from the thirteen or so taken to complete the thesis. In style and
form it differed greatly from the meticulous, detailed and documented
history of his thesis, its main thrust being painted in broad and bold
strokes. Yet curiously it represents a continuation of the central
investigation of his earlier work, albeit with a much heavier emphasis
on the development of the institutions and legal framework of industrial relations. Again what makes this work stand apart from the main
body of institutional labour in Canada is the concentration on the
dynamic interaction between the economy, the unions, the state and
technological change.

The 1960s were intensely productive years for Pentland. In addition
to his population research and Task Force report, he also pursued his
interest in technological change, producing three major reports for
both provincial and federal agencies on skills, training and technological change[34] plus a number of lesser reviews on related issues.

Unfortunately, this level of intellectual activity could not be maintained. Heart problems plagued the last ten years of his life, robbing
him of his stamina, a terrible frustration for a man so dedicated to his
teaching and his work. Yet despite this he shouldered a heavy administrative load in university affairs, as a member of Senate from 1963 to
1966 and again from 1969 to 1976, and as a member of the Board of
Governors representing the Senate from 1973 to 1976. He also served
on numerous university and Faculty Association committees[35] as well
as continuing to teach, write and research. Two articles (published
posthumously) and two reviews were the primary academic output of
the 1970s before his premature death on October 13, 1978.[36]

Pentland and the Study of the Labour Process

In recent years Pentland's work has received increasing attention,
partly as a result of a renewed interest in the nineteenth century, in
particular the transformation of Canada from a commercial to an
industrial economy which involves the question of the nature of the
emergent Canadian state (the "national question"); and partly because
of a renewed interest in the related question of the evolving nature of
the organization and character of work (the labour process), both
issues which are central to *Labour and Capital in Canada, 1650-1860*. It is
important to understand these key dimensions to his work in order to

appraise the contribution of this thesis to more recent research and scholarship.

The labour process, simply defined, is how labour is applied to raw materials, in conjunction with given tools and technology, to produce output of use and market value. While such a definition appears relatively straightforward, the real-life process is not. This is made clear by other definitions or approaches. One source defines labour process as "work as a system of power", thereby emphasizing relationships at the work place; while another emphasizes technology and technological change with the resulting consequences for the "nature of work, the composition (and differentiation) of the working class, the psychology of workers, the forms of working class organization and struggle, and so on."[37] As Pentland has noted:

> A fundamental problem of any society is the organization of its labour force for production. If the society is to survive as an entity, ways must be evolved or devised to maintain or increase the labour force, to determine the nature and extent of the division of its labour force (the techniques to be used), and to establish and enforce the system of incentives. The solutions of these problems fasten their appropriate behaviour patterns upon the various members of each society.[38]

Thus the organization of work far transcends the normal micro-concerns of technical economics with its preoccupation with technical efficiency, to include the social structure of society as an ongoing system. This is a basic key to understanding the importance of Clare Pentland's approach.

A second key to understanding this approach is to recognize that the labour process is not part of a stagnant equilibrium system but a constantly evolving and changing one, responding to both purposeful and unplanned change. This is the sense of what Thompson was trying to convey in his English study, the *making* of the working class. It is impossible to take a snapshot of working people at a point in time and then to objectively characterize a working class and its institutions at that point. They are in constant flux as are society and the economic forces acting on society. It is this concern with change that separates Pentland from the dominant American school of labour historians who tended to fix upon the institution of the (business) trade union as the apotheosis of labour activity, thereby neglecting all the other manifestations of working-class activity which reflect the workingman's response to his changing milieu.[39]

Is it possible to identify the critical elements in Pentland's world of work that interact to produce the changes he describes? The obvious starting point of his analysis is with supply and demand for labour. But this is not the mechanistic supply and demand for labour of the neo-classical equilibrium labour market. The supply and demand for labour do not just determine the level of wages and employment (although they do that, too); they determine the very system by which labour is organized to produce. That is, they determine the nature of the labour process.[40]

There are, of course, complicating (or perhaps one should say other interactive) factors, the most important of which are technology and the skill level of the workers. These are not independent. As long as technology was imbedded in the skills of the artisan, the work force exhibited a skilled-unskilled dichotomy that influenced the very basis of social organization and consequently the institutions of the work-ingman. This is a very important variable not only in the thesis but also in the Task Force report. As long as work processes were depen-dent on the skilled worker, the working class remained divided by a social and economic gulf that was reflected in both attitudes and insti-tutions. The period covered in the thesis coincided with the period when the differentiation between skilled and unskilled was most manifest, before the introduction of large-scale industrial capitalism with its routinized and homogenized labour force. Pentland was to take up this second transformation in the Task Force report.

His first concern was with the evolution of a capitalistic labour market in Canada — that is, the transition from personal (or feudal) labour relations where labour was personally bound to, and looked after by, an employer, to capitalistic labour relations where labour, skilled and unskilled, was sold on an impersonal labour market charac-terized by no personal obligations on either side for the welfare of the other. This represented a profound transformation in the labour process which required a major alteration in the nature of both the labour force and the employers. This is the substance of the last two chapters of this volume which Kealey has characterized as "the first and perhaps still, the fullest account of Canada's industrial revolu-tion".[41]

The further transition from a capitalistic labour market to indus-trial capitalism required two major transformations: first, the emer-gence of a coherent and integrated domestic market that could provide steady employment for a labouring class along with the intro-duction of machine processes that could serve such a market; and second, the creation of a disciplined labour force that could and would man the new processes. It required a remaking of the behaviour and attitudes of the workers themselves, a remaking that constituted a cultural as well as an economic transformation — a replacement of the habits of irregularity, ill-discipline and sloth and a preoccupation with the immediate, with habits of punctuality, regularity and order and a longer-term view, all of which were necessary to the working of an emerging capitalist order with its new scientific technology. The description in Chapter 6 of this book chronicles only the first phase of this transition of the labour process from pre-industrial man to the industrial worker, the phase that I have designated elsewhere as the artisanal period and which Leo Johnson designates as the period of the manufactory based on craft control of production rather than on the factory and machine control of production.[42] As Pentland notes:

> At the critical moment when industrial development was just getting underway, the artisan was the key man who held the new technology in his hands and brain and it was nowhere else.... Labour in the past had been regarded as a species of draft animal, and in the future would be regarded as a type of machinery; but, in the middle of the nineteenth century, skilled labour was an important part of the community.[43]

Unskilled labour, however, was another matter, particularly in the overstocked mid-century labour market. Yet this labour did learn discipline, often as the result of troop or police repression, and by learning to act together, developed a minimal measure of control over the degree of their exploitation.

Such a transformation of labour was not easy, not accomplished without intense worker opposition which, in turn, modified the very system that was making it. This view appears in the section on the Lachine strike, but it is more explicitly developed in the discussion of ethnic bodies and particularly of the Orange Order as the defensive organization of the artisan in this period of upheaval, although it should be noted that Pentland's identification of the Order with the artisan has since been questioned.[44]

What differentiates Pentland's approach from many of his contemporaries was his explicit use of class in the analysis of labour's transformation.

> The problem of production in all other [than primitive] societies is complicated by the division of these societies into a ruling class, which organizes the labour force (the rest of society) in its own interest, and the ruled or working group, whose satisfactions are a matter of expediency and of consistency with the demands of the rulers. The methods of replenishing and training the labour force, the choice of techniques and the organization of production, the system of incentives used, are now matters determined by the ruling group for those to whom they apply — the ruled group — as the interest of the rulers dictates. While the social shorthand of ritual retains a prominent role in such societies, change is facili-

tated by the fact that, from the viewpoint of the rulers, the conditions of production are impersonal forces. Only from this vantagepoint is the logic of such systems of production apparent. On the other hand, the ruled inevitably have a viewpoint also, different from, and more or less hostile to, that of the rulers, which also has to be taken into account.[45]

But Pentland was also well aware of the divisions within the dominant ruling class (between commercial and industrial capital), and within the working class (between the skilled and unskilled), a consequence of the choice of techniques and the organization of production. The result was that the demarcation between classes in terms of objective economic function was blurred in terms of consciousness such that industrial capital and its employees coalesced around the issue of protection. This was manifested in the political support given to the Conservative Party by the Orange Order.[46] Once protection was achieved by the coming together of the interests of commercial and industrial capital, class conflict emerged as a more dominant feature of society.[47]

Pentland's thesis carries the analysis of the transformation of the Canadian economy and its people through to the 1870s when large-scale industrial capitalism was still in its infancy. The factory system was just beginning to revolutionize once more the labour process that evolved by mid-century and this system was to destroy much of the basis for the artisan-labourer social and economic division that had been characteristic of the previous few decades. Nevertheless, the seeds of change were sown. The "personal" market had become a self-regulating market. Labour had become a commodity traded on a cash nexus.

The following period has since become fertile ground for a number of scholars, and recently published work is too extensive to review in detail here. Much of it has already been surveyed in a recent and generally comprehensive review article.[48] Nevertheless, it is perhaps worth singling out a few recent contributions from authors who acknowledge their debt to Pentland's pioneering work, and who extend the study to the period when industrial capitalism began to revolutionize production relations, destroying much of the basis for artisanal control of production with the introduction of machine control within the factory system, thereby deskilling the artisan class while opening up new semi-skilled — or perhaps more accurately, work-skilled — opportunities for the labourer.

Pentland discusses briefly some of the mechanisms by which this new labour force was to be made to fulfil the needs of a new industrialism: public education[49] and the Mechanics' Institutes, "an index of the diffusion of belief in progress, and of industrial ambition",[50] a diffusion more indicative of business aspirations than of working-class self-improvement.[51] He implies that industrial capitalism was already emerging by the 1850s, although it is questionable whether it is appropriate to think of mid-nineteenth-century Canada as industrial in any significant sense. Katz calls Hamilton, one of the centres of manufacturing output, a commercial city in the 1850s.[52] Langdon, in an ambitious attempt at synthesis, dates the origins from the 1860s, although there were, of course, isolated examples of mechanization threatening artisanal control of production in the 1840s and 1850s.[53] But as he notes, by 1867 there were no known local unions whose membership had reached 100 members, itself an indication of the relatively small scale of industrialism by Confederation.[54]

The most comprehensive and detailed studies of the evolution of large-scale industrial capitalism and a modern industrial labour force are the studies by Greg Kealey on Toronto and by Bryan Palmer on Hamilton, and their joint work on the history of the Knights of Labor.[55] To Palmer and Kealey, the 1860s mark the transitional decade from the period of primitive accumulation, merchant capital and capitalist handicraft production to that of modern industry, a transition accelerated by the American Civil War.[56] There was, however, no sharp discontinuity between the periods. As Palmer notes, "the later sixties would be something of an Indian Summer for the craft-based manufactory."[57] Although Toronto was highly industrialized by 1871, "numerous artisanal pursuits remained, in which production was still centred in small shops with very few employees and a minimal level of mechanization."[58] Still, the 1870s and 1880s saw the consolidation of industrial capitalism and with it the evolution of new types of working-class response, particularly in the form of the Knights of Labor.

Paralleling, even anticipating, the emergence of industrial capital was the increase in protectionist sentiment. Pentland was one of the first to recognize the common manufacturer-worker interest in protection and the resulting producer ideology championed by Isaac Buchanan,[59] at least until the 1880s when merchant and industrial capital coalesced around the National Policy.[60] But as Craven and Traves have recently demonstrated, Pentland's view (and similarly Palmer's and Kealey's) must be regarded as incomplete, largely ignoring the importance of the third major class "in, but not of, industrial Canada", the agrarian *petite bourgeoisie*.[61] The existence of three classes, they argue, led to an unstable series of alliances among the triad of classes from the 1870s to the depression of the 1930s.[62]

While Pentland's thesis ends with the opening of this second transformation of the labour process which destroyed the central position of the skilled worker just as the earlier transformation had created it, his study of the labour process and the industrial relations system did not. It is continued in the Task Force report. The central question of the report is the same as the thesis[63] despite differences in time frame and emphasis. Again, a central concern is the dynamic interaction between the economy, employers, workers and their institutions, the state, the organization of work and technological change. And again he used class as an analytic tool.

> Canadian experience has continually demonstrated another phenomenon observable elsewhere: that tensions in industrial relations increase or decrease as the "real" gap between employers and employed in capacity (as distinct from the constant gap assumed by the institutional structure) has narrowed or widened. Whenever workers have generally been advancing more rapidly than employers in sophistication, tensions have been more acute. When the real gap has widened, tensions have usually diminished.[64]

His conclusion about the major adaptation of the Canadian economy in the early decades of this century, a "profound re-orientation and transformation of Canadian industry which adjusted it to the market demands and technology of the twentieth-century as this applied most obviously to Canadian resources", and its effect in transforming a second time the labour force and its institutions,[65] parallels his analysis and conclusions about the mid-nineteenth-century transformation. The report also suggests that the post-World War II period was also one of those periods of intense change, in response to a new and evolving technology.[66]

It is our great misfortune that Pentland never had the opportunity to complete what I think was his ultimate goal, the complete history of Canadian labour (organized and unorganized) and to bring the original insights of his research together. Be that as it may, what he did complete is rich ground for students of working-class history.

Pentland and the National Question

Clare Pentland was a strong Canadian nationalist, sufficiently strong to reject French-Canadian aspirations which he considered a threat to Canadian nationhood. "Both history and logic (including Marxist logic) weigh against Canada's chance of survival as two (or ten) nations, against requiring exactly two cultures and 'founding races', and against insisting that 25 per cent of us must have a one-to-one relationship with the other 75 per cent, and so on."[67] His rejection of any notion of a bi-national state is also founded in his interpretation of the class basis of national policy rather than an ethnic basis, all of which is rooted in his interpretation of the economic nature of the original Canadian Confederation.[68] His conception of the birth of the Canadian nation, while not denying the importance of the Innis-Creighton imperialist conception (of the nation-building commercial class based on the staple trades), is based more on an analysis of the industrial roots. As such, his work bears directly on the debate on the "national question" engendered by the publication of *Capitalism and the National Question* in 1972, and more particularly the article by Tom Naylor contained in that collection, "The Rise and Fall of the Third Commercial Empire of the St. Lawrence", subsequently expanded on in his two-volume *History of Canadian Business 1867-1914*.[69] Naylor, following the conception of Creighton's *Commercial Empire of the St. Lawrence*, argued that Confederation and the policies which have come to be known collectively as the "national policy" were a continuation of the commercial dominance of the St. Lawrence to the exclusion of industrial capitalism.

> Canadian development policy in the National Policy era was essentially a policy of mercantilism, of consolidation and expansion led by a commercial capitalist class in which the state structure in the hands of that class played a critically active role.[70]

Pentland fundamentally disagreed. He would not even accept that no significant industrial working class had emerged prior to Confederation,[71] although he would not go as far as some of Naylor's critics, such as MacDonald, who treat the railways, designed to promote commercial activity, as industrial capital.[72] In fact, Pentland, along with Ryerson and Fowke, have from their differing perspectives, argued that Canadian Confederation and the national policy represent a *conjuncture* of capitalist interests, be they industrial, commercial, financial or transportation, French or English.[73] It was a capitalist imperialism of central Canada in relation to the resources and the markets of the west. Indeed, for Pentland, it was this conjuncture of commercial and industrial interests, once protection was achieved, that spelled the beginning of the end of the producer coalition of worker and employer and permitted the realignment of interests upon class lines.

Nevertheless, it would be a mistake to consider that this position is, somehow, a repudiation of the importance of commercial capital in the same period. Pentland was well aware of the impact of the needs of staple exploitation on the creation of conditions for the emergence of an industrially disciplined labour force, for it was in the creation of commercial infrastructure (transportation systems) that the labour force was forged. As he notes, "so long as economic life in Canada consisted of the looting of surface resources (furs, seals, timber), capital requirements were limited to the provision of short-term mercantile capital, circulating capital. [But] the canals and railway systems, designed to intensify staple production and intercontinental division of labour, had precisely opposite effects in the end."[71] Precisely! It was pursuance of commercial gains in the staple trades that required improved transportation systems which, in turn, created the demand for a capitalistic labour market and to which an elastic supply of labour was attracted. The filling in of the labour market with increased population and wage earners, and the product market with wage income, in turn created the demand conditions for the growth of industrial development. There were, of course, elements of a fundamental conflict of interest between merchants and manufacturers over tariffs and taxation to support transportation improvement. But, initially at least, this was an era of local industrial production — hand manufacture under artisanal control — in which transportation costs provided effective protection from competition from other centres where the same small-scale manufacture took place.

Railways, improved transportation and communications, and the growth of large-scale, mechanized factories in the United States began to erode and eventually destroy this protection, thereby creating in Canadian labour (both skilled and unskilled) and among rising capitalists a mutual interest in increased tariff protection. Some merchants — those in the export-import business and the maritime entrepôt traders — no doubt had a vested interest in opposing tariffs. Others, like the wholesalers serving agricultural markets, obviously acquired protectionist leanings, and increasingly integrated backward into manufacturing to supply their markets. (This whole process is described below in Chapter 5.) But it was the national question that ultimately resolved the conflict of interest as Pentland clearly recognizes.

> Soon after [Confederation], the prospect of exploiting the West now appeared.... The National Policy of 1879 was one weapon of Canadian imperialism. Its purpose was not to make Western Canada, also, a participant in a self-contained economy; for the West was fully expected to sell its staple product in world markets, as it did, and there appears to have been no regret that the new Canadian structure was thus made more vulnerable and dependent than its predecessor. The task of the National Policy of 1879 was rather to ensure that unprotected Western wheat was carried by protected Canadian railroads, handled by protected Canadian traders, and that the supplies of the West should come from protected manufacturers, in order that as much as possible of the gains to be made would be drawn to old Canada, and not to Britain or the United States. It was this imperial prospect that

brought merchants and railroads suddenly around in the 1870s, so that Eastern Canada spoke with an undivided protectionist voice.[75]

Thus Pentland cannot be said to be in unequivocal support of any interpretation that would give primacy to either mercantile capital or industrial capital in national strategy, including protection. Those who would relegate Pentland to the industrial capital side of the battle against the troops supporting mercantile capital, simply ignore what he has written. In fact, as he was to point out, the adoption of the national policy — whatever its industrial intentions — had the effect of increasing the staple dependence (and hence the commercial-financial strength) of the Canadian economy. "Now protection received for the first time the name of "National Policy", and ceased for the first time to be a national policy".[76] In *Unequal Union*, published seven years after Pentland's thesis, Ryerson appears to adopt a similar approach to the constitutional form of the national policy, Confederation.[77] In places Ryerson refers to "the men who shaped Confederation, the political spokesmen of the banking, trading and railway interests — the rising Canadian capitalist class..."[78] But elsewhere he says the pressure for Confederation was one of *"industrial capitalist development...* to unify and expand the home market."[79] However, in other places he tends to imply a contradiction between the interests and power of commercial and industrial capital.[80]

Thus, although Pentland and Ryerson are often linked as proponents of an industrial capitalist interpretation of the national policy, as opposed to the commercial capitalist view implied by the staple interpretation (best exemplified by Creighton and Naylor), it would perhaps be more correct to place Pentland on more middle ground. His position stressed rather the coalition of mercantile and industrial interests in the national policy, a position he shared with his fellow western economic historian, Vernon Fowke, albeit from a different regional perspective.[81]

This qualification is important to an understanding of Pentland's particular nationalism and his opposition to regionalism and Quebec's particularism. He attempted to integrate the conflicting interpretations of class divisions accentuated by the rise of industrial capitalism, and of regional divisions responding to the commercial exploitation of resource regions. One result of this is the tendency to criticize Pentland for accepting a form of western exceptionalism.[82] Rather, his view is tied to the differing class response in different regions to the tariff, and to the nature of capitalist development in the western and central Canadian regions.[83]

Labour and Capital in Canada, 1650-1860 — Twenty Years After

The two decades since the thesis was presented have produced a remarkable crop of labour and working-class studies in Canada, both within the traditional North American vein of institutional and political labour history and, more recently, in "the rescuing of Canadian workers from the margins of history", the social history of the "uncommon, common man".[84] The obvious question is the extent to which Pentland's observations and conclusions must be amended to incorporate the findings of more recent research. Although Pentland was one of the pioneers in the use of pre-Confederation documents in the Public Archives of Canada (many of which were not even catalogued at that time), he was nevertheless forced by gaps in the primary and secondary sources available to him to make some fairly speculative leaps in analysis and interpretation. Some of the recent scholarship relates directly to these gaps and should be incorporated into Pentland's analysis as no doubt he would have done, had he been able. At the same time, I would suggest that there is not a great deal in this recent work that contradicts the general thrust and argument while supplementing and, in places, correcting the detail.

Pentland's concern with slavery and "feudal" labour relations in Canada has been unique. What is perhaps more significant is that his economic analysis of pre-capitalist labour relations in Canada is supported and complemented by a more general theoretical analysis by the American economist, Evsey Domar, published a decade later.[85] Pentland's argument is based upon the supposition that labour scarcity relative to land encourages the establishment of a system of "unfree" labour — slave, serf, indentured servant or convict. As long as land is free (or at least cheap enough to be within the reach of the lower orders), labour will choose to cultivate its own holdings rather than generate a rental surplus to be appropriated by the large landholders or other employers in order to maintain themselves as a governing elite.

Domar uses the following formulation to develop the point. There will be economic benefit to a ruling elite to adopt and enforce unfree labour institutions where the lower productivity of unfree labour is more than offset by the lower wage costs of unfree labour; that is where:

$$P_f - P_s < W_f - W_s$$

where P_f = Productivity of free labour
P_s = Productivity of unfree labour
W_f = Labour cost of free labour
W_s = Labour cost of unfree labour

What determines the productivity and the wages or labour costs of free and unfree labour? Pentland's major contribution was to relate this question to the labour supply and demand conditions operative in Canada in the period before the development of large-scale industrial capitalism in the latter half of the nineteenth century.

Part of the thesis investigates the period when labour, while absolutely necessary to even minimal production — such as in iron production at the St. Maurice Forges — was not in sufficient year-round demand to maintain an unbound labour force. This study of the St. Maurice Forges remains the most insightful analysis to date of this early experiment in industrial development in a non-industrial environment.

Pentland made only brief reference to the role of the native population in the supply of labour to the European-oriented economy.[86] In part this reflects the fact that his main concern was the period of transition from a pre-capitalist to capitalist economy in central Canada. But it also reflects the fact that the investigation of the economics of the aboriginal participation in the European-based economy has only recently come under renewed academic scrutiny with results that remain in dispute. Was the Indian an "economic man" in the fur trade or was he a "political-military" man using the European fur traders as a strategic ally in the continuing battle for space in the northern reaches of the continent?

The recent literature is by no means unanimous. Some attempts have been made to argue that the aboriginal participation in the fur trade was basically non-economic in motivation. Eccles even suggests that the whole trade was non-economic in motivation, that from a European point of view it was an adjunct to military supremacy in the new world.[87] However, the most thorough exposition of the non-economic characteristics of native participation in the fur trade is given by Rotstein in his 1972 article.

> European traders were oriented towards profits which depended on fluctuating prices for their staples in Europe and tended to regard economic transactions as arms-length and impersonal activities. None of this was true for the Indian tribes which they encountered. Trade was a highly personal activity, an encounter of two political groups or their representatives (not of individuals) and followed established political patterns.[88]

Much of his argument is based on the reciprocal gift-giving that preceded any exchange and the accompanying ceremonial festivities and dances which likewise accompanied political missions and encounters.

Such a characterization of the nature of the fur trade relationship seems inconsistent with Pentland's characterization of the relationship as akin to the "merchant-employer system of commercial capitalism" in which the distribution of bargaining power was such that the Indians were forced to trade on "extremely disadvantages terms".[89]

On the other hand, work by Ray, and Ray and Freeman is much more supportive of Pentland's characterization.[90] The Indians appeared to behave sometimes in a non-economic fashion by reducing the number of furs traded when competition bid up the price (in terms of trade goods) or demanding a fixed price regardless of the state of the European market. This was eminently sensible economic behaviour to a migratory labour force which could not transport or accumulate surplus stores, but which had become dependent on European trade goods for their sustenance and their productivity.[91] Bargaining power thus was a function of the degree of competition among the buyers. In the pursuit of monopoly, therefore, the English and French or the Hudson's Bay Company and the North West Company were pursuing the goal of European commercial capitalism as would a later generation of Canadian commercial capitalists in their dealings with Canadian independent commodity producers, in particular with the staple farmers.

More difficult to assess is the apparent non-economic behaviour of the Indian in driving into virtual extinction many of the animals that were the mainstays of the fur trade. McManus has argued that this behaviour follows from the common property resources characteristic of fur-bearing animals.[92] But it is perhaps more general to argue that the natives' dependence on European trade goods — particularly if those goods had allowed the indigenous population to increase — forced them to deplete animal stocks merely to survive in the face of falling real prices for furs and pressure on the land resources by alternate economic uses. Such behaviour would be compatible with observed behaviour by other independent commodity producers.

On the other hand, it should at least be noted that recently the proposition has been advanced that the over-harvesting of fur-bearing animals was not economically motivated at all but occurred because the Indians associated animals with the spread of European diseases.[93] That proposition seems rather far-fetched.

In the original thesis, Pentland used the term "feudal" to describe the labour relations pertaining to the European labour force involved in the fur trade, the industries of New France and the timber industry. In the later revisions, this was altered to "personal" labour relations. The change is significant, a recognition that the concept of feudalism is too narrowly restrictive to a European-based *political* system of parcellized sovereignty, the applicability of which has been questioned even in the context of the seigneurial system of New France.[94] His central argument is economic, not political. Because of the insecurities of labour demand and the insufficiency of labour supply, a bond between employer and employee had to be established. But it was not a bond of market exchange. Rather, it was a bond of reciprocal if unequal obligation. That characterized by European feudalism, which may or may not have found comparable expression in the seigneurial system of New France, was only one political-economic form. The more generalized form, he designated *personal* labour relations, relations that were characterized by the employers' obligation to provide the "fixed costs" of sustenance and reproduction, in exchange for a loyal and reasonably efficient labour force, a form he explores in examples from the fur trade to the St. Maurice Forges.

Some contemporary readers may feel some unease at Pentland's discussion of paternalism in labour relations because of the apparent stereotyping of the French and Indian populations as "childlike" in their relations with employers.[95] Certainly one must reject the more excessive popular depictions of the aboriginal population as no more than backward children, despite the language (at least the English versions) of the Indian treaties and the language and customs of the fur trade ceremonies where the Indians were referred to as "children" and behaved in what might appear to be a childish manner.[96] But as Ray and Freeman conclude, "the pre-trade ceremonies were probably more akin to the exchanges of courtesies.or pleasantries which often precede business dealings in the modern Western world"[97] (including the prolific use of alcohol) and certainly no more childish than the goings-on at a modern Shriners' convention.[98] On the other hand, the docility of the French-Canadian worker has been remarked upon by other modern scholars.[99]

Nevertheless it is important to understand what Pentland is attempting to convey: that in an economy where the habitants and the Indians were dependent on employers, seigneurs or the trading companies for year-round sustenance and capital (the trade goods were in part capital goods to the fur trade), the relationship of the employer or trader needs to be paternalistic — implying a responsibility for the welfare of the habitant or native, a responsibility akin to that of a parent for a child.[100] It is true that the employer or trader was also dependent on the worker or Indian. But, as Adam Smith has pointed out in relation to the industrial worker and the industrial employer, the worker's needs are much more immediate and desperate.[101] In most cases, in pre-capitalist Canada, there was little or no alternative choice of employer.

In his discussion of pre-Confederation agriculture in Lower Canada, Pentland did not have the benefit of the research of the recent generation of French-Canadian historians, in particular the work of Ouellet and Hamelin, and Paquet and Wallot.[102] Ouellet and Hamelin generally support the argument taken by Pentland that French-Canadian agriculture, at least on the old seigneuries, did enter a period of crisis and retrenchment in the early years of the nineteenth century, a crisis of stagnation that was a factor in the Rebellion of 1837 and was at least in part caused by the habitants' inability and/or unwillingness to adopt more productive agricultural techniques. Paquet and Wallot challenge this view, arguing that the problems faced by Lower Canada were structural, arising from the inability of the agricultural economy to adjust quickly enough to changes in demand and prices in the period of instability that accompanied the Napoleonic Wars and the post-war readjustments. This interpretation does directly challenge Pentland's account. On the other hand, LeGoff's recent attempt to sum up the debate leans toward the Ouellet position, indirectly supporting Pentland, although he places his emphasis on population increase as the prime cause of agricultural stagnation.[103] At best, therefore, we must conclude that the jury is still out on the debate. But, whatever the outcome regarding the early years of the century, Pentland's more fundamental point regarding the inability of the Lower Canadian economy in the nineteenth century to absorb the natural increase and the consequent flood of emigration remains valid and is supported by Isbister in his recent article on the latter part of the century.[104]

Still, it is possible to question Pentland's conclusion that the French Canadian was only "a potential or marginal member, rather than an actual member, of the labour force."[105] The nature of labour market demand was hardly conducive to the encouragement of full-time participation by an unskilled, rural people. As Pentland has pointed out elsewhere, the artisanal and skilled jobs required a commitment to permanent occupation of the labour market as well as the requisite skills, a commitment that British industrial workers had only acquired from years of experience in Britain.[106] At the same time, factory employment comparable to the textile industry of New England was in its infancy and when it did develop, it did so with French-Canadian labour. The French Canadian did participate in the shipbuilding and timber industry, in the latter particularly as raftsmen.[107] It would appear that French-Canadian participation in the labour market was more akin to what Sacouman has called the semi-proletarianism of the domestic mode of primary production where subsistence farming provides for the fixed cost of maintaining and reproducing labour for seasonal, irregular, low-wage industries.[108] It was precisely the lack of steady demand for labour that would appear to have been the reason for lack of French participation in the labour force, a situation that Pentland contrasts with the immigrant Irish. But, again, the evidence is hardly that compelling. As he notes himself, the French Canadian was periodically in competition for jobs with the Irish[109] and, although competition for the heavy labouring jobs in public works was hardly keen, "Irish jealousy" served to make these jobs an ethnic preserve.[110] Given the wages and conditions on these jobs, the smaller physical stature of the French compared with the Irish, and the migratory nature of employment, it is hardly a wonder that the habitant stuck to his farm if he could. However depressed it might be, the alternative was hardly attractive. Besides, the French Canadian, at least in the earlier years of the nineteenth century, had not been oppressed and reduced to such economic destitution that he had come to consider agriculture an undesirable pursuit as Pentland argues the Catholic Irish had.[111]

One of the most original aspects of *Labour and Capital in Canada, 1650-1860* is its central concern with the role of the Irish immigrant in the making of the Canadian labour market, perhaps in part a reflection of Pentland's own Ulster heritage. Again, this is an area where more recent work has tended both to confirm his general argument and to raise some specific objections, particularly with his treatment of the Orange Order. Some readers may interpret the characterization of the Irish as "primitive", "intemperate", "improvident" and "gregarious" as ethnic stereotyping which should be amended in the light of more contemporary research. Yet much subsequent scholarship serves to reinforce Pentland's characterization rather than contest it. Duncan, for example, notes "violence and riots, disease, crime, drunkeness, and prostitution were rife" in the Irish Catholic working-class slums and goes on to show how the pattern of behaviour in Canada had its roots in Irish society.[112] Similar support for Pentland's view is given in the work of Michael Cross, particularly in his study of the "Shiners' War"[113] while the general fractiousness of the contending Irish religious groups is recorded by Palmer and Kealey.[114] The general conclusions of this later work extends Pentland's analysis but does not contradict it.

On the other hand, Kealey does challenge Pentland's interpretation of the specific role of the Orange Order as an institution that "represented the artisan"[115] while the unskilled labourers "used their ethnic community to organize their strikes and promote their political interests."[116] Kealey, by contrast, argues that the Orange Order in Toronto was "overwhelmingly working-class", but not limited to artisans. "No imaginative flights can transform Toronto Orangemen into a labour aristocracy."[117] His data on Toronto makes a persuasive argument that the Orange Order was, in fact, a fairly broad based, Protestant organization with strong working-class roots. The question that arises, however, is how general this working-class composition was. Cross's work, for instance, tends to emphasize the middle-class (Tory) agricultural basis of Orange activity around Ottawa which coincided with religious divisions.[118] It is probable that no single characterization of the Order is possible in all areas, at all times. Without doubt, the Orange Order did play a part in working-class organization, but its complex interrelationship with political and religious divisions suggests that no single characterization is possible. The answer to the question of the role of the Order, and indeed of the "green" organizations among Catholics, still awaits a comprehensive treatment, but it is probably safe to conclude that Pentland's original view is too simplistic.

Somewhat surprisingly, Pentland's treatment of the 1837 rebellions is somewhat fragmentary and cursory.[119] This is surprising if only because of the degree of attention since given them by other labour and social historians who were also interested in the development of the nineteenth-century Canadian political economy.[120] Ryerson's treatment, the most extensive, interprets the failure of both rebellions, which he defines as "bourgeois-democratic, class and national revolution(s)," as a reflection of the immature development of industrial capitalism and with it an undeveloped anti-mercantile bourgeoisie and a working class as yet unconscious of its existence as a separate class.[121] Lipton interprets the rebellions in even more class-specific terms, part of the "world fight for labour and democracy" and credits the defeats to the preference of business and capital for "collaboration with, and subordination to, foreign capitalists ... to alliance with workers and farmers in Canada, the 'mechanics' and 'yeomen'."[122] But Lipton's analysis seems somewhat superficial and simplistic.

It is Johnson who gives the most intensive and complex analysis, though limited to a relatively small area of Upper Canada, the County of Ontario. He sees the conflict as between the "Tory" vision of a stratified society ordered around a privileged social and economic elite,[123] and the reform vision, a democratic society ordered around the "sturdy yeoman, the independent producers — farmer, craftsman or [small] merchant."[124] The reform vision looked with increasing horror at the social implications of the industrial revolution and the creation of a degraded working class which was foreseen in the expansion of large-scale capitalism to Upper Canada. Though the rebellion was defeated by the British-backed establishment, Britain soon granted responsible government which swept away the Tory vision. But the reform vision would likewise perish, "prey to the new, rising class of entrepreneurs whose version of 'progress' was the accumulation of capital and the employment of men."[125]

These three interpretations, although based on the same type of historical approach to the development of labour and industrial capitalism, nevertheless are at variance to a greater or lesser extent with each other and with Pentland's treatment which is insufficiently developed to compare or contrast to the others in any detail. Yet what he does say suggests an interpretation blending some elements of both Ryerson and Johnson but different from both — a rebellion by an increasingly economically independent farmer-artisan alliance against subjugation to economic dependence on a transatlantic staple trade controlled by, and for the benefit of, the merchant class.[126] It would be wrong, however, to attempt to fit Pentland's interpretation to the procrustean bed of viewing the rebellions as an abortive "bourgeois revolution" or his artisans as an immature industrial working class.

The Pentland Legacy

When Pentland defended his thesis in 1961 he opened up an area of historical and economic studies which had previously been neglected by Canadian economic historians. He was not alone in his interest in the growth of manufacturing in Central Canada in the latter half of the nineteenth century,[127] but his concern with the broad historical aspects of economic development and social change made his work virtually unique. No one since has attempted a study of the industrial transformation of Canada of a similar depth or breadth, though, as I have indicated above, there has been considerable work on specific topics or regions included in the original thesis but not explored in detail. This has been particularly true of the nineteenth-century Quebec[128] and, to a lesser extent, of the emergence of organized labour[129] and the political and social history of the labouring classes.[130]

We are probably approaching the time when someone will attempt a new synthesis of this body of work, particularly now that the manuscript censuses are available for the critical transitional years. Nevertheless, despite the firm foundation of Pentland's work (supplemented by the variety of studies already referred to), the task is formidable, possibly even premature. Critical debates are still unresolved and may indeed remain unsolved in the largely pre-statistical era of the nineteenth century. Modern economic theory is a poor beacon to guide one in this period of systemic change, even for those who regard this theory as a valuable guide to contemporary economic processes (and clearly there are many in the field who do not). We have had more than a decade now of economic and quantitative history research, and the results, however, fruitful in enriching our knowledge of specific issues in economic history, have not brought us much closer to understanding broader issues of the type that Clare Pentland was trying to illuminate.[131] Probably the very nature of the questions quantitative history asks and the issues it is able to deal with preclude a real investigation of the industrial revolution in Canada and the accompanying transformation of the "uncommon, common man", the making of a modern working class.

Pentland's study was the first attempt at a general treatment of this social and economic transformation, though incomplete, as he was quite consciously aware. To date it remains the only comprehensive treatment and until now has remained unpublished — a situation that Watkins has found "incredible".[132] But responsibility for this lies with Clare Pentland, not with publishers or scholars ignoring his contribution. He was aware of the gaps, of the unresolved questions and of the deficiencies in evidence, which is why he was reluctant to submit it for publication. This was what excited him about the new generation of scholars probing the period, no matter that he disagreed at times with their reading of the evidence. But if and when a new synthesis is put together, the author of that work will owe a debt directly and indirectly to Pentland, and to the Social Science Federation of Canada whose financial support finally helped make readily available this seminal work to future generations of students and scholars.

P.P.

Notes to Introduction

[1] E.P. Thompson, *The Making of the English Working Class* (Harmondsworth: Penguin Books, 1963), pp. 12-13.

[2] H.C. Pentland, "The Lachine Strike of 1843", *Canadian Historical Review*, Sept. 1948, p. 255.

[3] In the notes that follow pertaining to the Pentland thesis, the reader is referred to the current volume.

[4] Kealey has noted that "Pentland's use of 'feudal' is unique. What he means by it is the 'labour organization that preceded the free market of industrial capitalism, that was not slavery, nor a putting-out system, nor the share system of early capitalist commerce' ("Labour and Industrial Capitalism", p. 54). Thus it is a name he uses for a preindustrial *but* capitalist form of labour organization and should not be confused with classic European feudalism" (G.S. Kealey, "H.C. Pentland and Working Class Studies", *Canadian Journal of Political and Social Theory*, Spring/Summer 1979, p. 92). Pentland was obviously not entirely comfortable with his original terminology, and in the revised manuscript "feudal labour" was changed to "personal labour relations". But the term retains the essence of the feudal bond, personal obligation, rather than the impersonality of the self-regulating capitalistic labour market.

[5] Staple studies are most associated with H.A. Innis, A.R.M. Lower, W.A. MacIntosh, V.C. Fowke and D.G. Creighton.

[6] H.C. Pentland, *Labour and Capital in Canada, 1650-1860* (Toronto: Lorimer, 1981), p. xlvi. See also H.C. Pentland, "The Role of Capital in Canadian Economic Development Before 1875", *Canadian Journal of Economics and Political Science*, Nov. 1950, pp. 83-84n.

[7] K.A.H. Buckley, "The Role of Staples in Canada's Economic Development", *Journal of Economic History*, Dec. 1958.

[8] See below, p. xlvii.

[9] It would appear that Pentland planned a more extensive revision and possibly an extension to the version presented here, but, in any case, this possibility was precluded by his illness and death.

10 Pentland, "The Lachine Strike"; "The Role of Capital"; and "The Development of a Capitalistic Labour Market in Canada", *Canadian Journal of Ecomonics and Political Science*, Nov. 1959.

11 These were *The Canadian Journal of Political and Social Theory*, Spring/Summer 1979; *Labour/Le Travailleur*, 1979; and *Studies in Political Economy*, Spring 1979.

12 Kealey, "H.C. Pentland and Working Class Studies", p. 79.

13 The local graveyard is named after the Pentland family, after the first person buried there, a Pentland infant.

14 The thesis was completed in 1942 but the degree was not conferred until March 1943.

15 For the biographical information on L.A. Wood, see Foster J.K. Griezic's introduction to *A History of Farmers' Movements in Canada* (Toronto: University of Toronto Press, 1975; originally published 1924), particularly pp. vi-viii.

16 *Loc. cit.* Wood used texts such as Gustavus Meyers', *History of Canadian Wealth* and O.D. Skelton's *Socialism: A Critical Analysis* (p.viii). Among the materials that Pentland collected during the war were a number of socialist and Marxist pamphlets, largely British.

17 *Ibid.*, p viii.

18 Pentland took the following courses at Toronto: economic history from Innis, economic theory from G.A. Elliot, labour economics from H.A. Logan, sociology from S.D. Clark, and industrial relations from F. Toombs. See Kealey, "H. C. Pentland and Working Class Studies", p. 80.

19 See in particular Daniel Drache, "Rediscovering Canadian Political Economy", *Journal of Canadian Studies*, August 1976, pp. 3-18.

20 Innis, frequently in his writing, was intensely critical of the received doctrines of economic theory without an historical and political dimension. For an interesting discussion, heavily weighted to Innis's views, see Ian Parker, "The National Policy, Neoclassical Economics, and the Political Economy of Tariffs", *Journal of Canadian Studies*, Fall 1979.

21 Indeed, Pentland considered himself as much an historian as an economist and was hostile to the concentration on economic theory courses in university economics programs. He was president of the Manitoba Historical Society from 1963 to 1965, a founding member and third president of the Manitoba Records Society and active in local history work.

22 It is interesting to note that, although it was some years after his thesis, Pentland took a side trip from the Montreal Learned Societies to view the remains of the St. Maurice Forges to which he had devoted considerable attention in his thesis.

23 This is not true of some of the people writing within the dependency school which may be considered to include the staple model as one of its varients. For a critical view of the dependency school see H. Veltmeyer, "A Central Issue in Dependency Theory", *Canadian Review of Sociology and Anthropology*, vol. 17, no. 3 (1980).

24 Kealey, "H.C. Pentland and Working Class Studies", p. 81.

25 Bryan Palmer, "Working-Class Canada: Recent Historical Writing", *Queen's Quarterly* 87 (1980): 613, fn. 45.

26 Leo Panitch made a similar distinction and point in reference to the work of C.B. McPherson in discussion at the Canadian Political Science Association meeting, "Papers in Honour of C.B. McPherson", Montreal, June 1980.

27 Pentland, "Capitalistic Labour Market", p. 2.

28 Pentland, "The Role of Capital", p. 457.

29 See below, p. xlvii.

30 For a discussion of his population studies see Paul Duprez, "Pentland's Scarcity of Labour and the Industrial Revolution", *Canadian Journal of Political and Social Theory*, Spring/Summer 1979.

31 Pentland first met Mathias when he was visiting in North America. But it was when Pentland was in England on Sabbatical that the contact and influence developed.

32 H.C. Pentland, "Population and Labour Growth in Britain in the Eighteenth Century", in D.E.C. Eversley, ed., *Proceedings of the Third International Conference of Economic History*, vol. 4 (Munich, 1965).

33 As with a number of other studies for the Task Force, it was not published in the background series apparently because it was not completed until February 1968, after the cut-off date established by the Task Force, although mimeographed copies were distributed to repository and university libraries.

34 "A Study of Labour Skills in Reference to Manitoba's Economic Future", for the Committee on Manitoba's Economic Future, 1962; "Implication of Automation for the Employment and Training of White-Collar Workers in Manitoba", for the Manitoba Economic Consultative Board; "Human Adjustment to Technological Change: The Case of Manitoba Rolling Mills", for the Department of Manpower and Immigration, 1968.

35 Pentland had been active in the early organization of the University of Manitoba Faculty Association and was one of its early presidents.

36 H.C. Pentland, "The Western Canadian Labour Movement, 1897-1919", *Canadian Journal of Political and Social Theory*, Spring/Summer 1979; "Industrial Relations Systems", *Labour/Le Travailleur*, 1979; "Were Canadian Capitalists Different? How the West was Won", *Canadian Forum*, Sept. 1972; "Marx and the Canadian Question", *Canadian Forum*, Jan. 1974.

37 *Canadian Dimension*, Dec. 1979, p. 18; Paul Baran and Paul Sweezy quoted in Harry Braverman, *Labor and Monopoly Capital* (New York: Monthly Review Press, 1974), p. ix.

38 See below, p. xlv.

39 R.G. Hann, et al., *Primary Sources in Canadian Working Class History 1860-1930* (Kitchener: Dumont Press, 1973), p. 10. See also Palmer, "Working-Class Canada", pp. 601-2.

40 See Pentland, "The Role of Capital", and "Capitalistic Labour Market".

41 Kealey, "H.C. Pentland and Working Class Studies", p. 83.

42 Paul Phillips, "Divide and Conquer: Class and Consciousness in Canadian Trade Unionism", *Socialist Studies*, no. 2, May 1980; Leo Johnson, "The Political Economy of Ontario Women in the Nineteenth Century", in Janice Acton, Renny Goldsmith, and Bonnie Shepard, eds., *Women at Work, Ontario 1850-1930* (Toronto: Canadian Women's Educational Press, 1974).

43 See below, p. 187.

44 See p. xxviii.

45 See below, pp. xlv-xlvi.

46 See below, p. 174.

[47] See below, p. 174.

[48] Palmer, "Working-Class Canada", pp. 594-616.

[49] See below, pp. 180-182.

[50] See below, p. 182.

[51] See below, p. 183.

[52] Michael Katz, *The People of Hamilton, Canada West* (Cambridge, Mass.: Harvard University Press, 1975), p. 7.

[53] Steven Langdon, "The Emergence of the Canadian Working Class Movement, 1845-1875", *Journal of Canadian Studies*, May 1973, p. 7.

[54] *Ibid.*, p. 11.

[55] Bryan Palmer, *A Culture in Conflict: Skilled Workers and Industrial Capitalism in Hamilton, Ontario, 1860-1914* (Montreal: McGill-Queen's University Press, 1979); Gregory Kealey, *Toronto Workers Respond to Industrial Capitalism, 1867-1892* (Toronto: University of Toronto Press, 1980); Bryan Palmer and Gregory Kealey, "Brotherhood, Sisterhood and the Bonds of Unity: The Knights of Labor in Ontario, 1880-1900". Paper to the Canadian Historical Association, Montreal, June 1980.

[56] Palmer, *A Culture in Conflict*, pp. 5, 15-18; Kealey, *Toronto Workers*, p. 18 ff.

[57] Palmer, *A Culture in Conflict*, p. 15.

[58] Kealey, *Toronto Workers*, p. 29.

[59] See below, pp. 171-173.

[60] See below, p. 173.

[61] Paul Craven and Tom Traves, "The Class Politics of the National Policy, 1872-1933", *Journal of Canadian Studies*, Fall 1979, p. 15.

[62] *Ibid.*, p. 16.

[63] Compare the first pages of the two studies.

[64] H.C. Pentland, "A Study of the Social, Economic and Political Background of the Canadian System of Industrial Relations", prepared for the Task Force on Labour Relations, 1968, p. 20. See also pp. 11-12.

[65] *Ibid.*, pp. 44-45. Pentland's analysis of the period between the 1870s and 1900 is sketchy. He was working on that period in the last few years of his life but, as with his demographic researches, illness and his premature death prevented his completing the task.

[66] *Ibid.*, pp. 226-62.

[67] H.C. Pentland, "Marx and the Canadian Question", *Canadian Forum*, January 1974, p. 28. Any lingering doubt that Pentland did not see the evolution of the labour market in class terms should be dispelled by this review. "Thus, capitalism has proceeded, as Marx predicted, to grind small independent owners out of existence, leaving ownership of the means of production concentrated in the hands of a small capitalist class, the rest of us dependent on sale of our labour, and Canadian society quite close to the simple two-class confrontation that socialists have expected — a conclusion that should surprise nobody, but that probably does need a lot of pointing out" (p. 26).

[68] See below, pp. 160-162.

[69] Gary Teeple, ed., *Capitalism and the National Question in Canada* (Toronto: University of Toronto Press, 1972); Tom Naylor, *The History of Canadian Business, 1867-1914*, 2 vols. (Toronto: Lorimer, 1975).

[70] Naylor, *Canadian Business*, vol 2, p. 186.

[71] Pentland, "Marx and the Canadian Question", pp. 26-27.

[72] L.R. MacDonald, "Merchants Against Industry: An Idea and its Origins", *Canadian Historical Review*, Sept. 1975, p. 264. MacDonald's contention that railways were industrial rather than commercial pursuits seems to me to be a red herring if not actually incorrect, however Marx characterized them. The railroads were advanced and built to support a staple commerce, not to support industrial production whatever their ultimate effect may have been. MacDonald also argues that V.C. Fowke, Alfred Dubuc and Stanley Ryerson as well as Pentland "have advanced the view that the development of trade did not require, and was even antagonistic to, the growth of production" (p. 264). This misrepresents both Pentland and Fowke in the period from Confederation.

[73] See below, p. 161.

[74] Pentland, "The Role of Capital", pp. 460, 463.

[75] See below, p. 173.

[76] *Loc. cit.*

[77] S.B. Ryerson, *Unequal Union* (Toronto: Progress Books, 1968).

[78] *Ibid.*, p. 355.

[79] *Ibid.*, p. 358.

[80] *Ibid.*, p. 428. Quoting Marx, "The independent development of merchants' capital... stands in inverse proportion to the general economic development of society."

[81] See Paul Phillips, "The Hinterland Perspective: The Political Economy of V.C. Fowke", *Canadian Journal of Political and Social Theory*, Spring/Summer 1978.

[82] Kealey, "H.C. Pentland and Working Class Studies", pp. 88-89.

[83] See Pentland's references to Dubofsky in "The Western Canadian Labour Movement".

[84] Kealey, "H.C. Pentland and Working Class Studies", p. 80. Palmer has designated these two schools as the first and second generations of Canadian labour historians. Interestingly, though perhaps misleadingly since Pentland bridged both, he associates Pentland with the second generation. See "Working-Class Canada", pp. 601-2.

[85] See below, pp. 1-4; E.D. Domar, "The Causes of Slavery or Serfdom: An Hypothesis", *Journal of Economic History*, March 1970. See also H.C. Pentland, "Feudal Europe: An Economy of Labour Scarcity", *Culture*, 1960, pp. 280-307. Pentland did not deal with slavery within the aboriginal societies but restricted his analysis to European society.

[86] See below, pp. 204-205; also p. 61.

[87] W.J. Eccles, "A Belated Review of Harold Adam Innis, The Fur Trade in Canada", *Canadian Historical Review*, 1979, p. 423.

[88] Abraham Rotstein, "Trade and Politics: An Institutional Approach", *Western Canadian Journal of Anthropology* 3 (1972).

[89] See below, p. 23.

[90] Arthur Ray, *Indians in the Fur Trade*, (Toronto: University of Toronto Press, 1974); Arthur Ray and Donald Freeman, *Gives us Good Measure*, (Toronto: University of Toronto Press, 1978).

[91] Ray and Freeman, *Give Us Good Measure*, pp. 161-62, 218-28.

[92] J. McManus, "An Economic Analysis of Indian Behaviour in the North American Fur Trade", *Journal of Economic History* 32 (1972).

[93] Calvin Martin, *Keepers of the Game* (Berkeley: University of California Press, 1978).

[94] Marcel Trudel, *The Seigneurial Regime*, Canadian Historical Association Booklets, no. 6 (Ottawa, 1967), p. 14. However, see Sigmond Diamond, "An Experiment in Feudalism: French Canada in the 17th Century", *William and Mary Quarterly*, 3rd Ser., XVIII (1961), pp. 3-34.

[95] See below, p. 55.

[96] Rotstein, "Trade and Politics", p. 8: Ray and Freeman, *Give Us Good Measure*, p. 66 ff.

[97] Ray and Freeman, *Give Us Good Measure*, p. 233.

[98] Cf. the description of the North West Company Partners party in G.C. Davidson, *The North West Company* (Berkeley: University of California Press, 1918), pp. 243-44.

[99] Michael Cross, "The Shiners' War", *Canadian Historical Review*, March 1973, p. 7.

[100] See below, p. 60.

[101] Adam Smith, *Wealth of Nations* (Homewood: Irwin, 1963; first published 1776), pp. 53-54.

[102] Fernand Ouellet, *Histoire économique et sociale du Québec, 1760-1850* (Montreal: Éditions Fides, 1966); Fernand Ouellet and J. Hamelin, "La Crise agricole dans le Bas-Canada", *Canadian Historical Association Report*, 1962, pp. 317-33; G. Paquet and J.P. Wallot, "Crise agricole et tensions socio-ethniques dans le Bas-Canada, 1802-1812: Éléments pour une ré-interprétation", *Revue d'histoire de l'Amérique française*, vol. 26, no. 2 (1972), pp. 185-237, and Paquet and Wallot, "International Circumstances of Lower Canada, 1786-1810", *Canadian Historical Review*, vol 52, no. 4 (1972), pp. 371-401. Additional references can be found in T.J.A. Le Goff, "The Agricultural Crisis in Lower Canada, 1802-1812: A Review of the Controversy", *Canadian Historical Review*, vol. 55, no. 1 (1974), pp. 1-31, and the response by Paquet and Wallot, "The Agricultural Crisis in Lower Canada, 1802-1812: mise au point", *Canadian Historical Review*, vol. 56, no 2 (1975), pp. 131-68.

[103] Le Goff, "Agricultural Crisis in Lower Canada".

[104] John Isbister, "Agriculture, Balanced Growth, and Social Change in Central Canada since 1850: An Interpretation", *Economic Development and Cultural Change*, July 1977, pp. 673-97.

[105] See below, p. 78.

[106] Pentland, "Capitalistic Labour Market". Even at that the paucity of demand for skilled labour meant that many were forced to emigrate to the United States. See below, p. 106.

[107] Ryerson, *Unequal Union*, p. 39 ff.

[108] R.J. Sacouman, "Semi-proletarianism and Rural Underdevelopment in the Maritimes", *Canadian Review of Sociology and Anthropology*, vol. 17, no. 3 (1980), p. 232 ff.

[109] Pentland, "Lachine Strike", pp. 266, 271. See also Paquet and Wallot, "The Agricultural Crisis in Lower Canada, 1802-12: mise au point", pp. 153-54; Ryerson, *Unequal Union*, p. 182; Cross, "The Shiners' War", p. 13.

[110] Pentland, "Lachine Strike", p. 259.

[111] Pentland, "Capitalistic Labour Market".

[112] Kenneth Duncan, "Irish Famine, Immigration and the Social Structure of Canada West", *Canadian Review of Sociology and Anthropology*, February 1965, p. 24.

[113] Cross, "The Shiners' War". See also M.J. Cross, "Stony Monday, 1849: The Rebellion Losses Riots in Bytown", *Ontario History*, September 1971, pp. 177-90.

[114] Palmer, *A Culture in Conflict*, pp. 43-46; Kealey, *Toronto Workers Respond to Industrial Capitalism*, ch. 7.

[115] See below, p. 129.

[116] *Ibid.*, p. 128.

[117] Kealey, *Toronto Workers Respond to Industrial Capitalism*, p. 107.

[118] Cross, "Stony Monday, 1849", p. 190. See also Leo Johnson, *History of the County of Ontario* (Whitby: County of Ontario, 1973), pp. 116-21.

[119] See below, pp. 144-145, 152.

[120] In particular see Ryerson, *Unequal Union*, part 1, pp. 29-132; Charles Lipton, *The Trade Union Movement of Canada, 1827-1959* (Montreal: Canadian Social Publications, 1967), pp.9-16; Johnson, *History of the County of Ontario*, ch. 7.

[121] Ryerson, *Unequal Union*, pp. 83, 132.

[122] Lipton, *Trade Union Movement of Canada*, pp. 13, 15.

[123] Johnson, *History of the County of Ontario*, p. 96.

[124] *Ibid.*, p. 99.

[125] *Ibid.*, p. 127.

[126] See below, p. 146, 152.

[127] G.W. Bertram, "Economic Growth in Canadian Industry, 1870-1915: The Staple Model and the Take-off Hypothesis", *Canadian Journal of Economics and Political Science*, vol. 29, no. 2 (1963), pp. 162-84; E.J. Chambers and G.W. Bertram, "Urbanization and Manufacturing in Central Canada, 1870-1915", in S. Ostry and T.K. Rymes, eds., *Conference on Statistics, 1964*, Canadian Political Science Association (Toronto: University of Toronto Press, 1964), pp. 225-58; J.M. Gilmore, *Spatial Evolution of Manufacturing: Southern Ontario, 1851-1891* (Toronto: University of Toronto, 1972); D.M. McDougall, "Canadian Manufactured Commodity Output, 1870-1915", *Canadian Journal of Economics*, vol. 4, no. 1 (1971), pp. 21-36.

[128] Among some recent sources are: J. Hamelin and Yves Roby, *Histoire économique du Québec, 1851-1896* (Montreal: Éditions Fides, 1971); J. Hamelin, et al, *Répertoire des grèves dans la Province du Québec au XIXe siècle* (Laval: Les Presses de l'Université Laval, 1970); Noel Bélanger, et al., *Les Travailleurs québécois, 1851-1896* (Quebec and Paris: Les Presses de l'Université de Paris, 1975).

[129] Among recent studies that may be cited in addition to sources previously mentioned, are: Eugene Forsey, *The Canadian Labour Movement, 1812-1902*, Canadian Historical Association Booklets, no. 27 (Ottawa, 1974); Forsey's major unpublished manuscript, "History of Canadian Trade Unionism", Public Archives of Canada, Reel M-2214; Forsey, "The Telegraphers' Strike of 1883", *Transactions of the Royal Society of Canada*, Series 4, vol. 9 (1971), pp. 245-60; Forsey, "The Toronto Trades Assembly, 1871-8", *Canadian Labour*, June-October issues, 1965; Robert Babcock, *Gompers in Canada: A Study in American Continentalism Before the First World War* (Toronto: University

of Toronto Press, 1974); James Rinehart, *The Tyranny of Work* (Toronto: Longman, 1975). More complete bibliographies are Palmer, "Working-Class Canada" and Gregory Kealey and Peter Warrian, eds., "Bibliographic Essay", *Essays in Canadian Working Class History* (Toronto: McClelland and Stewart, 1976).

130 Much of the best of this work has been published in the journal *Labour/Le Travailleur* since its inception in 1975. A second useful collection is Kealey and Warrian, eds., *Essays in Canadian Working Class History*. In addition to Katz's social demographic study on Hamilton, David Gagan and Herbert Mays have published on Peel County ("Historical Demography and Canadian Social History: Families and Land in Peel County, Ontario", *Canadian Historical Review*, March 1973, pp. 27-47), and Leo Johnson on the Home District ("Land Policy, Population Growth and Social Structure in the Home District", *Ontario History*, March 1971, pp. 41-60). Much of the political writing has centred on the political alliances of labour in the Macdonald period. See Bernard Ostry, "Conservatives, Liberals and Labour in the 1870s", *Canadian Historical Review*, vol. 41, June 1960, pp. 93-127; Ostry, "Conservatives, Liberals and Labour in the 1880s", *Canadian Journal of Economics and Political Science*, vol. 27, May 1961, pp. 141-61; Sally Zerker, "George Brown and the Printers Union", *Journal of Canadian Studies*, Fall 1975, pp. 42-48. On ethnic labour history see Lawrence Runnals, *The Irish on the Welland Canal* (Toronto, 1973), and Donald Avery, *"Dangerous Foreigners": European Immigrant Workers and Labour Radicalism in Canada 1896-1932* (Toronto: University of Toronto Press, 1979). Palmer's "Working Class Canada" also reviews the American work on the work process in this period.

131 See the appraisal by H.G.J. Aitken, "Myth and Measurement: The Innis Tradition in Economic History", *Journal of Canadian Studies*, Winter 1977, pp. 96-105.

132 Mel Watkins, "The Staple Theory Revisited", *Journal of Canadian Studies*, Winter 1977, p. 95.

Selected Bibliography

Aitken, H. C. H. "Myth and Measurement: The Innis Tradition in Economic History". *Journal of Canadian Studies*, Winter 1977, pp. 96-105.

Ankli, R. E. "The Reciprocity Treaty of 1854". *Canadian Journal of Economics*, February 1971, pp. 1-20.

Avery, Donald. *"Dangerous Foreigners": European Immigrant Workers and Labour Radicalism in Canada 1896-1932*. Toronto: University of Toronto Press, 1979.

Babcock, Robert. *Gompers in Canada: A Study in American Continentalism Before the First World War*. Toronto: University of Toronto Press, 1974.

Bélanger, Noel, et al. *Les Travailleurs québécois, 1851-1896*. Montreal, 1973.

Bertram, A. W. "Economic Growth in Canadian Industry, 1870-1915:

The Staple Model and the Take-Off Hypothesis". *Canadian Journal of Economics and Political Science*, vol. 29, no. 2 (1963), pp. 162-84.

Braverman, Harry. *Labor and Monopoly Capital*. New York: Monthly Review Press, 1974.

Buckley, K. A. H. "The Role of Staples in Canada's Economic Development". *Journal of Economic History*, December 1958, pp. 439-50.

Chambers, E. J., and Bertram, A. W. "Urbanization and Manufacturing in Central Canada, 1870-1915". In *Conference on Statistics 1964*, Canadian Political Science Association, edited by S. Ostry and T. K. Rymes, pp. 225-58. Toronto: University of Toronto Press, 1964.

Craven, Paul, and Traves, Tom. "The Class Politics of the National Policy, 1872-1933". *Journal of Canadian Studies*, Fall 1979, pp. 14-38.

Cross, Michael. "The Shiners' War". *Canadian Historical Review*, March 1973, pp. 1-25.

_____. "Stony Monday, 1849: The Rebellion Losses Riots in Bytown". *Ontario History*, September 1971, pp. 177-90.

Davidson, G. C. *The North West Company*. Berkeley: University of California Press, 1918.

Deprez, Paul. "Pentland's Scarcity of Labour and the Industrial Revolution". *Canadian Journal of Political and Social Theory*, Spring/Summer 1979, pp. 95-100.

Diamond, Sigmond. "An Experiment in Feudalism: French Canada in the 17th Century". *William and Mary Quarterly*, 3rd series, XVIII (1961), pp. 3-34.

Domar, E. D. "The Causes of Slavery or Serfdom: An Hypothesis". *Journal of Economic History*, March 1970, pp. 18-32.

Drache, Daniel. "Rediscovering Canadian Political Economy". *Journal of Canadian Studies*, August 1976, pp. 2-18.

Duncan, Kenneth. "Irish Famine Immigration and the Social Structure of Canada West". *Canadian Review of Sociology and Anthropology*, February 1965, pp. 19-40.

Eccles, W. J. "A Belated Review of Harold Adam Innis, The Fur Trade in Canada". *Canadian Historical Review*, December 1979, pp. 419-41.

Forsey, Eugene. *The Canadian Labour Movement, 1812-1902*. Canadian Historical Association Booklets, no. 27. Ottawa, 1974.

_____. "History of Canadian Trade Unionism". Public Archives of Canada, Reel M-2214.

_____. "The Telegraphers' Strike of 1883". *Transactions of the Royal Society of Canada*, Series 4, vol. 9 (1971), pp. 245-60.

_____. "The Toronto Trades Assembly, 1871-8". *Canadian Labour*, June to October Issues, 1965.

Gagan, David, and Mays, Herbert. "Historical Demography and Canadian Social History". *Canadian Historical Review*, March 1973, pp. 27-47.

Gilmour, J. M. *Spatial Evolution of Manufacturing: Southern Ontario, 1851-1891*. Toronto: University of Toronto Press, 1972.

Hamelin, J., et al. *Répertoire des grèves dans la Province du Québec au XIXe siècle*. Laval: Les Presses de l'Université Laval, 1970.

_____, and Roby, Yves. *Histoire économique du Québec, 1851-1896*. Montréal: Éditions Fides, 1971.

Hann, R. G., et al. *Primary Sources in Canadian Working Class History 1860-1930*. Kitchener, 1973.

Harris, R. C. *The Seigneurial System in Early Canada*. Madison, 1967.

Isbister, John. "Agriculture, Balanced Growth, and Social Change in Central Canada Since 1850: An Interpretation". *Economic Development and Cultural Change*, July 1977, pp. 673-97.

Johnson, Leo. *History of the County of Ontario, 1615-1875*. Whitby: County of Ontario, 1973.

_____. "Land Policy, Population Growth and Social Structure in the Home District". *Ontario History*, March 1971, pp. 41-60.

_____. "The Political Economy of Ontario Women in the Nineteenth Century". In *Women at Work, Ontario 1850-1930*, edited by J. Acton, R. Goldsmith, and B. Shepard, pp. 13-31. Toronto: Women's Educational Press, 1974.

Katz, Michael. *The People of Hamilton, Canada West*. Cambridge, Mass.: Harvard University Press, 1975.

Kealey, Gregory S. "H. C. Pentland and Working Class Studies". *Canadian Journal of Political and Social Theory*, Spring/Summer 1979, pp. 79-94.

_____. *Toronto Workers Respond to Industrial Capitalism, 1867-1892*. Toronto: University of Toronto Press, 1980.

_____, and Warrian, Peter, eds. *Essays in Canadian Working Class History*. Toronto: McClelland and Stewart 1976.

Labour/Le Travailleur, Annual 1975-1980.

Langdon, Steven. "The Emergence of the Canadian Working Class Movement, 1845-1875". *Journal of Canadian Studies*, May 1973, pp. 3-13; August 1973, pp. 8-26.

Le Goff, T. J. A. "The Agricultural Crisis in Lower Canada 1802-12: A Review of the Controversy". *Canadian Historical Review*, vol. 55, no. 1 (1974), pp. 1-31.

Lipton, Charles. *The Trade Union Movement of Canada 1827-1959*, Montreal: Canadian Social Publications, 1967.

MacDonald, L. R. "Merchants Against Industry: An Idea and its Origins". *Canadian Historical Review*, September 1975, pp. 263-81.

McDougall, D. M. "Canadian Manufactured Commodity Output, 1870-1915". *Canadian Journal of Economics*, vol. 4, no. 1 (1971), pp. 21-36.

McManus, J. "An Economic Analysis of Indian Behaviour in the North American Fur Trade". *Journal of Economic History*, vol. 32, no. 1 (1972), pp. 36-53.

Martin, Calvin. *The Keepers of the Game*. Berkeley: University of California Press, 1978.

Naylor, Tom. *The History of Canadian Business 1867-1914*. 2 vols. Toronto: Lorimer, 1975.

Officers, L. H., and Smith, L. B. "The Canadian-American Reciprocity Treaty of 1855 to 1866". *Journal of Economic History*, vol. 28, no. 4 (1968), pp. 598-623.

Ostry, Bernard. "Conservatives, Liberals and Labour in the 1870s". *Canadian Historical Review*, June 1960, pp. 93-127.

_____. "Conservatives, Liberals and Labour in the 1880s". *Canadian Journal of Economics and Political Science*, May 1961, pp. 141-61.

Ouellet, Fernand. *Histoire économique et sociale du Québec, 1760-1850*. Montreal: Éditions Fides, 1966.

_____, and Hamelin, J. "La Crise agricole dans le Bas-Canada." *Canadian Historical Association Report*, 1962, pp. 317-33.

Palmer, Bryan. *A Culture in Conflict: Skilled Workers and Industrial Capitalism in Hamilton, Ontario, 1860-1914*. Montreal: McGill-Queen's University Press, 1970.

_____. "Working-Class Canada: Recent Historical Writing". *Queen's Quarterly* 87 (1980): 594-616.

_____, and Kealey, Gregory S. "Brotherhood, Sisterhood, and the Bonds of Unity: The Knights of Labor in Ontario, 1880-1900". Paper to the Canadian Historical Association, Montreal, June 1980.

Paquet, G., and Wallot, J. P. "Crise agricole et tensions socio-ethniques dans le Bas-Canada, 1802-1812: Éléments pour une ré-interprétation". *Revue d'histoire de l'Amérique française*, vol. 26, no. 2 (1972), pp. 185-237.

_____. "International Circumstances of Lower Canada 1786-1910". *Canadian Historical Review*, December 1972, pp. 371-401.

Parker, Ian. "The National Policy, Neoclassical Economics, and the Political Economy of Tariffs". *Journal of Canadian Studies*, Fall 1979, pp. 95-110.

Pelling, Henry. *American Labor*. Chicago, 1960.

Pentland, H. C. "The Development of a Capitalistic Labour Market in Canada". *Canadian Journal of Economics and Political Science*, November 1959, pp. 450-61.

————. "Feudal Europe: An Economy of Labour Scarcity". *Culture*, September 1960, pp. 280-307.

————. "Human Adjustment to Technological Change: The Case of Manitoba Rolling Mills". Unpublished ms. Department of Manpower and Immigration, 1968.

————. "Industrial Relations Systems". *Labour/Le Travailleur*, 1979, pp. 9-23.

————. "Implication of Automation for the Employment and Training of White-Collar Workers in Manitoba". Unpublished ms. Manitoba Economic Consultative Board, 1965.

————. "Labour and the Development of Industrial Capitalism in Canada". Unpublished Ph.D. thesis, University of Toronto, 1961.

————. "The Lachine Strike of 1843". *Canadian Historical Review*, September 1948, pp. 255-77.

————. "Marx and the Canadian Question". *Canadian Forum*, January 1974, pp. 26-28.

————. "Population and Labour Growth in Britain in the Eighteenth Century". In *Proceedings of the Third International Conference of Economic History*, vol. 4, edited by D. E. C. Eversley, pp. 157-89. Munich, 1965.

————. "The Role of Capital in Canadian Economic Development Before 1875". *Canadian Journal of Economic and Political Science*, November 1950, pp. 457-74.

————. *A Study of the Social, Economic and Political Background of the Canadian System of Industrial Relations*. Prepared for the Task Force on Labour Relations, 1968.

————. "A Study of Labour Skills in Reference to Manitoba's Economic Future". Unpublished ms. Committee on Manitoba's Economic Future, 1962.

————. "Were Canadian Capitalists Different? How the West was Won". *Canadian Forum*, September 1972, pp. 6-9.

————. "The Western Canadian Labour Movement, 1847-1919". *Canadian Journal of Political and Social Theory*, Spring/Summer 1979, pp. 53-78.

Phillips, Paul. "Clare Pentland and the Labour Process". *Canadian Journal of Political and Social Theory*, Spring/Summer 1979, pp. 45-51.

————. "Divide and Conquer: Class and Consciousness in Canadian Trade Unionism". *Socialist Studies*, no. 2, May 1980, pp. 43-62.

————. "The Hinterland Perspective: The Political Economy of V. C.

Fowke". *Canadian Journal of Political and Social Theory,* Spring/Summer 1978, pp. 73-96.

Ray, Arthur. *Indians in the Fur Trade.* Toronto: University of Toronto Press, 1974.

————, and Freeman, Donald. *Give Us Good Measure.* Toronto: University of Toronto Press, 1978.

Rinehart, James. *The Tyranny of Work.* Toronto: Longman, 1975.

Rotstein, Abraham. "Trade and Politics: An Institutional Approach". *Western Canadian Journal of Anthropolgy,* vol. 3, no. 1 (1973), pp. 1-28.

Runnals, Lawrence. *The Irish on the Welland Canal.* Toronto: 1973.

Ryerson, Stanley. *Unequal Union.* Toronto: Progress Books, 1963.

Sacouman, R. J. "Semi-proletarianism and Rural Underdevelopment in the Maritimes". *Canadian Review of Sociology and Anthropology,* vol. 17, no. 3 (1980), pp. 233-45.

Saklins, Marshall. *Stone Age Economics.* New York: Aldine, 1972.

Teeple, Gary, ed. *Capitalism and the National Question in Canada,* Toronto: University of Toronto Press, 1972.

Thompson, E. P. *The Making of the English Working Class.* Harmondsworth: Penguin Books, 1963.

Trudel, Marcel. *The Seigneurial Regime.* Canadian Historical Booklets, no. 6. Ottawa, 1967.

Veltmeyer, Henry. "A Central Issue in Dependency Theory". *Canadian Review of Sociology and Anthropology,* vol. 17, no. 3 (1980), pp. 198-213.

Watkins, Mel, "The Staple Theory Revisited". *Journal of Canadian Studies,* Winter 1977, pp. 83-95.

Zerker, Sally. "George Brown and the Printers' Union". *Journal of Canadian Studies,* Fall 1975, pp. 42-48.

Preface

A fundamental problem of any society is the organization of its labour force for production. If the society is to survive as an entity, ways must be evolved or devised to maintain or increase the labour force, to determine the nature and extent of the division of its labour (the techniques to be used), and to establish and enforce the system of incentives. The solutions of these problems fasten their appropriate behaviour patterns upon the various members of each society. If a social system is to function well, the patterns of social behaviour must be self-consistent; and it follows that they are distinct from other sets of patterns that denote different social systems. However, the arrangements for production and the behaviour appropriate to them need not be entirely free of disharmonies and conflicts: the survival of production systems turns rather on whether the conflicts are within the limits of tolerance of the participants.

In primitive societies (and also, ideally, in socialist societies) the potential labour force consists of all the members of society, and the methods of production are those that these members conceive to yield the greatest mutual benefit. These societies, like others, must provide for mobilization of the labour force, for training in the chosen techniques, for ensuring the appropriate division and cooperation of labour, and for the distribution of the social product. For these purposes primitive societies appear to depend upon ritual and ceremony in a more exclusive and absolute sense than is true of other societies.

The problem of production in all other societies is complicated by the division of these societies into a ruling class, which organizes the labour force (the rest of the society) in its own interest, and the ruled or working group, whose satisfactions are a matter of expediency and of consistency with the demands of the rulers. The methods of replenishing and training the labour force, the choice of techniques and the organization of production, the system of incentives used, are now matters determined by the ruling group for those to whom they apply — the ruled group — as the interest of the rulers dictates. While the social shorthand of ritual retains a prominent role in such societies, change is facilitated by the fact that, from the viewpoint of the rulers, the conditions of production are impersonal forces. Only from this vantage-point is the logic of such systems of production apparent. On the other hand, the ruled inevitably have a viewpoint also, different from, and more or less hostile to, that of the rulers, which also has to be taken into account. Divided societies can be categorized into slave societies, feudal societies and capitalist societies, though there is obviously a good deal of variety within these general types. A large part of the interest of historians, in their studies, has been focused upon the manner in which one system of economic and social organization has superseded another in the transformation of particular societies.

The present study is concerned with the way that European society in Canada has evolved through the earlier portion of its existence up to the flowering of full industrial capitalism. It is conceived that this development involved the widespread use of a social organization that had much in common with feudal organization but which, to avoid confusion with the specific European political institution, will be designated in this study as the system of "personal labour relations"; that this type of organization was succeeded in one area after another by arrangements inspired by capitalist organization; and that these transitional forms in turn yielded to a true industrial capitalism. The particular thesis of the study is that the evolution of Canadian society depended heavily upon the state of the labour market: the quantity and types of labour available to employers. It will be of concern, therefore, to show why and how changes have occurred in the availability of labour; why and how the organizers of production in Canada adjusted themselves to new conditions in their labour market and in other respects; and why and how the workers themselves modified their patterns of behaviour.

European exploitation of Canada was begun, and the greater part of it proceeded, in the period when "commercial capitalism" was the dominant type of arrangement in Western Europe. Canadian development had necessarily to accommodate itself to that fact, and the typical system of the production of staple goods for export to metropolitan markets, which has characterized economic life in Canada and elsewhere, was the logical outcome. Commercial capitalism has been, however, a system that asked remarkably few questions about the method of production of the goods in which it dealt. It could readily integrate slave societies, feudal societies and incipient capitalist societies into its structure. It is not apparent that the European settlers of Canada were prohibited in any absolute sense from arranging production upon a basis of slavery, or upon a basis of small-scale capitalism: European settlers elsewhere had chosen such courses. But, in fact, it was a personal form of organization that was characteristic of Canada until well into the nineteenth century. The basis of the choice of method in Canada's case (and in the other cases) lay in the nature of the supply of, and demand for, labour. In Canada, there was a relative scarcity of labour, particularly of specific kinds of skilled labour, which ruled out dependence upon a capitalistic labour market. On the other hand, the nature of employment was so little adapted to slavery that that solution was never seriously attempted. In the nineteenth century a new condition, relative abundance of labour (unskilled and skilled), invited the replacement of personal arrangements by new ones that approximated more and more to the industrial capitalist methods developed elsewhere.

The tasks of this study are to delineate the personal organization of labour as it appeared in various Canadian employments up till about 1850, and to trace the transformation that produced a capitalistic labour market and a well-developed capitalistic economy, shortly after 1850. Before undertaking these studies, however, it is convenient to dispose of "Slavery in Canada".

CHAPTER I

Slavery in Canada

The Incidence of Slavery

Slavery was practised in Canada for nearly two centuries. The most striking thing about it, nevertheless, is how shallow-rooted and how limited in use enslavement was.

Slaves in Canada, in the classical meaning of perpetual bond-servants, were either negroes, brought from the West Indies and the American colonies, or "Panis" Indians. Possibly the first was a negro boy sold at Quebec in 1629.[1] By 1688 there had arisen a general demand at Quebec for the importation of negroes as a means of supplying the labour market.[2] In 1689, the colonial government received authority from France to permit the importation of negroes, and some slaves were probably brought in by 1692.[3] In 1721, again, as part of a program to promote hemp-growing in Canada, Intendant Begon sought the importation of negroes. The authorities in France approved and took some steps to provide the negroes, but it is not clear whether any actually arrived on this occasion.[4] Private persons, of course, brought in negro slaves sometimes, but the number was never large. In Quebec, it appears, there were only six negro slaves in 1716 and twenty-five in 1744.[5] Negro slaves, and the considerably larger body of Indian slaves, were used almost exclusively as domestic servants throughout Canada's history. Ownership was confined, therefore, to the few who could afford such a luxury.

The enslavement of Indians, much more common in New France than the enslavement of negroes, had appeared at Montreal by 1670.[6] "Panis", the name applied to these slaves, derived from the Pawnees, a distant and docile tribe, whose members were easily captured by the Foxes and brought by them for sale to the French.[7] However, members of a number of other tribes appeared among the Panis.[8] "There were a considerable number of panis in the colony, especially, naturally, in Montreal. A tenth of the names in the register of burials in Montreal in 1761 are those of Indian slaves."[9] The fact is the more

1

remarkable in that it was evidently the custom to release Panis after a few years of servitude, in the expectation that otherwise they would escape to the forest anyhow.[10]

The legality of negro slavery in New France, if there ever was any question about it, was settled apparently by the royal permission of 1689. The legality of Indian slavery, which was questioned at least once, was established by an ordinance of 1709.[11] The existence of freed Panis, and the additional consideration that Indians were sometimes employed as wage labour,[12] must have invited confusion concerning the status of Indians in New France. Some clarification may have been afforded by the requirement, from 1736, of a declaration before a notary when freeing Panis.[13] By the terms of the capitulation of Montreal in 1760, the inhabitants were confirmed in the possession of their slaves.[14]

Under British rule, there was a decline of Indian enslavement in Canada, but a sharp increase in the number of negro slaves. Many of the merchants and officials who came to Quebec and Montreal brought negroes with them for use as domestic servants.[15] Newcomers at Kingston, Niagara and Detroit (as in Nova Scotia) likewise often brought negroes with them. It has been asserted, also, that in the 1780s drovers brought slaves to Canada for sale, along with horses, cattle and sheep.[16] Indian tribes, too, did a certain amount of business in stolen or captured slaves.[17] A census of 1784 reported 304 (negro?) slaves in Lower Canada.[18]

The legal basis for slavery in the British period is not very clear. The confirmation of Montrealers in their possessions at least set a pattern, and an imperial act of 1790 authorized the issue of licences to new settlers in British North America to bring slaves with them.[19] Main dependence seems to have been placed, however, on the argument that Intendant Raudot's ordinance of 1709 was one of the laws and usages recognized by the Quebec Act.[20] In any case, there appears to have been no opposition to the institution until after 1790.

Slavery in Canada was disposed of quickly, nevertheless. The Upper Canadian Legislature in 1793 confirmed existing owners in possession of their slaves, but prohibited importation; provided that children born to slaves after 1793 should be freed at the age of twenty-five; and provided further that any children born to these offspring in turn in a state of slavery should nevertheless be free.[21] A negro boy in Upper Canada who in 1824 still had ten years of servitude before him as the child of a female slave was "probably...the last slave in Canada".[22] In Lower Canada, slavery was abolished, not by the Legislature, but by the courts. A bill to abolish slavery was introduced in 1793, but enjoyed little support. Various bills introduced from 1799 to 1803 have sometimes been represented also as intended to abolish slavery. In reality, they were acts to re-establish that institution, under the pretext of regulating and limiting it. For the fact was that the courts of Lower Canada refused to enforce enslavement from 1797, and that slaves quickly took advantage of this situation to end their own servitude. The slave-owners were left, as they said, without protection or recourse; but public opinion had moved far enough against them that the Legislature refused to correct the courts, so that slavery disappeared from Lower Canada about 1800 from lack of any means to enforce it.[23]

The Unsuitability of Slavery

While French and British settlers in Canada trifled uncertainly with slavery in the seventeenth and eighteenth centuries, French and British settlers in other parts of America were building the most extensive slave empires since classical times. The justification offered for the slave system in the southern colonies — the scarcity and expense of labour — was consistently echoed from Canada. There appears to have been little moral objection to slavery anywhere until about 1790, and where slavery proved useful to the ruling group, its popularity long survived that date. It is not apparent that the problems of obtaining slaves, and of retaining them, would have provided more difficulty in Canada than elsewhere. Yet the slave in Canada scarcely ever appeared beyond the kitchens of a few well-to-do households, and his departure aroused few regrets. Clearly, the Canadian environment was unsuited to the slave system of providing labour.

A ready reason why this was so, and a partially true one, can be found in the Canadian climate. The negro fared well enough in the severe Canadian climate, though the fear that he would not seems to have been the one ground for hesitation in French consideration of large-scale slave imports.[24] But the climate also confined Canada's major economic activities to a period of about six months in each year. The country's employments, therefore, fitted badly with the economics of slavery. For the economics of slavery are the economics of overhead costs: In return for the services anticipated, the slave-owner must accept a fixed burden in initial and in operating costs; these cannot be avoided without damaging or losing the asset. When the overhead costs are assumed, the services are available free of further charges, but must be regularly and fully applied to obtain the maximum benefit. The sugar planter certainly was in a better position to utilize the flow of services continually than was the typical Canadian employer — only in domestic service, indeed, was the continuous application of services readily made. It is suggestive that other colonies found slavery unhelpful the more their climates resembled the Canadian one.

It is clear that the Canadian employer faced handicaps as a slave-owner, as compared with his counterpart around the Caribbean. He had every reason to prefer a free labour market from which to draw labour as he willed, leaving to the labourer the problem of bearing overhead costs and surviving, somehow, the slack season. Climate (seasonality) seems to provide, indeed, all the reasons needed for the failure of slavery to flourish in Canada. Yet, for all that, it does not appear to have been the basic reason. However little employers might wish to bear the overhead costs of labour which they could apply only a portion of the year, they were compelled to bear those costs in leading Canadian employments until the appearance of a plentifully-supplied labour market in the nineteenth century. They were forced to undertake year-round costs precisely because labour was scarce and would not be available at the time and place wanted, if the labourer was not supported at the time he was not wanted.* Besides, there always were town handicraft employments of a year-round nature.

The inflexibility and lack of incentive tha characterize slavery seem to offer a better reason why it did not meet Canadian needs. To be sure, in the heart of a slave economy, selection and differential status and lack of other labour resources could lead to the use of slaves as craftsmen.[25] Slaves enjoyed a surprising variety of employments in New England, possibly because the ubiquity of the slave trade made it easy for a variety of people to have them.[26] Typically, however, the slave was utilized for field crops in ways that required a minimum of skill and permitted a maximum of supervision. Slave labour was notoriously unproductive — a common estimate was that the slave did one-third the work of a free labourer[27] — so, to make slavery pay, the slave had to be used where he stood at the least disadvantage. When he was used for work that required any degree of skill or initiative, the reason ordinarily was that free labour was unavailable. Even in the Southern States, free workmen replaced slaves in semi-skilled employments when free men were available.[28]

Now, whatever else might be said of Canada's employments — the work of the voyageur and bateauman, the lumberman and the raftsman, the artisan and, indeed, the farmer — they involved a considerable degree of skill and, still more, they required a high degree of initiative, independent judgment and mobility. Rarely could supervision substitute for the workman's own willingness to produce. Even with free labour, employers were continually pressed before the nineteenth century to establish mechanisms that would induce assiduity beyond range of the master's voice.

---

*Pentland's argument has been independently formalized by Domar. See Introduction, p. xxiii. Ed.

If slavery could ever have flourished in Canada, it should have flourished much more readily in Pennsylvania. But it is clear that Pennsylvania's abolition of hereditary slavery in 1780 rested upon the thorough unsuitability of the institution, as well as upon revolutionary and Quaker sentiment.[29] In New York — unsentimental New York — "many masters have found it for their advantage (1804) to promise their slaves liberty on condition of serving a few years under indenture — this the young are anxious to accept, and the hopes of freedom makes them better servants in the meantime."[30] New York already had enjoyed the premonition, if not yet the realization, of a well-stocked labour market. Canada might hope for one also before long. But even before a satisfactory labour market appeared, it seems plain that slavery was uneconomic in Canada.

The Aftermath of Negro Slavery

The abolitionist movement in North America at the end of the eighteenth century was, on the whole, a practical, unsentimental and local affair. Communities judged that slavery did not suit them, but did not profess to say how the case stood in other parts. But a new kind of abolitionist movement was already arising and was to become more and more powerful and vocal through the decades. This movement was concerned with the sins of others, and its zeal was in proportion to the distance of the evil from the observer. In England, abolitionist agitation made easy the destruction of the privileges of the West Indian planters. In the United States, it was a valuable aid in the drive for the tariff protection and free labour market conducive to industrial capitalism.

One consequence of abolitionist activity in the United States was the northward movement of escaped slaves, and of negroes in general. Particularly as slave-owners sought to recover their property from any part of the United States, slaves who escaped were inclined to seek sanctuary in Canada, usually by crossing the Detroit River frontier. Before 1835, there were very few negroes in Canada,[31] but the influx after that date provoked a violent anti-negro agitation, not unmixed with sympathy for slavery, in the 1840s and 1850s.

The negroes who came to Canada concentrated particularly in the Western District of Upper Canada, the region along the Detroit River. It was the part of Canada reached first — perhaps its climate was more agreeable than elsewhere in Canada — and it soon had a number of "colonies" of negroes that became the objectives of those in flight.[32] The negroes congregated, secondly, in the cities and towns from Toronto westward, and thirdly, in nearby rural areas.[33] How many of them there were is a matter of debate. The official count (taken at the behest of the agitated Western District) found that there were about

4,200 negroes in Canada West in 1842, and 5,500 in 1848.[34] Perhaps there were about 8,000 in 1852;[35] but much higher estimates were frequently given.[36]

Anti-negro sentiment flourished in proportion to negro numbers. It was most violent by far in the Western District, and its pervasiveness there is apparent by 1840.[37] During the next two decades, significant ways were found to express it. One was a periodic effort by magistrates and constables to return escaped slaves to the custody of their masters in the United States.[38] Another consisted in the imposition by magistrates of savage sentences on negroes for petty offences.[39] A third was the expulsion of negro children from the common schools. This occurred not only in the Western District,[40] but at St. Catharines in 1842[41] and at Hamilton in 1843.[42] The provincial school authorities saw no harm in meeting this difficulty by the provision of segregated schools,[43] a device upheld in the courts by a marvellous stretching of the law that authorized separate schools for Roman Catholics.[44]

Landon noted instances between 1849 and 1851 of the exclusion of negroes from juries, township meetings and the Sons of Temperance.[45] But there was nothing new about this sort of treatment.[46] What the publicity of the 1850s seems to mark is a greater public sympathy for the negro, and advances by the negroes to exploit it. Abolitionist agitation in the northern States, and the ease with which employment could be found in the 1850s, may explain the milder white attitude. By 1853, a protest from the Western District against the entry of negroes included the proviso that the petitioners abhorred slavery[47] — a sentiment conspicuously absent from many, if not all, earlier communications. By 1856, magistrates who tried to return an escaped slave to the United States were summarily dismissed,[48] whereas their counterparts a decade before had escaped with a reprimand.[49] Presumably the negroes sensed a more favourable climate: it was in the 1850s that they found courage to challenge segregated schools in the courts, though the segregation was of long standing. And in 1858 a negro mob, in the heart of the Western Disctrict, was bold enough to remove a slave forcibly from his American master.[50]

It could be said, and was said, that anti-negro feeling was strongest in the districts settled by Americans, and least in the districts settled from Great Britain.[51] The settlers of American origin, favouring a lower-middle-class equalitarian democracy, may have suspected a desire to reduce them to the level of negroes in an hierarchical society. The less affluent British settlers were likely to fear grasping Americans more than aristocratic Britons or, probably, negroes. But, of course, it was in American districts that the negroes settled. Perhaps anti-negro feeling would have been as violent, or more violent, in British districts, if the negroes had settled there.

The situation in the towns is clearer. The affluent and secure did not greatly care whether their inferiors were black or white. It was "the middling and lower classes of the whites" who drove the negroes from the common schools.[52] The greatest antagonism was found among "the French-Canadians and Irish"[53] — those who competed directly with the negroes for the poorest jobs — as was the case also in the United States.[54] But the relatively secure artisan class probably shared in some degree the feelings of contemporary American workers: no sympathy for negroes; indifference to the slavery question; detestation of upper-class philanthropists who excited themselves over the alleged miseries of far-off slaves, but who did nothing to relieve the wage-slavery on their own doorsteps.[55]

The Western District, with its extreme and unanimous hostility to negroes, invites particular attention. The District differed from others in the mildness of its climate and the nature of its crops: perhaps slavery would have suited it.[56] It felt, with justice, that labour was always more scarce and expensive in the District than anywhere else in Canada.[57] The inhabitants, then, might have been expected to welcome an ample supply of negro labourers. And there is something to support this view of them. The fact is that the negroes of the area were not commonly forced upon the labour market, but were insulated from it in the colonies established by missionaries and philanthropists. And the colonies drew a very large share of the abuse originating in the District; whereas, if the white inhabitants had merely been anxious to have negroes out of their sight, they should have welcomed the settlement system. But, on the other hand, negro employment was not much encouraged — indeed, farmers were inclined to agree among themselves to employ no negroes whatever.[58]

The inhabitants' case, as they stated it themselves, was that the negroes frightened off the British artisans whom they wished to settle among them. The negroes were a shiftless, immoral lot, addicted to petty crime, avoided as neighbours by immigrants. Above all, artisans were unwilling to settle where they faced the prospect of negro competition.[59] The arguments have limited merit, but they do suggest that the typical citizen was a simple yeoman who aspired to the ownership of a farm rather than of a plantation, and wanted a free (white) labour market.

The negroes regarded Canada as a temporary rather than a permanent place of residence, at least in the early years.[60] They showed a strong desire to abandon its unfriendly atmosphere in favour of a more hospitable environment.[61] The outcome of the American Civil War cut off negro immigration and led many negroes to return to the United States.[62] There was left only a negro residue, and a persisting hostility to it. But this encounter with slave society did serve to bring out the position of various groups in Canadian society, and to show the strength of the forces making for a free labour market, with a labour force undivided in respect to race.

Indentured Servants

The typical immigrant to the Caribbean region before 1800 was a slave, but the typical immigrant to more northerly colonies was an indentured servant. The indenture system played a predominant part in the early peopling of the United States.[63]* In the seventeenth century, it provided a similar service for New France, but its place in the eighteenth century was much reduced. The reason was that France experienced labour shortage in the eighteenth century (as against a surplus in the seventeenth), so that workers for Canada had either to be deported by force or beguiled with contracts generous far beyond the typical indenture.[64]

Barring kidnapping (of which there was probably a good deal), the indenture system required a European so despairing of the Old World and hopeful of the New that he would contract his labour for a period of future years in return for a passage to America, his keep and, perhaps, some stated amount of wages. It required a prospective employer willing to pay a substantial sum for this sort of servant. And it required, of course, a vessel master willing to do a business in immigrants. The indenture form resembled that of an apprentice, but the servant usually was an adult who made his own contract and possessed already whatever skill he would ever have.[65] Economically, the indentured servant accepted the position of a slave — a slave with a termination date to his servitude, but a slave, nevertheless.[66] The master likewise assumed the rights and duties of a slave-owner. He had to assume the initial and overhead costs of the servant, and it was his obligation to find employment for the servant that would make the investment pay. Nevertheless, demand was brisk in the eighteenth century, and the indenture traffic was taken over by enterprising specialists who provided the capital, enlisted the servants and auctioned their human stock in American ports.[67]

Despite its early popularity, the indenture system was falling into disfavour by 1800,[68] and the regular large-scale traffic in servants failed to survive the glutted market of 1819.[69] While this glut, which forced the indenture entrepreneurs to the desperate expedient of releasing the servants from their indentures, was the result of depression, it also signified something more fundamental and lasting: the

---

*Pelling, summarizing American evidence, estimates that one-half of immigration to pre-revolutionary America came as indentured servants. See Henry Pelling, *American Labor* (Chicago: University of Chicago Press, 1960), p. 7. Ed.

appearance of a well-stocked free labour market in America. The main factor in stocking that market was the sharp drop in ocean fares as shipowners sought return cargo for the rapidly-mounting volume of bulky goods shipped from America to Europe. The willingness of English ratepayers, Irish landlords, and friendly societies to pay the lowered fares for poor emigrants also contributed. Increasingly, the American employer found his convenience best served by hiring labour when he happened to want it, rather than retaining a supply that was often unusable. There remained, to be sure, one advantage to the indenture system that continued to fascinate a great many prospective employers: the ignorant and defenceless European could be induced to sign a contract at wage rates far below those in America. But as the free market and the area of settlement expanded, this apparent advantage increasingly proved to be illusory. The servant soon learned enough of wage rates in America to consider himself the victim of sharp practice. He could also observe that the settled area of America was amply large to lose oneself in. The servant had two effective courses open to him. One was to make himself thoroughly disagreeable. "Captain M. [Miles MacDonell] in returning from Scotland in 1790 — brought some indentured men — bound for 4 years — but they became discontented and so troublesome that he gave them up most of their time — they had nothing by their bargain to expect at the end of their time."[70] The other course was to desert. "Several hundred were brought over in this way in 1829 by the builders of the Chesapeake and Ohio Canal. No sooner did they land than many of them deserted and the agent went again to Liverpool in search of others."[71]

These hazards were enough to keep small employers from trying to import indentured labour. They preferred to agitate for more immigration, subsidized if possible from Great Britain. From the 1820s onwards, unskilled labour was usually in good, and often in excess, supply in Canada. If an employer was willing to pay good wages and to make his wants known,[72] he had a good prospect of obtaining a full (though not very dependable) labour force.[73] Indentures in the old style survived only in such a specialized field as the importation of pauper children as apprentices.[74] Yet the fascination of the system is illustrated by various grandiose schemes that were proposed as the nineteenth century struggled to clarify its thinking about colonies and emigration and labour, while the inappropriateness of the system in the new age also emerges from them.

The most comprehensive of the nineteenth-century schemes was the colonization plan of Edward Gibbon Wakefield. It was, among other things, a plan to introduce indentured labour into the colonies on a gigantic scale.[75] Wakefield knew, as others sometimes did not, that the indenture system continually failed from the employer's viewpoint because of the alternative opportunities available to the servant. Hence, he proposed the best-known feature of his scheme, pricing land beyond the reach of the poor, so as to leave the servant no choice but to remain with his employer — or, at any rate, some employer — for several years.[76] Wakefield did not explain, however, how his plan could be made to work in Canada, while cheap land remained readily available across the border in the United States.

There were many other plans. One was proposed by Robert Gourlay in 1822.[77] It depended on public works designed to improve the St. Lawrence, then the subject of considerable discussion. Gourlay proposed bringing out 5,000 men from the United Kingdom to dig the canals, under contracts or indentures by which the costs of their passages would be deducted from their wages, and another substantial part of the wages evidently withheld. But Gourlay proposed to pay going wages and to offer year-round employment. He calculated that his immigrants could be released after about eighteen months, with ten pounds each in their pockets, to settle or to seek other work. They would then be replaced by other labourers brought out in the same way.

The promoters of the Great Western Railway, in the course of the early difficulties of that railroad, put forward a plan of this sort in 1847.[78] It was suggested in part by the famine in Ireland. Spokesmen for the Great Western professed to be deeply concerned for the unhappy famine victims, and they said, truly enough, that only railroad construction could offer large-scale employment for them in Canada. All that the Company wanted was a loan of £800,000 sterling at no interest for the first two years. With this money, it would bring out 10,000 labourers with their families (50,000 people altogether) and purchase up to one million acres of land from the Province. The labourers would be imported under a contract providing that their wages, over and above subsistence, would be held and applied (1) to pay for their passages and (2) to buy land. The Company considered that the labourers would stick by their contracts rather than forfeit their land claims, and thus would be retained in Canada instead of going to the United States. Though the labourer would thus become "the consumer of British manufactures, and a producer of surplus agricultural products...", the British government (which was looked to for the loan) said "No."

All the same, the Colonial Office at this time had a lively interest in schemes involving Irish emigration and indentured labour. Despite Wakefield's opinion that Lord Grey was no friend of his projects,[79] Grey declared his sympathy for Wakefield's ideas;[80] and showed that he was interested in plans for large-scale migration. One plan discussed in 1848 was to build a Halifax-Quebec railroad employing no less than 24,000 "wild Irish of the Western Counties" enlisted in semi-military labour units for seven years.[81] What seems to have been a variant is "Colonel Tulloch's Plan for the formation of a Corps of Military Labourers in the Colonies" to be used in building roads and bridges.[82] These plans envisaged holding labourers to a low level of wages and inducing them to purchase land in Canada.

London demonstrated a remarkable attachment to the notion that immigrant labourers could be made to invest heavily in land. A great part of the investment in Canadian lands had, indeed, been made by immigrants with savings from their wages, and with obviously beneficial effects. If nearly all immigrants could be made to buy land, a much greater share of British emigration could be saved for the British colonies instead of ending in the United States. And the further thought keeps peeping out, that somehow through this mechanism, Canadian public works could be paid for by the reserves of unalienated land.[83] So the schemes involved paying off labourers with land instead of money, despite objections argued from Canada.[84] On the other hand, the Colonial Office appreciated that men indentured at inferior conditions in Canada would desert to the United States. This seems to have been the reason for planning military units: presumably it was expected that military discipline and sanctions could force labourers to observe unattractive contracts.[85] The expectation was not shared in Canada. The Canadian government (R.B. Sullivan) argued that if the men in these units could not escape, they would likely use violence and become dangerous to the country.[86] Whether or not the objections were appreciated, the plans were singularly unsuited to the labour market conditions that existed in 1848, and could hardly have produced anything but trouble if they had been tried.

What the labour market permitted and sometimes encouraged, however, was a very much milder form of the indenture system, the "contract labour" system. Just as the indenture system flourished while there was a real need for it — that is, while its purpose was to overcome a genuine shortage of labour, and not merely to cut wages — the contract labour system seems to have answered well when used to meet a real labour shortage. In the nineteenth century in Canada, periodic shortages of skilled labour could occur, even though labour in general was plentiful. Such shortages were likely when a new economic activity of unusual type or size was undertaken.

New France in the eighteenth century provides a striking example in miniature of the problem of skilled labour shortage and the use of contract labour to overcome it. It does not seem to have been very difficult to obtain unskilled labour in the colony. Skilled labour, however, was another matter. If the skilled labour was wanted for an employment not hitherto carried on in New France, it was entirely probable that not one practitioner of the craft would be found.[87] The lack of pressure in France, or of attraction to the colony, made it improbable also that skilled workers would emigrate on their own initiative, even if they knew of opportunities in New France.[88] The only way left to obtain skilled workers was to bring them from France, under contracts very favourable to the workmen. Whether it is the really substantial industries that were made to flourish in New France (the shipyards and ironworks), or ephemeral ventures like ropeworks and brickworks, the records tell of importations of skilled workers to start them. The industries that persisted required continued importations to keep them going, still under contracts at high wages.[89]

In the nineteenth century, it was railroads that were most likely to occasion a sudden large demand for labour, including skilled labour. A railroad company was in a position to undertake large-scale importation of labour to satisfy its own needs; and its monopsonistic position assured that the labour it brought out would very probably continue in its employment, given reasonable conditions.[90] Railroads were, in fact, very prominent importers of contract labour into Canada. In 1854, the Grand Trunk Railroad undertook to bring out to Canada no less than 4,000 construction workers, engine drivers, engine fitters and similar artisans, and labourers. "Constant employment has been guaranteed by the contractors to steady men for five years. The passage money of those who cannot pay it, as well as of their wives and children, is defrayed for them, on condition of the men being under stoppage of a shilling a day each, until the debt is liquidated."[91] The Great Western Railway earlier brought out men, apparently about one-third of its staff, but possibly more — probably under a similar arrangement.[92] The Canadian Pacific Railway took the same course to staff its Angus Shops in Montreal in the 1880s.[93]

Though these arrangements involved the importation of workmen under contract, as did the indenture system, the obligations imposed on employer and on employee were very much lighter. While the employers perhaps used the system to lower wages, or to prevent their rising, they do not appear to have been high-handed in this respect. Actual (absolute) shortages of skilled labour provided their primary motivation.[94] The men brought out by the railroads seem to have thought themselves well treated, and the system worked satisfactorily.

There may, on the other hand, have been a good deal of importation of contract labour by smaller employers in which the motive was to get very cheap labour. Canadian wage earners, at any rate, expressed that opinion. Of course, they did not in any case wish to have their bargaining position weakened by new competitors. Organized labour in Canada, from the earliest opportunity, stated its strong opposition to contract labour, as it did also to immigration subsidized by the federal or provincial governments. The employer interest in Canada was, however, much too strong to allow any success to the demand that the importation of contract labour be made illegal, and this was still one of the unrealized goals of the labour movement in 1902.[95]

Convict Labour

Edward Gibbon Wakefield was acutely conscious of the fact that modern slavery arose from the shortage of labour in new countries.[96] He classified modern slaves as consisting of (1) negroes, (2) indentured servants and (3) convict labourers.[97] He was particularly concerned over the use of convict labour in Australia, and gave a striking description of the slave economy which had been able to develop in Tasmania because of the availability of convicts.[98] He noted that those who profited by the deportation of convicts from Great Britain sought the continuance of their labour supply, while other colonies not yet favoured begged to have convict slaves sent to them.[99]

Canada, during the French period, had considerable acquaintance with deported convict labour. This developed from the circumstance that, in the eighteenth century, France was short of labour relative to demand, so that the voluntary emigration to New France of the preceding century was not continued on any scale.[100] After 1715, the failure of indentured servant schemes to provide much labour for New France was counteracted increasingly by the deportation of convicts. In the 1720s, the convicts were mostly petty criminals. They were to be bound as indentured servants to masters in New France for five years. However, these newcomers proved to be far more adept at crime than at work, and the outcry that resulted led to the substitution of other types of criminal. Already in 1728 there had been sent prisoners convicted of "minor crimes such as poaching and wife beating".[101] And from 1729 to 1743, prime reliance was put upon the *faux saunier*, the person who had been convicted of evading the salt tax, but who, for all that, was likely to be sober and industrious. This type of convict labourer was eagerly sought in New France in the early 1730s, when the country experienced a decided boom. The smaller numbers imported after 1733 appear to reflect crop failures and poorer business conditions rather than dissatisfaction with these convict labourers. War ended this traffic from 1744. The supply of *faux sauniers*, about six hundred altogether, could not have had a very pronounced effect on the labour market. Moreover, many of the convicts, apparently under the slightest of supervision, soon left the labour market in one way or another; and these immigrants were rarely capable of meeting New France's really acute need for skilled labour.[102]

Convict immigrants played no part in Canada's labour market after

1760. Nor, for eighty years, did local convicts have any notable share in it. There were indeed criminals, but the tone and circumstances of the age were inimical to the use of their labour for commercial purposes. Though gaols were crowded, they could not be said to be heavily-populated, so that the desirability of regular routine and activity did not suggest itself. Neither did employment seem appropriate for those who suddenly and temporarily overflowed the gaols after riots and political disturbances. The public's attitude was one of detach- ment. The citizen was reluctant to pay for the care of convicts, and he was indifferent to their miseries; but he was not interested, on the other hand, in imposing any particular order upon them. As for gaol- ers, they thought they had their hands full preventing escapes, and saw no reason for multiplying their difficulties by unnecessary inter- ference with the prisoners' existence.[103]

In the second quarter of the nineteenth century there was in Can- ada, as elsewhere, a drastic change in the prison order. It developed both from new public attitudes and from matter-of-fact material cir- cumstances. With respect to the latter, the growth of population and of cities involved a rapid increase in the number of convicts. The gaols were more and more overcrowded. And as time went on, most of them were more and more tumble-down and easy to escape from. Additional prison room had to be provided somewhere. But the heavi- ly crowded condition of the gaols drew the attention of gaolers and officials, at least, to something else: the desirability of some means of segregating different classes of inmates. Female prisoners should, of course, be segregated from males.[104] Insane persons in the prisons created exceptional difficulties and, from the 1830s, there were ef- forts in Upper Canada to segregate them in an "asylum" under medi- cal care.[105] Finally, there was appreciation at least from the 1830s of the distinction between minor offenders on the one hand and "hardened criminals" on the other. To segregate these, and relieve pres- sure on the gaols,[106] Upper Canada accepted the necessity for a peni- tentiary. The penitentiary was built near Kingston in 1834 (though a good deal of the original plan was not followed for reasons of econ- omy), and the first prisoners were received on June 1, 1835.[107] By 1836, there were over eighty convicts, and about the same number appeared in each of the next few years, except for the crowding occa- sioned by the Rebellion of 1837.[108] After the Union of the provinces, the penitentiary received convicts from the eastern province also (1842), so that it was crowded to its capacity of 150. A program of en- largement was pursued through the 1840s.[109]

The penitentiary was more than a product of administrative neces- sity, however. It was a mark of a new spirit of puritanism and efficien- cy associated with capitalist progress. Even before the construction of the penitentiary, that *bona fide* representative of the pushing new age, William Lyon Mackenzie, was demanding that the prisoners of the Toronto Gaol be segregated and that they be put to useful work.[110] From the beginning, those appointed to supervise the penitentiary were keen students of "prison discipline" — penology is an approximate equivalent — which implied that prisoners should be, and could be, reformed. They followed contemporary discussion and experiment in the United States with close attention.[111]

From the doctrines of prison discipline, but even more, from the difficulty of getting money from the Legislature to run the penitentiary,[112] the Commissioners were anxious from the beginning to put their charges to remunerative work. Workshops, part of the original plan of the penitentiary hitherto neglected for reasons of "economy", were built in 1839 by convict labour,[113] and from this time on some measure of craft training was provided in the prison shops. In 1839, the penitentiary was prepared to hire out to private employers an imposing variety of workers: blacksmiths, tinsmiths, stone cutters, masons, lathers, carpenters, painters, tailors, shoemakers, ropemakers, labourers and quarrymen, seamstresses, cooks, and nurses.[114] In practice, the most important commercial product of prison labour in 1839 was shoes.[115] In the same year, the penitentiary embarked on rope making,[116] an internal manufacture carried on pretty regularly thereafter.

What were the artisans of Canada, and of Kingston in particular, likely to think of this competition offered by convict labour? Elsewhere the energetic commercialization of convict labour brought quick protests from free workers, and a long vendatta followed. In the United States, by 1842, "the competition of prison labour had reached the point where, for the first time, an organized protest of considerable weight was made." This, a veritable labour movement in its own right, started in Buffalo and reached its peak in 1847.[117] If workmen had not been greatly disturbed in the same period over the competition of new machines and of cheap immigrant labour, the movement against convict labour might have been still more powerful. But a petition of Kingston mechanics (carpenters and joiners) of 1842 is against the displacement of artisans by machinery, not against convicts.[118] At the end of 1844, Kingston shoemakers petitioned against the competition of American shoe imports, "the larger portion of which as they believe is made up in the Penitentiaries of that Country, by the labor of the Convicts confined therein...", but Canadian convict labour was not mentioned.[119] A similar petition at the same time from Belleville shoemakers against American leather imports failed to mention convict labour in either country.[120] These petitions were, in fact, part of the general demand for protection and encouragement of native industry in Canada in the 1840s, and they were successful in winning discriminatory duties against American leather products.[121] It was not until 1850, fifteen years after the opening of the penitentiary, that the Kingston craftsmen formally complained of the competition in their local market of convict-made shoes, clothing, chairs, ironworks and other items.[122]

This long delay might suggest that the Canadian mechanic did not fear the competition of convict labour, or that he was singularly obtuse. But the facts reveal a different explanation for this long silence. For fifteen years the (more or less Conservative) management of the penitentiary had quietly arranged to minimize the amount of actual convict competition on the labour market in return for the tacit agreement of the artisans to accept what competition there was.

Back in 1836, when the penitentiary had barely opened, James Nickalls was recommended as President of the Board of Penitentiary Commissioners (and was subsequently appointed) because:

1. He is in a manner the Head of the Mechanics — i.e. they all look up to him for advice in their affairs.

2. Having a Mechanical turn he would be exceedingly useful to the Board, in devising proper modes of employing the Convicts so as to render their labour productive, and at the same time, as little offensive and injurious to the mechanics as possible.[123]

In the same year an investigation was undertaken into the feasibility of moving the penitentiary to the elaborate iron and other works erected at Marmora in the 1820s and then abandoned. No doubt there was a wish to realize something from the equipment and buildings at Marmora, and to make Upper Canada an iron producer, but the major motive was frankly stated: "the intense anxiety of feeling which prevails among the Mechanics of the Province, whose interests it is apprehended will be unfavourably affected by the present Penitentiary system, if continued."[124] The difficulties of the move to Marmora were considered too formidable, however, and the project was abandoned.

The inauguration of rope manufacture within the penitentiary in 1839 had the same object of avoiding direct competition with free artisans.[125] Unfortunately, the rope produced proved to be hard to sell.[126] The real solution found for the problem seems to have been internal construction in the penitentiary using convict labour. Despite the other kinds of employment mentioned, it appears that most of the prison labour was used in this way in 1839.[127] In 1842 also, when the arrival of Lower Canada's convicts invited expansion, the labour was used largely for internal construction.[128] In 1847, the convicts were still building away.[129] It looks very much as if the penitentiary authorities kept their charges to internal construction work throughout the 1840s, and that they judiciously avoided hiring out more than a token force of convicts to compete with free labour.

This policy satisfied Canada's free workmen and probably contributed a good deal to the solidity of the Conservative hold on the Kingston region. No doubt those in charge of the penitentiary could have gone on indefinitely devising uses for convict labour that did not conflict with free workmen. Their policy was overturned, however, by the rise of a powerful "reform" sentiment: a general "reform" movement in Canada and elsewhere associated with the development of industrial capitalism, and the specific Clear Grit movement of Canada West.

Increasingly through the 1830s, 1840s and 1850s, Canadians were adopting the pushing puritanical attitudes appropriate to industrial capitalism. The demand of middle-class groups for exclusive political power, the interest in mechanical inventions, the spreading gospel of efficiency, were aspects of this development. Another, and a good barometer of the spread of industrial capitalist attitudes, was the rise of the temperance movement.* Only an occasional Temperance Society appeared in the 1830s,[130] but in the 1840s there was a widespread campaign to render the labouring classes more industrious and obedient by removing liquor from their reach.[131] When a profitable cash market for grain developed in the 1850s, there appeared a really powerful temperance movement, reaching peaks (if one may judge from the avalanches of petitions directed to the province's administration) in 1852 and 1856.[132] Temperance agitators were at pains throughout to claim a close relationship between intemperance and crime and the large prison population.[133]

Prison administration was not exempt from the new interest. In the earliest days of the penitentiary there had been "reform" discussion among the officials. Methods of "discipline" were viewed in terms of their effectiveness in making prisoners, upon their release, into honest and industrious persons. In 1839, the Penitentiary Inspectors observed, acutely, that among philanthropists in Europe and America the subject of discipline had got on much faster than that of reform. They thought reform was likely to be effective only with a probation system to watch over prisoners after their discharge. To promote the attractiveness of the straight and narrow path, they suggested withholding part of the earnings of prisoners, and paying the money in

---

*For a more extensive treatment of the temperance agitation in Canada West in this period, see Leo Johnson, *History of the County of Ontario, 1615-1875* (Whitby: County of Ontario, 1973), pp. 215-22. Ed.

instalments after discharge. Following the prison chaplain, it was proposed that sentences not shorter than three years nor longer than seven years were most likely to make prisoners receptive to reformation.[134] The Inspectors reflected the new age in their view that men's behaviour was conditioned by their environment, that prisoners could be reformed; and in their desire for efficiency and economy, by avoiding recidivism. But these were rather private and practical discussions among relatively informed and responsible men.

But now another, public sort of interest in prison reform arose. There came first, in 1846, an outcry against the whipping of women prisoners. The Warden explained that women were rarely beaten with a rawhide whip, being punished usually by confinement in a dark cellar on bread and water for 24 or 48 hours; and that male prisoners were usually whipped with a rawhide whip, rarely with the cat.[135] But the Executive Council appeared to doubt that the public would be satisfied with this explanation, for in January 1847, it limited and regulated the methods of corporal punishment at the penitentiary.[136] At the same time, more attention was paid to the prison diet, and a hospital within the penitentiary was planned.[137] The unhappy prison officials, accepting the proposition that the lash "is repugnant to the feelings of the Public", substituted "the Box", a form of confinement originated in the penal establishments of Tasmania, despite the doubts expressed by Governor and Council. The officials pleaded that they had to have *some* method of punishment.[138]

The "public" next, in 1849, subjected the Executive to a barrage of petitions for the abolition of capital punishment, as ineffectual, demoralizing, possibly applied to the wrong persons, and "opposed to the spirit of Christianity".[139] The sentimentality in these attacks, another symptom of the age, should not obscure the practical goal of efficiency in keeping prisoners alive, healthy and well-fed, and free from physical abuse, not only so that they might be efficient workmen to supply the labour market upon their release, but also so that they might make a similar contribution during their time of servitude.

Until 1849, these attacks had made little impression on the position of the entrenched prison authorities; but in that year the most ardent spokesmen of the new era realized their ambition to give the penitentiary a thorough overhauling. A special Commission of Inquiry was established, with George Brown as secretary. Its report sorrowfully pointed out that segregation of different classes of criminals had still not proceeded far in Canada; that as a consequence, young offenders who might have been saved, were led by evil companions to a life of crime; and that the penitentiary, the "one penal Institution of which the aim is reform", had not in practice contributed much to that cause. The reason for this was that the penitentiary relied upon the negative powers of severe punishment, rather than on measures to lead the prisoners to a better attitude.[140]

There was good sense in this. The financial saving from reducing recidivism was particularly likely to appeal to bustling Reformers. But the humanitarianism expressed seems forced and artificial. Beneath the facade, more practical and immediate objects indeed were in view. One of them was to blast the Conservative officials out of the penitentiary, and generally upset the happy family relationships by which the Kingston area was kept a Conservative stronghold.[141] But an even more important object, it may be suspected, was to force prison labour on the market to the greatest possible extent. If the penitentiary could not be made to pay, at least the burden on the taxpayer would be lightened; and, at the same time, wages would be forced downward on the free labour market. These advantages could hardly fail to recommend "reform" to George Brown. Early in 1850, Brown was able to write an editorial congratulating the new Commissioners upon their efforts to introduce the "new spirit" and their success in having cut costs by 35 per cent.

> The Inspectors are entitled to some credit also for the introduction and establishment of the contract labour system, often and again, but unsuccessfully, tried to be commenced by their predecessors. A contract for 100 convict shoemakers is now in operation, and another for 50 cabinet-makers, for whose use costly machinery has been erected by the contractors. A further contract for 50 tailors has just been signed, and one for 50 blacksmiths is now negotiating... [142]

Presumably this exhausted the able-bodied convicts, and Brown informs us that the new administration was also seeking employment for the semi-invalids. The same editorial expressed pleasure at the decrease in the use of corporal punishment, which, of course, might have interfered with employment.

Reform had accomplished its object — it is significant that the "public" sent in no more petitions about prisons, though it found plenty of other causes to uphold in the 1850s — and contract convict labour was to be a feature of Canadian life for many decades. Nobody thought it necessary to ask what the convicts thought of the system. But the new order was not established without an opposing petition from several hundred "Inhabitants of Kingston". The petition was directed "against the teaching of trades at the Provincial Penitentiary"; and since this teaching had been undertaken in the name of reform, the petitioners felt it necessary to declare that the "reformation" anticipated from trades' training was "one of the philanthropic delusions of the day".

But what the Kingston mechanics particularly objected to was that

> the Convict labour being hired out to contractors for the manu-
> facture of such articles as are in more general domestic use, estab-
> lishments for the disposal by retail, within this City, of Convict
> made Boots and Shoes, ready made Clothing, Chairs, Cabinet-
> makers ware, Blacksmiths Work, and various other articles are
> now in full operation; this has had the effect of driving a large
> number of our Mechanics from the City and neighborhood, by
> lowering the price of articles manufactured ... to a price at which
> they cannot be produced by an honest Mechanic, having a family
> to support ... [143]

The petition was, of course, ineffectual. Brown had assured the
continuance of the new system by entering into contracts for prison
labour for five years; and in 1851 the main contract was extended to
1860. The Conservatives found that the extension had been effected
without inviting others to bid; they calculated that the contract was
too favourable to the contractor by $150,000; and they wondered how
much George Brown had been paid by the beneficiary.[144] But it is
doubtful whether they were anxious to abandon the new system itself. It is doubtful, too, whether in the booming years of the 1850s
even the Kingston mechanics felt convict competition very severely.
But the same boom permitted mechanics to express their feelings with
relative impunity. In the fall of 1853, masons building a bridge across
the Humber River for the "Guelph railroad" struck against the use of
stone worked by convict labour.[145] It is interesting that others besides mechanics could find their interest affected by convict labour. In
1856, steam power for manufacturing was installed in the penitentiary. One result was a set of petitions, circulated in various parts of
Canada West and signed by merchants and manufacturers who
owned steam engines. The petitioners asserted that penitentiary
competition with their businesses was unfair, and stated that such
competition was prohibited in the United States.[146]

When Canadian workingmen achieved a sufficient degree of organization to provide a means of expressing their common sentiments, they
took up the convict labour question vigorously. Almost the first act of
the first meeting of the Canadian Labour Union, in 1873, was the
appointment of a Committee on Prison Labour.[147] The Committee
denounced the policy of the Inspector of Prisons: "simply to make the
prisoner self-sustaining during the period of his incarceration". They
protested especially against teaching trades to convicts "as in a great
measure placing a premium on crime, which system unscrupulous men
are not slow in taking advantage of for the aggrandizement of their
own pockets and to the detriment of the honest labourer."[148] At each

succeeding meeting of the Canadian Labour Union — that is, annually to 1877 — similar resolutions against convict labour were passed, a special vehemence being contributed sometimes by Kingston delegates.[149] The successor organization, the Canadian Trades and Labour Congress, took up the work in its turn. Whereas it had been proposed in the 1870s to confine convicts to road work, or to require articles made by them to be sold at the outside price, the Congress suggested use of the union label to direct consumers away from convict-made goods, and exclusive employment of convicts to make articles required by government offices.[150] However, as the perennial resolutions indicate, organized labour was entirely unsuccessful in curtailing the commercialization of convict labour in the nineteenth century.

But there is one modification required to this record of failure, and one other episode in the history of convict labour that merits recital. It concerns the Toronto Trades Assembly, organized in 1871, and mainspring of the Canadian Labour Union which the Assembly organized in 1873. Nearly a year before the first meeting of the CLU, the Toronto Assembly dispatched a delegation to remonstrate with the Ontario government over a decision to contract the labour of the province's Central Prison to the Canada Car Company for fifteen years. Meetings with the Attorney General failed to alter the plan to hire out Ontario's prisoners, and the Assembly planned a mass meeting of protest.[151] But then, something odd happened. The meeting was postponed; and when it reappeared, it became a mass meeting to discuss a Mechanics' Lien Law, to discuss the Master and Servant Act, to discuss nearly everything except convict labour, which was given a very minor place.[152] Moreover, from this time forward the Assembly conspicuously avoided mention of convict labour. The Toronto delegates joined, of course, in the annual resolutions on this subject of the Canadian Labour Union, and in January 1878, in the last months of the Assembly's existence, a protest was entered against a proposal that the City of Toronto use prison labour for some of its work.[153] But in five to six years of the Assembly's minutes, there is not another single reference to convicts. It looks very much as if a deal had been quietly arranged between the practical men of the Toronto Assembly and the practical men of the Ontario government. The trade unionists made no more fuss about convict labour. In return, the Ontario government provided a Mechanics' Lien Act in 1873 and amended the Master and Servant Act in 1876.[154] These were important gains; and they indicate, if the reconstruction of events offered here is correct, that the Ontario workingman of the nineteenth century did not come off badly in the contest over convict labour, after all.

A Note on the Role of the Aborigines*

The Europeans who settled in Canada did not, like those in Australia and most parts of the United States, endeavour to exterminate the aboriginal inhabitants of the country. Neither did they undertake to reduce the aborigines to slavery or serfdom, as was done in Latin America: Canadian efforts to enslave, like those of New Englanders, were early and half-hearted. But the Canadian settlers were no more sentimental about their natives than were Europeans elsewhere. The Indians were preserved in the interest of the fur trade, and there was a marked reluctance to maintain them any longer when that trade passed. That Indians were not formally enslaved, then, depended on the fact that they provided a far more satisfactory labour force for the fur trade on a "free" basis. The Indian was skilled in taking pelts and certainly required no instruction from the European. The industry, by its nature, was highly dispersed. Native dependence on trade goods provided an unrivalled incentive mechanism. Finally, the producer was skilled in supporting himself from the sparse but unappropriated resources of the country, so that bargains extraordinarily favourable to the European could be driven without destroying the labour force. Even so, the European sometimes passed this limit and a particular segment of the labour force perished.

The relationship established between the fur traders and their aboriginal labour force, then, certainly was not one of slavery, though it has sometimes been loosely characterized as such. Slavery involves a high degree of direction and of care of the slave, both of which are lacking here. The relationship, again, obviously was not that of industrial capitalism — the trader did not buy labour time. A paternalistic aspect of the relation between traders and Indians, very often present, might suggest a feudal relationship, with the Indians as serfs. But there are so many ways in which relations between whites and Indians differed from those which are usually called "feudal" that this description seems unhelpful. It may be observed that if the fur trade *had* been organized on a basis of slavery, or of serfdom, or of wage labour, the costs of production would almost certainly (in every case) have been much higher and the profits correspondingly lower.

Of the types of organization familiar to European society, the merchant-employer system of commercial capitalism is closest to the relationship established between merchants and the Canadian aborigines. Of the relationships favoured by Europeans overseas, the closest parallel is with the tribute system imposed on Oriental peoples, and with trading systems for commodities in the Orient that approached closely to the collection of tribute. It is significant that these relationships were built up during Europe's mercantile period. The tribute

---

* For a survey of recent work on the economic relations of the Indians and Europeans in the fur trade, see Introduction, pp. xxiv-xxv. Ed.

system could be considered a merchant-employer system in which the bargaining advantage is so overwhelmingly on one side that goods are obtained with no payment whatever in return. The Russians conducted a good deal of their fur business on this basis. But costs of collection are likely to be quite high in a tribute system. A smaller outlay on force directed not to obtain unrequited tribute, but to induce "voluntary" trading on extremely disadvantageous terms, seems usually to have proved more profitable. In general, this is the arrangement the European peoples sought to impose on their neighbours in the mercantile period, and it is the arrangement they were successful in imposing on Oriental peoples and on Canada's Indians.

CHAPTER II

The Pre-Industrial Pattern: Personal Labour Relationships

Slavery provided a means to overcome an actual or apprehended shortage of labour. In Canada it played a minor part in the organization of the labour force. But it was only in the nineteenth century, and in the latter half of it rather than the first, that Canada approached the classical (abundant) labour market of industrial capitalism. Historically, that labour market depended upon the creation and continuance of a large enough labour force that employers could always hire the number of workers wanted, whenever they were wanted. This condition, in turn, required such obstacles to the acquisition of property that the labour market could not be unduly depleted by the escape of wage earners to remunerative self-employment. These conditions were met after 1850, and existed in parts of the Canadian economy earlier. They permitted employers to force workmen to bear their own overhead costs, and yet to be available for use, even for very short periods. The advantage of this arrangement to the employer was particularly momentous in Canada, whose climatic variations promoted peaks of employment in the year lasting a few months or even a few weeks.[1]

But there was a long period in Canadian development during which the labour supply (especially of most kinds of skilled workers) was not abundant enough to permit a free labour market. The employer who thought to escape the overhead costs of labour by pulling a worker from a pool, just at the moment and for the moment that he happened to want him, was taking a gamble in which the odds were against success. The chances of failure mounted if the employer had to obtain a particular quality or kind of labour, and if he required his labour in a remote location. The disability undoubtedly discouraged economic activities that would have been undertaken if labour had been more plentiful. But it also led employers, when the prospects of profit justified the course, to act positively to recruit and retain an appropriate labour force. To meet their requirements, they might have acquired slaves; but it has been observed that they did not choose this solution. They had another way, a well-integrated system of organizing labour, that was a feature of several important employments till about 1850.

In this system, the employer had first of all to catch his labourer, perhaps by recruiting in Europe, with consequent expense and delay. The essence of the system lay, however, in the measures adopted to hold the labourer when he was caught. The employer had to offer continuous employment, or regular employment over a substantial part of the year: that is, he assumed the overhead costs of the labourer. There might be periods during which the employer could not use his workman; still, the workman had to be kept when not wanted, so that he would be available when wanted.[2] But there was a countervailing advantage to the employer: the same imperfect labour market that forced the employer to cling to a particular employee, similarly forced the employee to cling to his employer, for lack of alternative ones.

The economic relationship was thus one of mutual dependence, with limited alternatives on both sides. But unlike the notoriously impersonal relationships of industrial capitalism — following from unlimited alternatives — personal and institutional relationships also developed in this system. The employee, tied to one employer, usually made himself submissive and agreeable. The employer, deprived of the sanction of dismissal, substituted positive incentives to induce conscientious work. These incentives were developed through the techniques of personal leadership on which reliance for results is still placed in most non-economic institutions. To win enthusiastic support, the employer-leader endeavoured to display superior energy, intelligence and fairness. He cultivated desired attitudes by an abundance of personal (superior-inferior) contacts; by expressing and demonstrating his paternal interest in the welfare of his charges, especially in their lifetime employment and care in old age; and by appropriate festivities, favours and rewards. He catered to the foibles of his subordinates and sought to turn them to account. He could turn to expulsion as a last resort, but his success lay in winning positive loyal service.

This system of labour organization has marked similarities to that found in classical feudalism. It was, apparently, the same difficulty of labour shortage that induced the feudal organization of labour through serfdom in medieval Europe, in place of classical slavery based upon abundant labour supplies.[3] It is arguable that the organization of craft labour in medieval Europe was similarly predicated upon labour shortage; provided a sensible way to develop appropriate quantities and qualities of labour under conditions of scarcity; and disintegrated when an abundance of craft workers destroyed its basic assumption.[4] The long survival of feudal traditions, by which those in low stations owed submission and respect, and those in high ones owed paternal care, not only provides analogies for Canadian practices in earlier centuries but probably facilitated their development. On the other hand, the analogies are admittedly limited, and the variations of practice in Canada are almost as extensive as the variations grouped under the title of "feudalism" in Europe. But there was a kind of labour organization that preceded the free labour market of industrial capitalism, that was not slavery, nor a putting-out system, nor the share system of early capitalist commerce. Its distinctive characteristics will become evident in the examples examined below.

New France

There are obvious feudal aspects to the organization of New France.* Kings of France thought of the colony as a fief, bestowed paternal care and generous material provisions for the colony's welfare, and anticipated loving obedience. Royal initiative and royal capital were the indispensable agents for the colony's development: the industries that flourished in New France were the ones that the government made flourish.[5] The colony's agriculture rested upon a basis of feudal tenure.[6] Urban industry consisted of small, independent craft establishments, not unlike those of medieval towns. The attitudes of the colonists were marked by an aversion for regular work and an immature appetite for profit-seeking in a Sombartian sense[†] — that is, they were pre-capitalist attiudes, not untypical of feudal Europe.

Nevertheless, New France did not employ feudal methods with anything like the purity and consistency that might easily be supposed. With respect to urban industry, it appears that New France may have had a few guilds,[7] and it has been complained that the "old guild system" was prolonged to the disadvantage of industrial development.[8] Craft traditions probably militated against new methods of production, in New France as in other places. But the facts seem to support the sweeping assertion that the colony never had a guild system at any time, and therefore, that whatever delayed development, it was not guilds.[9]

The colony did, however, have a chronic condition of labour short-

---

*On the issue of whether New France was "feudal" or not, see Introduction, pp. xxv-xxvi. Ed.

†Werner Sombart (1863-1941), a late member of the German Historical School of Economists. The "Sombartian sense" is revealed in this quotation from his contribution, "Capitalism", to the 1930 edition of the *Encyclopaedia of the Social Sciences* (New York: Macmillan). "The purpose of economic activity under capitalism is acquisition, and more specifically acquisition in terms of money. The idea of increasing the sum of money on hand is the exact opposite of the idea of earning a livelihood which dominated all pre-capitalistic systems, particularly the feudal-handicraft economy." Ed.

age, and what is most striking about its industry is the failure of employers, for the most part, to accept fully the limitations imposed by this fact and to meet them by establishing suitable personal relationships. It is true that workmen's overhead costs were assumed by the employer: unskilled labour was commonly hired for one or more years, and skilled labour from France came on contracts, usually of three years.[10] But employers seem to have assumed usually that nothing further was required of them. They knew that skilled labour was scarce, that the workmen knew they were scarce, and that both parties understood that dismissal was a remote and desperate sanction. Yet little effort was made to supplement wage payments by other incentives. Instead, matters were let run on until something went wrong, whereupon the workmen were denounced and berated for their moral deficiencies.

A reason for this behaviour was the prevailing conviction that wages paid in Canada — about twice those in France — were outrageously high. It seems to have been assumed that workmen, in return for this generous remuneration, should perform an uncommon quantity and quality of work, and expect no pampering.[11] The view that Canadian wage rates were immoral because they were above French wage rates rested, of course, on a moral approach (looking from the top downward) to the question of wage levels, rather than an economic one. But it is a viewpoint not only common among employers in later centuries, but one which writers on New France seem unable to escape, so that its popularity in New France itself should not be surprising. The views of the workmen can only be inferred, but the inference is that their first acquaintance with the colony convinced them — perhaps also on moral grounds — that their wages were too low for such harsh situations.[12] In 1671, the carpenters recently imported to establish a shipbuilding industry staged a strike at work — a slowdown — to win living expenses in addition to their stipulated wages.[13] They did win, but the authorities were outraged. Similarly, the workmen brought out in the 1730s to establish the St. Maurice Forges were at first quite insubordinate.[14] Labour morale and discipline improved greatly after these unpleasant initiations. Many workmen seem to have decided that Canada was not a bad place after all, and presumably employers gained skill in handling them. Still, a cleavage on the question of the morality of the Canadian wage level remained as an obstacle to the provision of effective incentives for labour.

In addition, some of the colony's employments were urban — notably Quebec shipbuilding — whereas personal methods are best suited to isolated work forces. Moreover, the colony's major industries were operated by the government. A government director might, of course, have outstanding capacity to lead and inspire his workmen: perhaps

Levasseur, in the Quebec shipyards, had this talent. But, in New France, there were too many officials exclaiming over the dreadful expense; there was frequent and unwise interference; there was manipulation and graft. In consequence, whatever may have been feudal in the colony, its labour force was not always given a personal organization in the sense set out at the beginning of this chapter. Rather, New France often tried to apply the methods of the free market without enjoying the sanctions of that market.

The Fur Trade[15]

The matured Montreal fur trade required, in addition to the labour of aboriginal primary producers, two distinct types of employee: traders and canoemen. The foundations of its labour organization, as of the trade itself, were laid in the French period. Early development depended upon the incentives provided by free trading. Pioneering activity is generally better suited to profit-sharing than to a wage system, so that large incentives may be offered and so that risks may be spread. In any case, New France lacked an effective mechanism to compel a wage system — that is, to prevent access to the trading opportunities that constituted the means of production. In the days of spectacular extension of the trade, near the end of the seventeenth century, there does not seem to have been a very clear distinction between traders and canoemen: the *coureurs de bois*, who operated in small parties, were, in effect, both. But before 1760, the French had worked out nearly all the machinery of the later trade: food depots, strategically-located posts in the northwest and elsewhere, and the supply system dependent on two sets of canoemen who met at Grande Portage or Michilimackinac. What was most precious in this heritage was the remarkable investment in skilled labour to produce a set of highly-skilled traders and another set of highly-skilled canoemen.

The difficulty of developing a race of competent fur traders has been stressed,[16] and the French ones were supremely competent. Yet other nations managed to produce traders of reasonable shrewdness and tenacity; and the incentive system applied to traders and their clerks seems to find its logic in the strategic position of the trader, permitting him to make or break the trade, rather than in the rarity of competent men. On the other hand, the cadre of skilled canoemen which New France developed was, and remained, unique. Others had either to hire them or to make do with inferior substitute methods of transportation.[17] The differentiation of this labour force from the traders, and the development of the traditions that marked it, remain obscure processes. The men seem to have been drawn largely from New France's farm population which, for this purpose, served as the labour reserve of the fur trade. In the French period the number involved may have been about 15 per cent of the colony's total labour force.[18] Men were drawn to the trade because of its superior attraction as against alternative opportunities. Wages appear to have been about 250 livres, with provisions, and perhaps some petty trading privileges.[19] This remuneration was about the same as the annual income of skilled workers in New France.[20] Whether the canoeman went for a three-month summer run, or wintered in the interior (at higher pay), it could be said that the trade provided for his annual overhead costs, while affording him the leisure which his undisciplined nature sought.* These advantages reconciled the men to their heavy and dangerous work, and to their unattractive diet. The techniques of the canoe brigades probably developed between 1700 and 1750, during which period canoes grew from three- to eight-man size, and came to travel typically in brigades.[21] At the head was the conductor (originally the trader). Engagés were placed in the canoes and paid according to their skill, an arrangement that provided both a status system and a training program. The isolation, the risks, the semi-military nature of organization, facilitated dependence on the leader and development of *esprit de corps*. The nature of the mature extended trade put a premium on paternalistic labour techniques in every respect. Only by winning the approval of his engagés could the trader expect to have skilful, willing men at the times and places required. It was necessary to select wisely, to pay well and to offer effective leadership. The men, though not without alternative means of employment, had a heavy investment in a specialized skill and an obligation to work effectively as a means of self-preservation.

The Albany traders who descended upon Montreal after the Conquest appreciated the value of the French fur-trading machine.[22] But their desire for quick profits and the acute competition of the early years militated against its preservation. Canoemen's wages may have been high enough — they started off at about the old French level, and rose considerably (seemingly in real as well as nominal terms) by 1800.[23] But deficiencies of leadership or organization, probably heightened by a lack of sympathy between old and new subjects, resulted in an extraordinary and unhealthy amount of unrest and labour turnover in the last quarter of the eighteenth century.[24] Nor were the measures taken by the traders to overcome their difficulties very suit-

---

*It is a well recognized pattern of behaviour of pre-industrial societies that the work effort was very intermittent. The literature on the disciplining of the labour force in the industrialization process to enforce regular and persistent effort has become extensive in recent years. As Sahlins notes, the concept of regular labour is "bourgeois ethnocentrism", adopted also by contemporary Marxism. "That sentence of 'life at hard labour' was passed uniquely upon us." Marshall Sahlins, *Stone Age Economics* (New York: Aldine, 1972), pp.3-4. Ed.

able. One was the application of (more or less illegal) force, which can be very effective in a feudal labour situation if applied judiciously.[25] But the traders seem to have depended more on requiring engagés to bear certificates from their priests, and on demanding vigorous support from magistrates.[26] These expedients confess failure. The traders were corroding their superb labour force by their petty spirit and their failures of leadership.[27]

Nevertheless, the Montreal traders prosecuted their business vigorously, and their grasp of labour and other techniques improved as they moved towards combination with an eye to the development of the northwest trade. The Northwest Company was distinctive in the generous incentives which it offered traders in the field, and in their energetic response. The method was a logical one for a partnership that grew from many small firms, but it was also the effective method for dispersed operations. The Company profited further, in some degree, from the skill of French traders left by the old regime. With respect to canoemen, the Company seems to have taken little pains to maintain the morale of the Montreal brigades. Their position was scarcely strategic, and the partners may have concluded that these brigades would survive the need for them. But for the northwest brigades, the Company developed a thoroughly paternalistic organization.

The Company's problem was the replica of one the French had faced a century before: expansion into new territory. In the northwest there had to be new food resources, smaller canoes, harder runs and tougher men. The men had to go to far places under difficult conditions. They could not be replaced quickly, and their zeal was important. The solution also followed French precedent: the earlier expansion had produced the paternalistic labour force of the trade, and the Northwest Company re-emphasized its techniques. Isolated employees may have been more at the Company's mercy than the Company was at theirs; but it was sound economy to curtail labour turnover and promote loyalty by offering conditions that encouraged lifetime service. Some attractions could be offered. Fur trade employees enjoyed a superior status among the Indians and could keep Indian women more or less at Company expense. There were long intervals when there was little to do. There was a guaranteed annual wage and a good prospect of lifetime employment. A fund existed to provide care for disabled men — provided, to be sure, from their own contributions. But the Company did not depend on these inducements alone. It successfully promoted an *esprit de corps* based upon personal loyalties, upon rivalries among brigades and groups, and upon a difference in status between eastern and western canoemen.[28]

It may have been the power of its non-wage incentives that permitted the Northwest Company to retrench on wages after 1804, when monopoly was re-established, without any noticeable harm to morale.[29] Neither is it apparent that truck pay harmed morale, even though, at times, the Company could cut its wage costs nearly in half by this device.[30] In fact, truck pay is a practice with two faces. It has seemed often to constitute a blatant form of exploitation that has produced violent hostility among employees. But in these cases, employees have had physical access, or expected to have access, to other suppliers besides the company store. In remote locations, on the other hand, the company store may be accepted gratefully by employees as a great benefit to themselves. Engagés in the northwest could take satisfaction in the knowledge that no supplier other than the Company was within their reach. Perhaps they considered, too, that the wages as well as the prices were high by distant standards.

The Hudson's Bay Company, though it had a different economic situation and a different type of employee from Montreal organizations, was no less bound by the exigencies of personal labour organization. To an even greater degree than the Northwest Company, it operated in remote areas from which little local labour could be hoped for. Replacements of and additions to its labour force were brought across the Atlantic during a short shipping season. Consequently, they were even more awkward and expensive for the Bay Company than for its rival. To get men in the first place, the Company was bound to hire on a long-term basis. Once men were got, it was to the Company's advantage to keep them indefinitely and to encourage competence and spirit among them. The Company's dependence on its men was great; but the dependence of the men, in their isolated posts, upon the good faith and wisdom of the Company, was still greater. For this labour problem there was no conceivable solution except a paternalistic one, beginning with the Company's underwriting of labour's overhead costs. The only question is, how effectively the Company used paternalistic techniques in building its labour force.

Opinion on this question has varied. Innis, in his *The Fur Trade in Canada,* drew contrasts between both the economic positions and the methods of organizing labour of the Bay and Montreal companies. In position, the advantage lay with the chartered company, for its nearness to the sources of fur supply allowed it, for a long time, to avoid the trouble of penetrating the continent, without its suffering disastrous losses in consequence. Offsetting this, however, was the advantage possessed by Montreal of a more lively and aggressive work force. The different calibres of labour reflected different systems of motivation and, ultimately, the different economic positions. Precisely because its competition was coast-bound and passive, the Hudson's Bay Company conceived that it had no need for zeal, and it retained a wage system which provided the slightest of monetary rewards for effort. The result was an apathetic and dissatisfied work force.[31] When the Company entered the inland trade, it turned to a system of incentives based on profit-sharing, with some other rewards for good service.[32] Considering that the Company did practise some measure of paternalism, that men were engaged for long terms and spent their lives in the Company's service, the pay system adopted after 1810 would appear to have rounded out a strong system of personal labour organization. Nevertheless, the impression is left that the Company showed little capacity to provide or develop positive leadership and relied far too heavily on negative sanctions unsuited to its situation; hence, it could not match the flair and tone of the Montreal brigades, and won its victories in spite of its disspirited work force, not because of it.

Dr. Glover also contrasts the Bay and Montreal companies, but with sharp revisions on both sides.[33] He finds that the Montreal traders were pretty unsatisfactory employers, not only from a humanitarian viewpoint but also in terms of their own self-interest. Glover considers that the Hudson's Bay organization, on the other hand, was a good one: good both in the quality of its leadership and the morale of its men. He thus directs attention to the merits of that particular military type of personal labour organization upon which the Bay Company depended. Rigid class and military differentiation has not very well suited either the Canadian environment or the Canadian temperament, and Canadian writers may have judged them too harshly. Given a class of recruits to whom a military structure was not inherently distasteful, and officers of some capacity, a military system relying on negative and non-monetary sanctions for its discipline might still achieve a reasonable level of morale and effectiveness. It may be, on the other hand, that Glover gives the Hudson's Bay system too much credit. It cannot be without significance that it was found useful to add monetary incentives to the system, or that these were highly inconsistent with the previous system; and it is difficult to avoid the conclusion that the Company had a muscle-bound organization.

What may have hardened views against the Company, and brings out clearly some of the essentials of personal labour relationships, was the Company's record after the merger of 1821. Governor George Simpson, of the combined companies, set out to apply the *dicta* of capitalism to employees and Indians alike.[34] A redundant labour force, the fruit of monopoly, seemed to make this a possible policy in dealing with labour. Labour reserves were settled at Red River, from which settlement men could be hired for short periods. Various other sources of labour for transport were explored on a temporary and parsimonious basis. If there had been a true labour surplus, barred from a livelihood except by sufferance of the Company, Simpson's capitalistic labour policy might have been permanently successful. However, the bars against alternative means of livelihood were not tight enough: the advantages to employers of a capitalistic market were sought before the market really existed. What Simpson actually did was to produce profits out of the unrecognized corrosion of the magnificant labour force bequeathed him by the Montreal company. When the demoralization of that labour force had become complete, about 1850, the Company could no longer command labour at the times and places desired and had to content itself with a very inferior quality of employee.[35]

Calvin & Company

The timber and lumber trades generally do not afford instances of highly-developed paternalistic labour organization. However, one firm engaged in producing and shipping timber appears, from its perceptive biography,[36] to have been a striking exception. This is the firm of Calvin, Cook & Company, founded by D.D. Calvin, with its successor and subsidiary companies.

It is significant that the Calvin firm was not just another competing timber firm. It specialized in the production of hardwood instead of the usual softwood, in exploiting the St. Lawrence rather than the Ottawa watershed, and especially in forwarding timber down the St. Lawrence. It was also unlike other firms in its possession of a permanent, convenient, yet isolated base in Garden Island, near Kingston, and in the breadth of its activities, which included shipbuilding. In short, the firm sought positions of monopoly from which it stretched out to conduct related activities that served to reinforce the monopoly. To be successful, such an enterprise required unusual quantities and qualities of labour in unusual places. If it did not require exceptional loyalty from its employees, it certainly fared better for having it. To obtain these advantages, the Calvins developed a personal organization of their labour force, which appears to have numbered 600 to 700 men.

The key employees of the firm were those highly skilled in raft construction, raft pilotage and shipbuilding. All of them were housed at Garden Island. To these retainers and their families:

> "the Governor" and his son felt a responsibility similar to that which they felt for their customers. "You can always depend on us" might equally well have been said to employee as to customer. When the timber market broke in the panic of 1873, timber-making was of course curtailed, but "the Governor" had no thought of "shutting down altogether" as so many firms and employers did in that grim time... (The employees were told) we must all carry a little piece of (the load) — the firm will lose money — wages must be cut, and we must all "live close" — it will be hard

> going but we'll fight it out altogether... The building of the
> ocean-going barque *Garden Island* was the peak of "the Gover-
> nor's" effort (and sacrifice) to keep the men employed... and a
> very costly effort it was... [37]

This organization did not depend upon generous wages or short
hours of work, but rather expected its other advantages to substitute
for them.[38] *Esprit de corps* built up by security and paternalism in the
communal society of Garden Island[39] was evidently an effective incentive, as demonstrated in the faithful service to the firm of three generations of sail-makers.[40] For certain superior classes of employees to
which these devices could hardly be applied, such as wood bosses and
ship captains, reliance was put upon the incentive of profit-sharing.[41]
There was even a paternal aspect to the firm's relations with its temporary employees, especially Indian raftsmen, although these men
were notoriously undependable and exasperating.[42]

The Calvin organization survived a good deal of technological
change, but it is significant that the change enhanced the productivity
of existing skilled trades for the most part, rather than promoting a
substitution of unskilled labour.[43] Integration facilitated the shifting
and economic application of the fixed labour force with lifetime tenure
that was the product of the firm's labour policy, and spread the burden
and risk of the overhead costs of the workmen. The eventual decline
of the firm was inevitable in view of the obsolescence of the economic
activities in which it engaged. But it seems doubtful whether the Calvin organization could, in any case, have survived the development in
Canada of a capitalist labour market, with the consequent power of
other employers to hire on a temporary basis and avoid the overhead
costs of their labour.

The St. Maurice Forges

Production of iron at the St. Maurice Forges, near Three Rivers,
began tentatively in 1732, and seriously in 1738. It continued until
1883. The Forges, then, operated for one hundred and fifty years — a
record matched by few North American enterprises. These years
spanned the French regime, the English regime and Confederation; the
ancient economy of the eighteenth century and the modern one of the
nineteenth. In their day, the Forges were almost the only large industrial plant in Canada, their scale sharply out of proportion to that of
other enterprises. Their products dominated the Canadian market for
iron goods for a century. For about the same period, the Forges seem
to have been the most advanced ironworks in America, and they were
favourably compared with European works. Yet they have not attracted
much attention, and some general history of this enterprise is given

here to provide a perspective for discussion of its labour force.[44]

There had been explorations for iron and talk of ironworks in New France for about seventy years before 1730, mostly with the thought that local iron production would guard against the uncertainties of ocean communications in wartime. However, the St. Maurice Forges had the rare distinction in New France that they were not initiated by government action. Entrepreneurship came from the family of merchant-seigneurs that held St. Maurice seigneurie, which had slight prospects for farming, but a wealth of minerals. The timing was good: in 1729, New France was beginning its period of greatest prosperity. François Poulin, who promoted the project, possessed a capital of 30,000 livres — not much, but more than most colonists had. The authorities readily granted a twenty-year monopoly of iron mining and manufacture in the Three Rivers area, with right to mine not only in St. Maurice but in neighbouring seigneuries.

In terms of the techniques of iron production used before the nineteenth century, St. Maurice was probably the best location in Canada. Within ten miles of the proposed iron works there was so much bog iron ore — rich, easily worked and at the surface of the ground — that the Forges scarcely ever found it necessary to look further. There were, however, other deposits in the neighbourhood which eventually supported other ironworks. The region was also rich in forests for charcoal-making. Limestone for flux was conveniently at hand. The stream selected to provide waterpower proved to be very satisfactory for that purpose. The Forges had access to Three Rivers (about seven miles distant) by road and by the St. Maurice river. Three Rivers was well placed for the distribution of finished goods to the Canadian market.

The original plan was a simple one. A couple of workmen (miners) were imported from France in 1731, but chief dependence was put upon a local blacksmith who was sent off in 1732 to study the operation of contemporary New England forges. These were primitive affairs, which made an indifferent job of refining a small output of iron. But they had the immense advantage of economizing both on capital and labour.[45] Their plan seemed to fit the needs of a cautious small capitalist, like Poulin. Under the blacksmith's direction, a New England-style forge was built and some iron, apparently good in quality, was turned out. But operations were pursued only for a few months, the method used being judged unsatisfactory and unprofitable. From the economic viewpoint, it seems unfortunate that the simple, inexpensive system of operation was not given a thorough trial.

However, the syndicate that had been formed to operate the Forges in 1733, and then reorganized after Poulin's death in that year, suspended operations until skilled advice and aid could come from France.

A French ironmaster (Vezain) arrived in 1735, and, indeed, a second one (Simonnet) in 1736. Vezain proposed a big blast furnace with two forges, the requisite dams for waterpower, and supplementary buildings. A hundred men worked from mid-1736 to the fall of 1737 to complete the establishment, which included an elaborate chateau intended for living quarters and offices. The cost was over 200,000 livres, as against Vezain's estimate of 36,000 livres. The company had received a government loan of 100,000 livres, but required additional support; it owed nearly 200,000 livres by 1738.[46] Strenuous efforts had been made to recruit skilled labour from France, and about sixty men were on hand in 1738, most of them arrivals of that summer.[47]

Some reconstruction proved necessary to correct faults in the works, and an additional forge was added in 1739 because — ridiculous though the fact may appear in the light of the earlier shortage of labour — the large force of skilled workers assembled could not be fully employed by the existing equipment. But, though substantially rebuilt on two or three later occasions, the Forges of 1736 remained unaltered to 1883. The high furnace refined by use of a cold air blast, a method evidently suited to the local ore. In general, methods were the best in use in France, and French methods had been much advanced in the half-century before 1740 by copying Swedish examples, following pilgrimages of French ironmasters to Sweden promoted by Colbert.[48] The iron produced was usually declared better than American or English, and sometimes was held to be as good as Swedish.[49]

The capacity of the Forges was supposed to be 300 tons of iron per year, but what they proved able to produce (after 1739) was 200 tons.[50] The difference may be accounted for by the fact that winter operation proved difficult, and the custom was established, and ever after observed, of closing down the furnace for about five months each year. From May to December, the furnace operated 24 hours each day, with two sets of workmen taking alternative six-hour shifts. At first, the Forges produced only forged iron. However, moulding was added in 1741, evidently with the aid of additional skilled workmen imported in 1740. Pots and pans were produced from 1741, and stoves and bullets from 1744.[51] In 1747, the Forges were equipped to produce round bar iron, and the casting of cannon and production of steel were attempted the same year, though without enough success to encourage much further effort.[52] Cast iron products seem to have made up 5 to 10 per cent of the firm's output, by weight, during the French regime.[53] Forged iron output exceeded the colony's needs, and perhaps half was exported to France, for use in French naval shipyards, at somewhat inflated prices.[54] On the other hand, the Forges suffered when French ironworks dumped their products in Canada in 1748.[55]

The syndicate that had built the Forges did not survive long. The great expense of the works caused difficulty, but the real trouble was the bankruptcy of François Étienne Cugnet, the chief partner, occasioned by losses on his far-flung operations of which the Forges was only a part. No other private operator was found for the works during the French regime.[56] To save the royal investment and to maintain a military resource, government operation was undertaken from 1742. The ironworks did not make good returns under royal control, but their record is not so bad as some comments imply. Up to 1757, total profits were about 130,000 livres — two-thirds of the original royal investment.[57] However, profits were inflated by artificial pricing. What contemporary observers stressed was not the nominal profit record but the bad management that prevented appearance of very substantial profits, which they thought the Forges could produce.[58] The marked profitability of the works after 1767 suggests that this criticism was just. The beneficiaries of the Forges during the French period were officials at the Forges and at Quebec who prospered on the plant's indifferent balance sheet. The very effort to overcome this difficulty, by dividing authority among officials who reported separately on their trusteeships, prevented efficient managers, if there ever were any, from tightening up the Forges' organization.[59] A good deal of weight should be given also, and was given by contemporary ministers, to the pioneer aspect of the venture. It was inevitable that there should be expensive mistakes in exploring the potentialities of the ironworks. Finally, the numerous wars of the period had a depressing effect. Generally, there is little evidence to support the proposition that the St. Maurice Forges was a marginal and premature enterprise, at a comparative disadvantage to farming and the fur trade.[60]

The Forges fared badly during the Seven Years' War. Its labour force was partially dispersed, and its equipment was all but useless by 1760. British military authorities had the Forges worked on a small scale in 1761 and 1762. They passed formally to the British crown in 1763, but stood idle for several years. Then, in 1767, a syndicate of Quebec merchants leased the property from the government for sixteen years at an annual rental of £25 currency, but with the important provision that £4,500 was to be invested to repair the dilapidated works. The leading figure in the syndicate was Christophe Pelissier, though most of the partners were old subjects. Reconstruction was vigorous and effective, and a sawmill (and perhaps a gristmill) was added to the establishment.[61] Three skilled Welsh ironworkers and perhaps some other British experts were added to the labour force. It is hard to be sure of the company's success,[62] but it would seem that profits were high after the first years of reconstruction. Gross sales in the 1770s were claimed to be £10,000 to £15,000 per year, with one-third of this as profit.[63] The sales figures are reasonable, and the profit not unlikely.

Discussion of this Forges' regime has centred on the sympathy of the ironworks' directors for the American Revolution. In 1775 and 1776, Pelissier not only supplied the American armies in Canada with iron products and munitions, but showed decided sympathy by his personal behaviour. When the American forces retired from Canada in 1776, Pelissier thought it best to leave also, with the cash assets of the Forges. Laterrière, his assistant, had been imprisoned in 1776 as an American sympathizer, but was released after a month. When Pelissier fled, Laterrière appropriated both the direction of the Forges and Mme Pelissier. The rivalry thus implied with Pelissier may have protected him from arrest.[64] Eventually, in 1779, Laterrière too was imprisoned for his obvious encouragement of a second American invasion. These events have led one author to discuss the mildness of Haldimand's* rule,[65] another to seek an explanation of peculiar behaviour in the personalities of the men involved.[66] There is an additional aspect of these episodes that deserves attention.

Canadians, in 1775, were decidedly sympathetic to the American Revolution[67] despite a good deal of subsequent writing designed to create a contrary impression, and the Forges' directors might very well have shared this sentiment. Moreover, industrialists do not ordinarily oppose conquering armies, least of all when they offer liberal contracts. In addition, those whose fortunes depended on a colonial manufacture had unusual reason to desire independence. The Forges catered to the local market and would benefit from its growth, whereas mercantile doctrine envisioned the colony as an attenuated supplier of raw materials. The Forges had much to fear from British hostility to colonial manufactures expressed through political restrictions and through unrestricted economic competition. It has been argued that Canada's fur traders wished to preserve the imperial connection because of their business interest in it.[68] The same ground of economic interest furnished the Forges' directors with reasons for revolution.

The term of the Pelissier company was completed with another partner, Dumas, directing the Forges. In 1783, a new lease for sixteen years, still at £25 per year, was given to Conrad Gugy. Since the ironworks was now a profitable concern, this lease can only be regarded as a novel method of rewarding an old soldier. In 1787, the lease was taken over by the Quebec firm of A. Davidson and Lee, at a valuation

---

*Sir Frederick Haldimand was governor of the Trois Rivières district from 1762 to 1765, including the St. Maurice Forges. After a posting in the American colonies and a British military command in the revolutionary war, he returned to Canada as Governor-in-Chief from 1778 to 1786. Ed.

of £2,300, apparently in settlement of Gugy's debts.[69] In 1793, the small remainder of the lease passed for £1,500 to the more or less related firm of G. Davidson, Munro, and Bell, which became Munro and Bell in 1794. Thus dawned the Bell regime at St. Maurice.

Matthew Bell came to Canada from Scotland as a youth. He was only twenty years old when he joined the firm of Davidson and Lee (and perhaps acquired an interest in the Forges) and about twenty-nine when he took direct charge of the works in 1798.[70] Bell remained as manager, and a sort of feudal sovereign besides, until 1845, when his age was seventy-six or seventy-seven. He was long in partnership with David Munro and may have had other effective connections, but he was the key man throughout. His was the longest and most remarkable tenure of the Forges.

From 1800 till 1806, Bell leased the Forges from the government at £850 per year, with the additional commitment to spend £1,500 on improvements.[71] That the rent was so far above the amounts paid before was party due to· the fact that other interests, associated with the Batiscan Forges, offered competitive bids for St. Maurice. Even so, Bell received special consideration on the ground of his heavy investments and his efficient management.[72] Profits could hardly have been less than £1,500 per year, and may have been higher.[73] The Forges was in a strong position. It had been remodelled so that two-thirds of the output consisted of the more remunerative cast products.[74] Bell had undertaken from the start to meet English prices,[75] the quality of his products was said to be superior to that of English imports,[76] and sales agencies were maintained in the principal centres of the colony.[77] The firm catered also to new demands: iron for the Quebec shipbuilding industry, potash kettles, and machinery for early steam boats.[78] The Canadian market for iron was expanding and prices were rising: Selkirk put sales in 1804 at £10,000 to £12,000, though they had "formerly" been only £7,000.[79]

In view of the investments required of Forges' grantees, a long lease for twenty-one years was offered in 1806 by public auction. But only Munro and Bell bid, to receive the lease at the ridiculous rental of £60 per year. The collusion necessary for this result appears to have depended on political considerations. Bell's pocket borough of Three Rivers seems to have had special value in 1806 to Thomas Dunn, a principal partner in the rival Batiscan ironworks.[80] This iniquitous arrangement caught the eye of the Colonial Office, and Bell was required to raise his rental to £500 per year in 1810. However, his lease was extended to 1831. Various extensions and renewals carried Bell through to 1845, all at £500 per year, though sometimes additional woodland was included in the lease. This effective monopoly could be justified on the ground that a capable operator who made the Forges prosper was a benefit to the country. The argument was made, and the disintegration of the Forges after 1845 indicates its soundness. But Bell did not depend on this theoretical support. He offered lavish entertainment to Governors and others in positions of power. This "expense of doing business" is alleged to have taken the largest share of Bell's generous profits, though it is difficult to believe that the director did not enjoy this obligation.[81] These circumstances demonstrate the diverse demands that might be put upon entrepreneurship in the early nineteenth century. However, the prosperity of the Forges offered compensation for Bell's efforts.

Output was valued at £24,000 to £30,000 per year in the early 1830s, about half of this being accounted for by stoves. The establishment was valued at £12,500 in 1833, while its inventory of materials, supplies and finished goods was put at no less than £48,000.[82]

In 1830, Bell had long ago overcome the competition of other ironworks. He successfully maintained his favoured position in government circles. He was celebrated for the "certain and uniform employment" which he offered.[83] But he lacked an effective defence against a new threat, the rise of popular and nationalist agitation in French Canada. From the 1830s to the 1850s orators in Lower Canada, and particularly the deputies from Three Rivers, found an inexhaustible store of ammunition in attack on the Forges' monopoly. The agitation constituted, as one writer has said, an anti-trust movement.[84] It was spiced by Bell's connections with unpopular government and by the fact that the workmen at St. Maurice were French, whereas the masters and foremen were English, though it may be doubted that there was any support for this agitation from the workmen at the Forges. The ostensible ground of attack was that the extensive woodlands held by the Forges blocked agricultural settlement up the St. Maurice.[85] The period was marked by agricultural over-population in French Canada, and by a search for new land to colonize, but the Forges land was not an answer to this problem. A considerable amount of it was offered for settlement in 1843, but there were no takers.[86] A few years later, when the Forges was entirely stripped of land, settlers still did not come, and Sulte wrote, "Even now the Forges land is almost without inhabitants."[87] As Bell had insisted, these lands were badly suited to agriculture.

Several writers have taken the colonization argument seriously; yet the main and different grounds of attack on the Forges were made clear at public inquiries.[88] The complainants were not intended settlers but the merchants of Three Rivers. They did not argue that lack of settlement harmed landless habitants but that it hampered the development of Three Rivers.[89] A special grievance was that the Forges' people did not spend their wages in Three Rivers but, as always, at their local (company) store.[90] Nor was iron sold at Three Rivers, any more than at other places. The merchants wanted ironworkers to reside and spend money in Three Rivers, and implied that an iron staple should be located there, to which buyers would have to resort.

To the aristocratic paternalism of the Forges was opposed the logic of the capitalist market, as understood by local lower-middle-class representatives. They proposed to lease ore lands to many people, and thus to encourage competition. Competition in producing ore for the Forges was workable and eventually existed so far as collusion among the habitants allowed.[91] But competition among ironworks also was envisioned: with a magnificent dismissal of the distinction between an ironworks and a blacksmith's shop, it was even pointed out that several forges already existed in Three Rivers.[92] That this argument rested on a monstrous misapplication of the new economic ideas of the competitive market may have been appreciated by those who used it. What is significant is the use of these new ideas by the pushing middle class to enforce and rationalize their grip upon the state. The merchants expressly disclaimed any responsibility for the welfare or future of the Forges' employees.[93] Their resort to the mores of the capitalist labour market implied that such a labour market was appearing.

By 1845, Bell was at last ready to yield up the Forges. No secret was made of the fact that the plant had been allowed to run down very badly in anticipation of a change in tenure. Forges and lands were sold (separately) to the highest bidder, Henry Stuart, for £11,475. The importance of management was now illustrated. Stuart undertook an ambitious program of renovation, but he evidently made costly mistakes, and his efforts to operate were not very successful. Within two years, the Forges passed to the Ferrier family of Montreal. These were hard, economical managers who are supposed to have made large profits.[94] In 1851, the Ferriers sold the Forges (without land) to Andrew Stuart and John Porter, of Montreal. These owners paid little attention to the works, allowed them to deteriorate, and were dispossessed in 1861 when the Canadian government foreclosed the mortgage held against them.[95] The works were again sold and eventually passed in 1863 to John MacDougall & Sons of Three Rivers.[96] The MacDougalls were energetic ironmasters and made profits while the Forges experienced a last blaze of glory. Finally, in 1883, the Forges were closed forever, because of the increasing difficulty of competing with more modern ironworks in an era of low prices.

Long before 1883, the St. Maurice Forges was an anachronism, its equipment outmoded by technical advances, its paternalistic labour organization by the growth of a free labour market. But its early technical superiority, the good management it frequently enjoyed, and the support of an established labour organization, provided a momentum that permitted its survival almost to the twentieth century.

The St. Maurice Workmen

The heart of St. Maurice was its furnace, forges and foundry. Thirty to forty men were required to operate these. Some of them were labourers, but a dozen or more were key craftsmen, without whom the plant could not operate. Other permanent employees included book-keepers, carpenters, boatmen, prospectors for ore, sawmill employees and a large force of carters, making a total of about 120. Finally, there were three or four hundred temporary employees, hired in season each year to cut wood, make charcoal, dig and haul ore and make roads. These were drawn predominantly from among the habitants of the Three Rivers area. The structure of this work force and its total of 400 to 500 workers persisted through the whole life of the Forges, though there were temporary fluctuations. As many as 800 were employed when production was pushed in the 1770s,[97] and perhaps at some other times.

The permanent employees lived at the Forges with their families, making a village of 400 to 500 persons. The site was described as pleasant and healthful. From the beginning, the firm provided the housing for its workmen. The great chateau had been intended to house employees, though it never accommodated a large part of them. There were also many cottages, scattered haphazardly with their gardens, usually described as clean and comfortable.[98] In the early nineteenth century, there were fifty-five of these cottages.[99] In addition, there appears to have been a good deal of accommodation of a temporary bunkhouse nature, which may have been used for temporary employees. Physically, as socially, the village was a world apart.

The villagers of St. Maurice developed a solid, complacent and introverted community, but not without strenuous conflicts to begin with. The seventy-five skilled workmen imported from France between 1736 and 1740 — mostly in 1738 — came without enthusiasm, despite the high wages that drew them. Matters were made worse by unskilful management and by the fact that the works were not ready to employ the men fully for some time. The workmen were "not amenable to discipline for they knew that they could not easily be replaced"; they "exacted exorbitant wages and dawdled over their work"; some were "unreliable, insubordinate and quarrelsome".[100] A major issue in the first years was the truck system. There was then a company store at St. Maurice, as there was ever after. The remoteness of the village would seem to justify this. But the workmen believed in the early days, no doubt correctly, that their wages were sharply discounted through store pay; and, on the other hand, they bought too much brandy. Efforts to restrict abuses from 1739 by providing for cash wages on demand, and by measures to curtail the consumption of brandy, may have had some success.[101] At any rate, there is nothing to indicate discontent with store pay in later years. The workmen were incensed also by delay in paying wages due about 1741, and by the failure of wages to advance with prices during the war-bred inflation of 1744.[102] Probably a greater cause of discontent still was the unstable and ineffectual management, and the inability of the workmen to produce the quantity and quality of output expected of them. Colonial authorities tried to solve the Forges' labour problems by applying legal authority and force. In 1745, after a visit to St. Maurice, Hocquart issued

> an ordinance limiting strictly the amount of liquor sold to the workmen and forbidding any man to absent himself from the Forges without permission on pain of losing his wages the first time and the additional penalty of corporal punishment for any other offence. The directors were enjoined to punish "scandals and public debauches" and to proclaim the ordinance every three months by way of a reminder.[103]

Repression often worsens labour conflicts, but it may have improved discipline in this instance, especially if other sources of dispute were overcome about the same time. Another control device used was the location of a priest at the Forges.[104]

Even in the 1740s, however, the work force became coherent, moderately efficient and reasonably satisfied. Management seems to have improved. Then, as contracts expired, men did not often return to France, despite talk of this recourse; but they did take the opportunity to negotiate terms — called "exorbitant" by the authorities — that suited them much better. But the main device to hold workmen and modify their opinions of Canada — Canada's secret labour weapon — seems to have escaped notice, though it is implicit in the genealogy with which Sulte adorned his work on the Forges. Workmen had scarcely arrived at Three Rivers when they began marrying Canadiennes. Soon they were baptizing children — the next generation of ironworkers. Sulte's parish records also reveal the development of blood relationships among the Forges families, and of a local caste system. The most highly skilled workers tended to be interrelated, and conversely, an outsider who married into this group was likely to be worked into a responsible position.[105] It seems reasonable to suppose that family and community ties soon played a vital part in keeping workers at St. Maurice.

The Forges group had distinctive social traits which their isolation preserved to the end. Unlike most settlers in New France, they came predominantly from the iron regions of Burgundy, particularly the

neighbourhood of Dijon.[106] They preserved the pride of craftsmanship associated with the master craftsmen of medieval Europe.[107]
They had peculiar customs and dress and an unusual wealth of super-
stition.[108] Sulte said they were notorious for swearing, quarreling and
bad manners; but notable for cleanliness and very moral (meaning
that illegitimate births were rare).[109] In keeping with their other characteristics, the ironworkers were sticklers for tradition. The operation of the blast furnace had always been directed by the ringing of a
bell. Late in the nineteenth century, MacDougall proposed to substitute clocks and speaking tubes, but the workmen would have none of
this: "On ne change pas de religion, voyez-vous; ce qui était bon pour
nos pères est bon pour nous..."[110]

Peculiarity was fortified by the paternalism that became the tradition of the Forges. Early difficulties emphasized the dependence of
management upon the key workers, but the workers were hardly less
dependent on management, for other employment suited to their
skills was not to be found easily. The more successful operators, at
least, were solicitous for the welfare of their employees. They had the
advantage that they could offer lifetime employment. Dating probably from Hocquart's intervention, the Forges' director was a social and
moral arbiter of the community ,as well as an employer. In the 1770s,
and perhaps long after, workers were not to entertain outsiders or
keep them in their homes without permission. There was much gaiety
and entertainment at the Forges in the winter, but this came under
the director's control.[111]

The feudal pattern was sustained by the fact that succeeding generations of workers grew up at the Forges. In 1852, St. Maurice was described as made up of "several hundred families", all born at the village,
and of "more than a thousand persons accustomed from their childhood
to a particular branch of labor" and contented with their lot.[112] Many
were descendants of the earliest workmen there. A few British ironworkers were introduced during the last decades of the eighteenth
century to overcome the pronounced shortage of key workmen of the
1760s. But these never numbered more than a dozen, they were absorbed by marriage and familiarity into the French Catholic community,
and they fitted readily into a high niche in the hierarchical structure.[113]
Workmen for the other end of the hierarchy, for the jobs outside the
forges themselves, had been drawn to a large extent from local French-
Canadians, and these men sometimes advanced to key positions. But
they, too, absorbed the social and cultural traditions of the French
ironworkers.

Methods of work and pay tended to isolate the group and to
strengthen its feudal structure. The furnace operated continuously,
and furnace and forge men were divided into two shifts which worked

alternate six-hour periods. However, moulders, as well as the other employees, worked the usual hours of dawn till dusk.[114] Housing and fuel were supplied to workmen throughout. Other supplies had to be bought, but were obtained ordinarily at the company store. Fixed annual salaries were paid at first, but eventually nearly all wages were paid by piece rates.[115] Most of the workers were said to have been paid 400 livres per year in 1754.[116] In 1796, the men were reported to receive ¾ of a piastre per day, which might amount to 800 livres per year.[117] Lord Selkirk reported in 1804 that the forgemen made £50 per year (1,000 livres) by piece rates, and the moulders received £40 (800 livres) in the same way.[118] Even allowing a good deal for free housing and fuel, and five months of winter leisure, these wages do not appear to be remarkable.[119] It seems probable that the immobility of the St. Maurice labour force after the first few years, the stickiness of wages in periods of rising prices, and the influence of profit-motivated directors, combined to keep wage rates at modest levels. An object of paternalistic management is to keep down wages, without reaping lowered morale in consequence, by substituting non-economic rewards. The communal life of the Forges probably was attractive to most workmen. Especially so must have been the long winter of festivities, in an age that did not find the sole end of existence in work. The heavy overhead costs of the Forges, untypical of industrial enterprises of their age, enforced steady operation. This provided a regularity of employment that offset whatever deficiencies daily pay rates were thought to have, and invited paternalistic mechanisms intended to minimize labour mobility.

Sharply distinguished from the permanent employees at St. Maurice were the three hundred men, more or less, hired temporarily each year. In the French period, especially around 1750, there was difficulty in getting enough of them. Conscription of local habitants by force, sometimes resorted to, met hostility and threats to move from the district. Companies of the Troupes de la Marine were stationed at Three Rivers in the hope that the soldiers would be willing to work at the Forges, but these men were not much used because they demanded "exorbitant" wages.[120] It is plain that the Forges' directors were unwilling to pay wages appropriate to their labour market and relied, with limited success, upon force to obtain labour on their own terms. However, these events also illustrate the hazard of seeking temporary labour from a market without a surplus, or insufficiently stocked to be a free labour market. The British military government of 1760 also relied upon authority to operate the Forges. In 1762, militia units were peremptorily ordered out to cut wood for the ironworks.[121] The private operators who followed had no power to enlist workmen in this summary fashion, but there is no evidence that any of them suffered on that account. Presumably, they offered wages sufficient to attract labour, which was anyhow in more generous supply. The seasonal employees necessarily came into some contact with the St. Maurice community, but the two groups do not seem to have affected each other very much. The relation of the temporary employees to their employer was largely confined to the "cash nexus" of the free labour market.

The early difficulties of the Forges in assembling sufficient temporary employees demonstrates the importance of measures taken to retain the far more valuable skilled workers. These measures met the social wants of the workmen well, and the economic wants acceptably. They militated against movement and against any agency that would induce movement. Once established, therefore, the paternalistic organization of the work force tended to be self-preserving. The system was not devoid of flexibility, for while it lacked the power to vary the numbers of the work force (downwards), there was some power to vary economic rewards so long as social advantages were preserved. The system also contained effective incentives in the *esprit de corps* of the labour force and of its status groups, and in the paternalistic activities of the employer. Since new workers were generally the children of old workers, they fitted in without disturbing the Forges' folkways. The system was decidedly advantageous to the employer economically (and possibly in other ways) in the first century of the Forges' operation. Indeed, only the effectiveness of this labour system would seem to offer a satisfactory explanation for the long survival of the Forges in the face of new methods of ironworking. On the other hand, drastic technical changes probably were impossible within this framework. Even apart from the question of technological advance, it seems doubtful whether an employer assuming the overhead costs of his labour force could survive indefinitely the competition freed from this obligation by the filling up of a free labour market.

Marmora

There were several ironworks besides St. Maurice in Canada during the eighteenth and nineteenth centuries. The Batiscan Forges operated from about 1794 to 1813. The physical characteristics of their site paralleled those of St. Maurice, and the works were patterned after the earlier ones. The proprietors were English-speaking mechants and officials from the topmost circles. Skilled workmen for the key positions were brought from England, but most of the workers were French, and the English workmen were absorbed into the French Catholic majority. Batiscan is said to have suffered heavy losses, but information is too meagre to say why.[122]

The Islet Forges were established in 1856 on the left bank of the St.

Maurice, and survived to 1876. Further up river, the Radnor Forges built in 1854 constituted an extensive and ambitious venture. Its chief difficulty was its remoteness from transportation lines, but it continued in operation on a small scale until 1908. French-Canadians built the Islet Forges and shared in the Radnor ones, and some at least of the men involved probably came from St. Maurice.[123] The later ironworks were fifty years too late, and all newcomers were handicapped by the competition of the established St. Maurice works. Perhaps, too, there was insufficient time and inappropriate management to develop the type of labour organization that supported St. Maurice. A labour problem of this sort seems to have been one factor in the failure of the Marmora ironworks.

The establishment of Marmora was undertaken in 1820 by a wealthy Irish immigrant, Charles Hayes. He was encouraged by the offer of a contract to supply iron to the naval dockyard at Kingston, and by a grant of 11,000 acres of provincial lands.[124] Nevertheless, Hayes's venture was decidedly incautious. Marmora had iron and waterpower and wood, but it was thirteen miles from the Trent river which provided a long and indifferent route, but the only possible route, to markets. Hayes cleared roads and built vigorously in 1821 and 1822, spending £30,000 before he had finished in 1823.[125] His establishment was extremely ambitious. He had two furnaces of four tons daily capacity and apparently coal-fired, a casting house, foundry buildings, coal houses, a sawmill, a grist mill, a tannery, houses for the staff and various other buildings.[126] Iron was produced by 1822, and was satisfactory in quality, but the transportation problem was never solved. Hayes was in financial difficulties by 1824, and Marmora passed in 1825 to his creditors and, in a short time, to Peter McGill.[127] The plant operated for some time in 1826, and McGill seems to have thought it had prospects for profitable operation, but it remained closed thereafter, nevertheless.[128]

Labour difficulties were not the fundamental cause of Hayes's failure, but they did complicate his problems. To operate Marmora required about forty-five workmen, whose wages made up the bulk of operating expenses which exceeded £100 per week.[129] Hayes appears to have considered that drunkenness was the basis of unsatisfactory behaviour among workmen. He took great pains to maintain the sobriety of his employees by preventing their access to liquor supplies — it is the one matter on which he exhibited firmness and power of decision — and his efforts appear to have achieved success.[130] In spite of this, or because of it, labour raised serious problems. In May 1822, work was greatly delayed by "Wicked Combinations among my work people — whom I have no power to control".[131] Later in that year, the workmen were orderly, but some of them were declared to be ineffi-

cient.[132] Shortly after, Hayes was "much plagued about workmen" —
probably this time the problem was to retain them — and the work
was badly delayed again.[133] These events suggest the initial difficulties at St. Maurice; they are perhaps the inevitable accompaniments of
building a labour machine from workmen unfamiliar with each other
and with their workplace. The rawness and remoteness of Marmora,
and the hostility exhibited by Indians and settlers, were further disturbing factors.[134] To provide a smooth and economical labour machine for Marmora, to make the mutual dependence of employer and
employees in their isolated position operate to the benefit of the employer, would seem to require the development of the kind of personal
labour relationships, and the contented and immobile labour force,
that were found at St. Maurice. But it is doubtful whether Hayes,
accustomed to situations in which workmen were at the mercy of
their employer, conceived of his problem this way; and still more
doubtful whether he possessed the generalship to cajole his workmen
into willing dependence. In any case, the time that would have been
necessary to cement new relationships, even with brilliant leadership,
was denied him.

Military Labour

Even now, military personnel are sometimes used to perform civilian
labour; nor, indeed, can a line be drawn precisely between military and
non-military employment. But the use of men organized in military
units for labouring work was a great deal more common before 1850
than afterwards. The practice reflects those inadequacies of the early
labour market that inspired other forms of pre-industrial capitalist
labour organization. Military units were already organized on feudal
lines, with an hierarchical leadership and the provision of non-
economic incentives. The frequent wars that affected Canada in the
seventeenth and eighteenth centuries invited a conception of permanent and total war, by which civilians could be regarded as soldiers
temporarily out of service, economic activity as the support of the
machinery of war, and the use of regular soldiers in its aid as consistent
with the military objectives of the state.

New France applied its total manpower resources almost indiscrimi-
nately to military and civilian tasks. The militia — the adult males not
in permanent military units — was called out periodically to fight, and
much more regularly for the King's corvée. In this duty they worked
under centralized direction on roads and bridges (wanted for both
military and civilian use) and on fortifications, for a varying number of
days each year. The system resembled the statute labour of English-
speaking communities and rested, like it, upon the undeveloped state
of the labour market and of other markets. Unemployed or surplus

labour was not concentrated in particular individuals, but spread more or less generally over all the available manpower. The unrationalized nature of the economy, which prevented the individual from making economic use of all his time, prevented him also from acquiring money with which he could conveniently pay for someone else to perform his public duties. However, French practice differed from statute labour in the indeterminate amount of time that might be claimed and in the central direction of work for predominantly military ends. Special corvées organized to build large fortifications or to provide transport services in support of troops emphasize the distinction.[135]

The conscription of civilian labour for military tasks is a fringe of the subject. Another fringe is represented in the settlement of military veterans on farms, in the expectation that they would support themselves by farming but remain available for military service. Settlement of the Richelieu Valley — an exposed frontier in 1666 — by about 400 men from the Carignan-Salières regiment was New France's most successful effort of this sort.[136] But members of the colony's regular troops, the Troupes de la Marine, were encouraged to settle, and did so, creating a chronic reinforcement problem for the French government.[137] Like the troops induced to settle by British authorities after 1783, with the same objectives, French veterans proved to be better frontier guards than they were farmers.

The real test, however, consists in the use of regular forces. In New France, these were the Troupes de la Marine, which may have numbered about a thousand men at most times.[138] Though not highly trained, they were enrolled in regular military units and were paid and supported as such. They, along with other regular units, were occasionally used for the semi-military task of building fortifications.[139] But what is surprising is the freedom with which they were turned to distinctly civilian employment. As military parties, they replaced rebellious carpenters in the shipyards,[140] were sent out regularly to manufacture tar, resin and turpentine,[141] and were employed in mining Cape Breton coal.[142] Besides these specialized employments, arising in each case from a deficiency of other workmen to perform tasks important for state policy, the Troupes were made a labour reserve serving the general labour market. "Discipline was so slack as practically to be non-existent, and the soldiers were free to take employment to supplement their pay."[143] The men were not only permitted but encouraged to serve the civilian market: companies were moved in 1749 to Three Rivers in the hope that men could be induced to take employment at the St. Maurice Forges.[144] The overhead cost of these workers was, of course, borne by the state, and the supplementary wage paid them was presumably less than an economic wage. Nevertheless, the labour of these men did not prove to be very cheap, at least for officials producing tar for the shipyards, or for the St. Maurice Forges,[145] and it may be doubted that the provision of cheap labour was the primary object of the authorities in making these men available. Rather, they sought to overcome the lack of reserves in their pre-capitalist labour market, which told particularly against sudden or short-term demands for labour and against demands from remote locations.

During the Seven Years' War, British commanders had to accommodate themselves in labour matters to the economy of the British seaboard provinces. They were compelled to hire civilian workmen required in support of their troops in the open market — a seller's market — under conditions which they found extremely irksome and at wages rates which they considered extortionate.[146] They were happy to find a different social order in Canada, and the military governors who took control from 1760 did their best to perpetuate the congenial French practices. Militia captains were again called upon to enforce the corvée for the maintenance of roads and bridges;[147] civilians were conscripted to build ships;[148] while special calls were made upon the militia to cut wood for the St. Maurice Forges[149] and to provide bateaumen for the movement of military stores.[150] A stubborn resistance to these demands was sometimes displayed by the habitants.[151] However, while the new regime was ready to conscript local men to get the work done, it did not have the disposition which the French regime had displayed, to make regular troops available to the civilian labour market. Rather, soldiers were utilized for such semi-military tasks as building forts.[152] But there were at least some occasions when British soldiers found discipline slack enough that they could take private employment near their posts.[153]

The undeveloped nature and special problems of Upper Canada after the American Revolution induced Governor Simcoe to recommend, and his superiors to approve, the establishment for that area of a special military force, the Queen's Rangers. This unit, of around 400 men, was established about 1792 and maintained until 1802. Its three functions were the same which the Troupes de la Marine had fulfilled in New France: to defend the country; to provide loyal settlers having military experience; and to build roads, bridges, barracks and forts, clear land, and navigate vessels.[154] That the Rangers should be wanted as road builders followed naturally from the existing scarcity of labour and from the desirability of roads in the new province, for military as well as civilian use. Regular British troops evidently were not very satisfactory as road makers: those employed to open the Lake Shore Road did so poorly that a compassionate militia officer offered to substitute his own, more competent men.[155] But the Rangers were of North American origin and experience, and besides building the fortifications of York, they can be credited with forming the road system that radiated from that place.[156] They worked first on Dundas Street, which they completed from Burlington Bay to the Grand River by 1794.[157] They were on Yonge Street in 1796, when they completed the road "as far as the navigable water on the Holland River", and a work party was there again in 1799. And in 1798, the Rangers worked on "that part of the Western road lying between Burlington Bay and Woodstock".[158]

The Rangers made their contribution at a time when, and place where, no substantial number of civilian workmen existed. Eastward, however, with the advance of settlement, a civilian labour supply was appearing. It is significant that Asa Danforth, who cut the road from York to Kingston in 1799 and 1800, seems to have had no trouble in getting enough civilian labour.[159] Perhaps soldiers were not needed very much for civilian tasks after 1800. All the same, large bodies of men could not be assembled without suggesting to many that their potential labour should be turned to account; and every augmentation of the supply of military labour in the years that followed brought fresh demands for road builders. This was especially true of militia units, made up after all of civilians and assembled at the expense of the civilian labour market. Thus, the militia "was used extensively on the roads" during the War of 1812-14, executing improvements at the Long Sault, Gananoque and Glengarry, and around York.[160] The military forces assembled in and after 1837 also made their non-military contributions, as will be noted below.

While military forces were being used systematically for civilian work in western Upper Canada, a reverse case, of civilians employed by a military authority under quasi-military conditions, can be found at Kingston. The problem here was not a scarcity of labour in general, but a scarcity of skilled labour; and if British administration at the beginning of the nineteenth century had not had a feudal tone already, the necessity to maintain a supply of artificers for the Kingston dockyard might well have inspired one. As it was, the Navy's way of handling and holding artificers was authoritarian but paternalistic. The Navy maintained its "company town" at Point Frederick, in which there were about 200 workmen, and an additional 170 members of their families, in 1816. The Navy did not hesitate to expel people from this community or to move them within it. But it went to considerable pains to promote the workmen's welfare and satisfaction. It provided housing, permitted artificers to buy goods from the naval stores at prices below those in Kingston, and encouraged the maintenance and expansion of a school that served the workmen's children.[161] Dockyard paternalism also extended, in 1830, to a battle to preserve the right of artificers to free passage over Cataraqui Bridge.[162] Aside from these benefits, the Navy's workmen long shared in the customary reward of a grant of land upon completion of satisfactory service.[163]

The construction of the Rideau Canal (1826-32) introduced another type of military labour for non-military purposes into Canada. Two companies of Royal Sappers and Miners were raised in England specifically for this work, and arrived on the Rideau in 1827.[164] The companies may have been raised on the supposition that skilled construction labour would not otherwise be available in the remote and raw territory of the canal, though no lack of labour seems to have been experienced in fact.[165] Cheapness may also have been a consideration. If it was, disappointment ensued, for the Sappers and Miners were not cheap, since they had to be paid whether they worked or not, instead of being sloughed off in slack times as civilian workmen were.[166] Even so, it was found politic to offer the Sappers and Miners land grants at the completion of their terms to counteract the corrosion of desertion, the curse of contemporary British military forces in America.[167] In practice, the Sappers and Miners were not used extensively in construction, but were rather employed in coordinating the work, in guarding stores, in testing projects to estimate a price for contractors, in emergency repair work, and above all, in intimidating the civilian labourers employed on the canals in order to suppress unrest and strikes and to keep down wages.[168] This example for decades afterwards inspired programs for stationing troops on public works to suppress strikes or other manifestations of protest from construction workers. It probably also inspired James FitzGibbon's "Plan proposed for making and repairing Roads in Upper Canada. Let a Regiment be raised to be composed of (1,000 men plus officers and NCO's)." FitzGibbon's soldiers were to be skilled men: "I contemplate employing the privates to superintend squads of Day laborers — these Privates not only guiding the laborers but also working at the same time with them." FitzGibbon thought that his force could be maintained at the same cost as the regular troops then in Upper Canada (£30,000 per year), that it would do more good than either regular troops or civilian labourers, and that suitable incentives, including the inevitable land grants upon the completion of service, would obviate "the great loss by desertion".[169]

The Rebellion of 1837 set off a new burst of military activity in Canada, and the continued threat of raids from the United States, together with the possibility of war with that country, induced the establishment of permanent units drawn from the militia but paid for by Britain.[170] These units were not called on very often for actual combat, and it was natural that their use for other purposes should be considered. Adam Fergusson in 1838 advanced a plan designed to turn the need for defence to the advantage of local employers. He preferred that defence forces be raised in Great Britain and Ireland rather than in Canada.[171] The troops he proposed to raise would have been kept in service from five to seven years and would have been paid even less than regular troops, with a portion of their pay forcibly withheld, but would have been compensated by free land grants. They were, nevertheless, to be kept free of subversive political views so as to prevent "that base tampering which unhappily seduces so many of our Soldiers to desert their Colours..." However, Fergusson's main point was: "Should the Military Services of these Regiments not be in constant requisition, the men would prove a valuable addition to the supply of labour in the Province, and tend to moderate, what is found to be the heaviest drawback upon the Cultivation & improvement of the Province...(i.e., wages)."[172] Such pressure from employers may have inclined the authorities to make military labour available for civilian uses. Moreover, though the labour market had been glutted in 1837 and 1838, there was a labour shortage in 1839.[173] It may have been considered also that employment and additional income would aid morale. In any case, troops (mostly if not entirely negro units) were offered to some of the road commissioners of the province in 1839 and after. The (extra) wage demanded for these men was one shilling a day, for privates, as against a civilian rate for labourers of two shillings and sixpence, or better. Not unnaturally, road commissioners considered this offer a bargain, and demand for military labour for road work exceeded supply.[174] Road work was carried on in this way at least until 1842.[175]

The military units of the 1840s were used also to suppress unrest amongst the labourers on various public works. Military commanders considered that the regular policing of public works (as distinct from emergency attendance at riots) was a civil function improperly thrust upon them by the parsimony of civil authorities. Nevertheless, military units were stationed for long periods on the main public works of this period. The Welland Canal was given particular attention: troops were retained there to control the canal labourers from 1842 until 1850.[176] Negro troops were used for this service — apparently, the same ones that had done road work until 1842. This employment of negroes to suppress Irishmen was hardly calculated to soften the antipathy which Irish workmen typically felt for negroes, and probably contributed towards the widespread anti-negro agitation in Canada West of the 1840s.

From about 1840, the Canadian labour market was well-stocked and possessed the reserves of a developed capitalist market. There was, therefore, little reason to call upon military personnel to perform civilian functions, except in emergencies. For the same reason, the custom of drawing upon militiamen for various kinds of supporting labour tended to fall into disuse. The earlier practice offers, however, examples of personal labour organization and illustrates its nature.

Since the overhead costs of military personnel were assumed by governments on defence grounds, their labour became available at small additional cost. It was particularly likely to be called upon for short-term tasks, and for work in remote places, which the reserve-less labour market was especially unable to provide. In contrast to contemporary arrangements in most civilian employments, discipline was maintained in military work parties by a minimum of attractions and a maximum of force and punitive threats. However, there was the same *esprit de corps*, the same hierarchical if not paternal leadership, and the same reliance upon non-economic incentives.

Paternalism and Other Forms of Personal Labour Relations

In the early nineteenth century, Canada had many more examples of paternalistic labour organization. It also exhibited substantially different types of employment. Comparisons among the various arrangements that existed serve to delineate the limits, and the essential nature, of personal labour organization.

Paternal methods occur frequently in the wood and water employments that loomed so large in early Canada. Among the Quebec timber handlers, it has been asserted, "There was an odd old-world paternalistic (not at all democratic) relationship between the cove men and their employers. Father and son worked for the same merchant, the merchant and his son looked after their men, down the years..."[177] Bateaumen, when the bateau was the basis of St. Lawrence transportation, were organized in much the same manner as the canoemen of the fur trade. The men worked about seven months each year, so that they enjoyed a relatively continuous and dependable kind of employment; they should therefore have found this work attractive. Employers, on the other side, had reason to attract and retain reliable men because of the regular service expected and the high degree of skill required to operate bateaux on the St. Lawrence. Bateaumen worked in brigades with appropriate hierarchies, like canoemen, and descriptions of them suggest that they responded warmly to leadership and readily developed an *esprit de corps*. They were paid by the trip, at rates amounting to $15 to $20 per month — high wages for the period. Nor, according to Selkirk, were these rates reduced by truck pay.[178] French-Canadian bateaumen were enlisted for the operation of an early ferry service at Niagara. The conditions of their employment were evidently much the same as on the St. Lawrence, with additional payment for their transport from Montreal and back each year.[179] In contrast, raftsmen on the St. Lawrence were handled with much less delicacy. Their employment was less regular, and their reliability may have been a less vital matter to the employer. Their nominal wages in 1804 were $15 to $16 per month, but employers reduced their actual expense to half this amount by encouraging raftsmen to take their pay in liquor.[180]

Bateaumen, like the canoemen of the fur trade, were of French origin. Most of the raftsmen on the St. Lawrence were Caughnawaga Indians. Both these groups were cheerful and docile employees; or, as an observer put it, "by habit (they) are taught to be decent and respectful to employers."[181] It is worth recalling Selkirk's opinion of docility among the canoemen: "The aptness of the Canadians for this employment depends perhaps more on this circumstance than any other"; and his comment upon a related characteristic: "The Canadians require a good deal of temper and attention in their leaders to be completely managed."[182] It seems proper to conclude that French and Indians exhibited some child-like qualities that encouraged their organization into paternalistic labour systems.* However, ethnic peculiarities were a subsidiary aspect of the matter and must not be pressed too far. It was not only Calvin and other employers like him who handled English-speaking workmen in a similar way when they desired to retain labour under conditions of scarcity. Early shipowners on Lake Ontario, for instance, found it expedient to retain their seamen on a permanent basis, though they could employ them on the lake for only six months of the year.[183] Similarly, in early years in Upper Canada, carpenters were retained on an annual, or, at the least, on a full season, basis.[184] At the same time many workmen, including French ones, were employed under different conditions.

An industry that displayed an exceptional diversity of employment conditions is agriculture. It was first developed, of course, with the French system of agricultural settlement, which has been justly celebrated as an outstanding instance of feudal organization.† French-Canadian agriculture was feudal in its possession of seigneurs and seigneuries, of strip farms occupied by peasants who owed annual payments of rent and produce in return for their holdings, and of corvées (though these, indeed, were royal rather than seigneurial).‡ But the system was more than that. Munro held that it was a truer form of feudalism than seventeenth-century Europe knew, because it approached closer to the mutual accommodation which feudal agricul-

---

*On this characterization, see Introduction, p. xxvi. Ed.

†The most recent authoritative and comprehensive treatment of the seigneurial system is R.C. Harris, *The Seigneurial System in Early Canada* (Madison: University of Wisconsin Press, 1967). Ed.

‡According to Trudel, there were also seigneurial corvées, governed by contract, of three or four days a year. Marcel Trudel, *The Seigneurial Regime*, Canadian Historical Association Booklets, no. 6 (Ottawa, 1967), pp. 12-13. Ed.

ture originally denoted in Europe. The seigneurs were not simply rent collectors, but made a crucial contribution to production by assuming heavy burdens of capital investment and leadership. In return, the habitant worked the soil and provided annual income, as well as respect. The system was based upon the mutual needs of its parties: the seigneurs' problem was the scarcity of labour, and the habitants' problem was the scarcity of capital. With each side dependent on the other, a balanced and equitable system developed, in which the seigneurs gave much, lived close to their people and were well-regarded by them.[185] It is significant that the habitant clung to his system and resisted freehold tenure even after the old order had been abused and distorted by two generations of English-speaking seigneurs.[186]

British agricultural settlement was not directed towards the creation of a feudal system, and, in general, did not produce one. It furnished some examples of feudal-style organization, nevertheless. Philemon Wright's settlement at Hull, established in 1800 in much the same patriarchal way that chieftains long ago spread settlement in Europe, was a microcosm of a self-sufficient feudal economy. Wright's essential problem, as he freely acknowledged, was to get and hold labour. To keep men, he had to offer continuous employment; otherwise, "it would have been impossible for me to obtain men in the spring, when I most wanted them, as the distance from any settlement was so great."[187] Scarcity of labour forced Wright, therefore, to assume a continuous responsibility for the livelihood of his men. Having done so, he found himself with a surplus of labour in winter; and it was a desire to recover something from this surplus that put him in the lumber and timber business.[188]

There were other patriarchal settlements, like those of McNab on the Ottawa and of Thomas Talbot on Lake Erie.[189] These did not achieve, or even aim at, the same economic coherence that Wright's settlement had. However, in isolated settlements like Talbot's, there was a characteristic scarcity of capital and of alternative opportunities that made men dependent on a patriarch, and a scarcity of labour that made the patriarch dependent on the men; and consequent development of a society built around mutual economic and social responsibilities more elaborate than those found in capitalistic society. Indeed, each of the more affluent among early immigrants from the United States and Great Britain was inclined to set himself up in the same way as a local lordling; and until about 1830, the scarcity on the frontier of men on one side and of capital and alternative employers on the other, crowned many of these ventures with some little success.[190]

In spite of these flirtations with patriarchy, it was perfectly apparent before 1830 that the agricultural organization of English-speaking Canadians was headed in a different direction, towards a mass of family-sized farms operated by independent farmers. The ease with which land could be acquired, the ambition of most men to own a farm, and their consequent reluctance to work for someone else, supported this trend. Proposals to reverse it, by putting the ownership of land beyond the reach of poor men, so as to promote the growth of large estates and of a pool of landless labourers with no alternative but to work on them, may have made some appeal to the near-squires and would-be squires in Canada. Nevertheless, Wakefield's system, and other plans like it, were never taken very seriously. A fundamental objection to them was that as long as poor men could acquire farms in the United States, attempts to deprive them of this privilege in Canada would only drive them out of the country.[191]

On the other hand, the ease with which small men became land-owners in Canada, their unwillingness to work for others, and even the multiplication of potential employers that followed, do not provide the full explanation of the failure of hierarchical agriculture or of the complexity of agriculture's labour problem. If they did, it might be expected when the land was all taken up, that a class of landless labourers available for hire would appear. Farmers with the most land and the most capital would have the best chance to hire them, and to keep them if others tried to hire them away. If the big farms proved to be the more efficient (perhaps, even if they did not), the agricultural economy would gradually become an organization of squires and labourers.

The fact that nothing of this sort occurred depended on a failure of labour supply only indirectly, in the sense that supply cannot develop far without an effective demand to support it. The essential trouble was with demand, and less with the volume of demand than with its unevenness over the year. Demand for agricultural labour mounted to a great peak at harvest time. If it was economic to hire labour through the rest of the year — as it was on large farms — it was nevertheless true that several times as much labour as usual was required for the harvest month. The small farmer's solution to this problem was to employ himself fully for one month and to underemploy himself during the other eleven months. The big farmer appeared to have two choices open, though neither turned out to be very satisfactory. He could hire on an annual basis the number of men required for harvest and try to find something for them to do during the other months — a pre-capitalistic solution. Or, trusting in the capitalistic market, he might hope that the extra harvest hands he needed would turn up from some source.

The best supplier of extra harvest labour was the immigrant stream. After 1815, there was a good supply of immigrants in most years, especially in the fall when they were most needed. Still, this supply depended on factors beyond local control and might fail.[192] A second possible source was the very large surplus of labour available from small farmers who needed cash income. But this labour was blocked from a straightforward response to market demand by whatever attentions the settlers devoted to their own farms, by family obligations, and by the distance between the intended labourers and employers.[193] Efforts were made in the 1830s to create a special variant of this type of labour, more satisfactory for big farmers. Sir John Colborne and the emigration agents tried placing immigrant families on five-acre lots, preferably among well-to-do farmers. These smallholders, like their ancient English prototypes, would be so dependent on the labour market that they would take employment readily, but would be supported enough by their holdings that they might survive long periods of unemployment. These settlements were defended as reasonably satisfactory. However, they were admitted to have only a limited application, and even their proponents acknowledged that demand for these cottage labourers was very irregular.[194]

A more systematic alternative for the large farmer was to hire men on an annual basis, for an annual wage which would balance the worker's satisfaction with continuous employment against the employer's satisfaction with having a man for harvest time. If employers had been few and collusive, annual engagements might have worked well. However, even at the beginning of the nineteenth century, this condition was not met. "Frenchmen were frequently employed (by farmers near Kingston), yet they could not be depended upon to remain during the whole season. At harvest time, when large wages would be offered, the hired man would often, without hesitation, leave his employer to go to another who would give for a while larger wages..."[195] At the same time, "American" farm labourers, though more competent, were hard to manage, "being scarce and knowing their own importance", and would leave upon the least complaint or difficulty.[196] In general, it was "impossible to procure trusty servants, clerks &c as in Europe — Every well doing man so easily finds means of setting up for himself that none but the refuse will remain in the employment of others."[197] Small industrial enterprises, like farms, were unable to escape this instability of labour supply, or would not pay the price to overcome it.[198] One consequence was a popularity of share systems in the conduct of farms and mills by which the owner offered a share of the profit rather than a wage to the operator.[199]

The final handicap to large-scale agriculture was the satisfaction which the mass of small farmers derived from the ownership of their homesteads. For the sake of his homestead, and the independence which it permitted, the small farmer would accept a smaller income, if necessary, than he could win in the market. The competition posed by this preference, added to the difficulty of commanding a satisfactory labour supply for the large farm, determined that no hierarchical labour organization would persist in Canadian agriculture. If it can be said that there ever was a solution to agriculture's labour problem, it lay in the motivation which pride of ownership provided for a numerous yeomanry. This solution was made more tolerable, and was more firmly entrenched, by the development of farm machinery that economized on labour, especially at harvest time.

The position of the small farmer shaded into, and was most starkly represented by, the position of the Atlantic coast fishermen. The fishermen were notoriously dependent upon merchant-employers who employed them under a truck system, under which fishermen were chronically in debt to merchants, and merchants put their own prices both upon the articles they sold and the fish they bought.[200] Small farmers in English-speaking Canada had similar relationships, in greater or less degree, with country merchants — themselves links in a chain of indebtedness stretching through Montreal to London. While the local merchant was in debt himself to others, his power over the farmer was much like that of the fish merchant over the fisherman: because he was likely to be the only merchant available, and because the farmer was typically in his debt, he could put his own valuation on the supplies he sold and on the produce he bought from the farmer.[201]

Ostensibly, the relationships in these cases revolved around "prices", in the modern sense in which prices are distinguished from wages. The cases are variants, in fact, of the ubiquitous merchant-employer system. They differed from paternalistic labour relationships, not because prices were involved, but because the bargaining advantage was overwhelmingly on the employer's side. This imbalance depended, for one thing, upon the scarcity of and lack of competition between merchants; but paternalistic relationships were also founded upon a scarcity of employers. The essential difference of the merchant-employer system from feudalism arose on the other side, from the inability of fishermen and farmers to escape, or to raise any other significant threat to, the merchant's power. Custom, law, the hope of preserving a little property and the lack of convenient alternatives all contributed to make the producers immobile and, therefore, defenceless. The impotence of workers and the undisputed power of a merchant-employer seem to have reached their possible limit on the Saguenay, where the Price interests were able to operate a peculiarly vicious form of debt slavery among the early settlers who cut wood for them.[202] Such situations display neither the paternalistic virtues imposed by a scarcity of workmen and employers, nor the capitalistic benefits that appear when there are many workmen and many employers. In the merchant-employer system, the merchant is a monopolist, who can treat workmen as if they were plentiful (whether they are or not) because they have no way to escape or retaliate. Hence, the merchant-employer can transfer all risks to the workmen and all profits to himself.

The conditions of personal labour organization may now be summarized. There must be a sufficient scarcity of labour (labour in a particular place, labour of a particular skill), so that workmen enjoy about as much bargaining advantage as their employers. Workmen must also retain their bargaining power, by preserving a power to leave or to damage the employer by bad work. Employers must enjoy a degree of monopoly in their labour market, so that they may expect to retain their workmen indefinitely and be able to recover their outlays on the workmen's overhead costs. Employers must assume these overhead costs in order to get and hold scarce workmen; and since dismissal is thereby made very costly to employers, the employers must find some effective incentive other than the threat to fire. These are all essential conditions. Some other circumstances appear to have facilitated the appearance and persistence of personal organization. The system was more likely if the labour required was skilled. It appeared more readily in employments that were permanent or were carried on during the greater part of each year. It appears to have been deemed especially suitable when the employees required were French. However, these circumstances were not essential to the system. Though skill, permanence and Frenchmen survived, the system passed away when workmen or employers ceased to be scarce.

CHAPTER III

Canada's Labour Force: Population Growth and Migration

Canada's inadequately-supplied labour market of the eighteenth century gave place, around 1850, to a capitalist labour market with permanent reserves. The transition was accompanied and facilitated by a rapid growth of the labour force. Mere numbers, however, could not produce a capitalist labour market: changes in the nature and organization of the work force were required for that. The business of this chapter and of the succeeding one is to examine how so fundamental a development came about.

Though thousands of Canada's aboriginal inhabitants were, in some sense, members of the labour force, their patterns of behaviour were but slightly related to those that give shape and meaning to a European-style labour market. Discussion is therefore confined to the European population resident in Canada. Canada enjoys an unusual wealth of census records for its European population.[1] They permit close estimates to be made of the population for any date, from which the potential membership of the labour force may be inferred.[2] The actual paticipation of this potential membership in the labour market is more difficult to state. Different ethnic groups played distinctive roles in the development of the labour market; the character of the groups, and the relations among them, are therefore important parts of the discussion to follow.

Canada's Population

There were about 70,000 Europeans in Canada in 1765, nearly all of French origin. By the processes of natural increase and of immigration, the population of the two Canadas rose to about half a million by 1820, a little over a million and a quarter in 1844, and was reported by the censuses to be 1,842,000 in 1851-52, 2,508,000 in 1861 and 2,812,000 in 1871.

The original French-Canadian population, despite the lack of reinforcement by immigration after 1760, succeeded by a rate of natural increase of about 3 per cent per year in numbering close to a quarter of a million in 1815, half a million about 1840 and a million in 1870. It has always constituted the bulk of the population of the lower province, though almost outnumbered in the cities in the middle of the nineteenth century, and provided over half of Canada's European population until about 1835. In the second half of the nineteenth century, its numbers were depleted by emigration of some hundreds of thousands to the United States. Nevertheless, it was not until sometime in the 1850s that English-speaking Canadians contributed a larger portion than the French of Canada's native-born.[3]

English-speaking Canadians multiplied particularly by immigration after 1760. The basic stock was provided by American settlers. The estimate that there were 25,000 persons of English speech in the Canadas in 1784[4] seems to put the number too high, but the great majority of whatever number was present were of American origin. The same must have been true of the population of 150,000 more or less of 1815. After that date, the American element was increasingly mixed with, and indistinguishable from, English-speaking immigrants from overseas. The overseas immigrants are of the greatest importance, not only because of their substantial contribution to the numbers of Canada's population, but also because their character and technical equipment played a great part in changing Canada's industrial structure. Until 1815, overseas immigrants were largely Highland Scots, and there may have been no more than 5,000 of them. But immigration at Quebec set in seriously from 1816, at a rate of about 10,000 per year in the 1820s, but rising to 50,000 a year at the peak in the early 1830s. These immigrants were predominantly Irish, at first almost exclusively Protestant, but with Roman Catholics appearing in significant numbers around 1830, and becoming predominant in the fantastic famine migrations of the late 1840s. Meanwhile, English and Scottish immigrants contributed steadily though unobtrusively to the totals, and in the 1850s English immigration seems to have become as predominant as the Irish had been earlier.[5] From 1842 onwards, censuses report the numbers of these immigrant stocks in Canada. In the 1842-44 census, there were 230,000 of British birth in Canada distributed in the ratio, English — 2, Scottish — 2, Irish — 7. In 1851-52, the British-born numbered 410,000, in the same ratio. By 1861, the English had gained and the Irish lost significantly, and in the two Canadas there were 127,000 who had been born in England and Wales, 112,000 born in Scotland and 242,000 born in Ireland: 480,000 British-born altogether, or about one-fifth of the total population.

French, Americans, English, Scottish and Irish together made up almost the whole of Canada's labour force in the nineteenth century. Each of these ethnic and cultural divisions had its own peculiar character and its distinctive role. The present chapter discusses the part played in the development of the Canadian labour market by the French, by immigrants from the colonies that formed the United States, and by overseas immigrants from England and Scotland. Immigrants from Ireland will be dealt with separately in Chapter IV, in consideration of the complexity and importance of the impact which the Irish had on the labour market, and the space required to deal with that subject.

The French-Canadians

It was the ambition of the French administration to make an agricultural community of New France, but the success achieved can easily be exaggerated. Towards the end of the French regime, the fur trade occupied at least 2,000 able-bodied men,[6] and possibly twice as many. Fishing and forestry, shipbuilding, iron-making and various handicrafts, must have occupied some thousands more. Perhaps no more than half the labour force of New France were full-time agriculturalists. On the other hand, the bulk of those in non-agricultural pursuits retained a foothold on the farm, from which they came and to which they returned. Only Quebec and Montreal, with about 15,000 inhabitants between them, had truly urban populations.

French Canada has been a perennially frustrated society. Lunn offers the view that the growing population of the colony was in 1760 "on the verge of overflowing the valley of the St. Lawrence into the abundant Niagara peninsula", where a firm base for the colony could have been established.[7] A century later, Ontario was to destroy the dream of a second Quebec in Western Canada. If New France neglected agriculture, French society after 1760 was driven back upon the land. Commerce fell into the hands of newcomers, the fur trade declined, and the artisans and labourers of Lower Canada were drawn in the nineteenth century from British immigrants. The canals and railways of the province were built and operated by others, and the French won only a marginal position in the timber industry. Census returns do not offer a precise measure of the habitant population, but they indicate, and other authorities agree, that a very large proportion of the French population — perhaps 80 per cent — was in agriculture throughout the nineteenth century.[8] Concentration on one industry is not necessarily detrimental; but final frustration lay in the fact that the habitant agriculture in which the French were confined failed to support its practitioners. Their standard of life, their satisfaction and probably their physical condition, deteriorated.

TABLE I

POPULATION, CANADA: CENSUSES AND ESTIMATES

Date	Lower Canada	Upper Canada	Both Canadas	Native-Born French	Native-Born English	British-Born English	British-Born Scottish	Irish	'Effectively' Irish Roman Catholic including those born in Canada
1784	113,012	10,000 est.	123,000	100,000 plus	25,000[1] est.				
1815[2]	335,000	95,000	430,000	247,000 est.					
1825	479,288	157,923	637,211	332,000 est.					10,000
1844[3]	697,084	567,000	1,264,000	541,000	340,000	54,000	55,000	125,000	85,000
1851-52	890,261	952,004	1,842,265	695,945[4]	650,799	93,929	90,376	227,462	200,000
1861	1,111,586	1,396,091	2,507,677	880,902	1,037,541	127,469	111,996	241,568	280,000
1871	1,191,516	1,620,851	2,812,367	1,005,200	1,278,029	136,433	102,067	188,828	260,000

[1] English-speaking, mostly Americans, in Canada. The estimate (from the Census, 1870-1) seems too high. In terms of natural increase, there should have been about 107,000 French in Canada at this date.

[2] The estimates are Bouchette's, except for the French population, which is my estimate.

[3] Since the Upper Canadian census was taken in 1842 and the Lower Canadian one in 1844, interpolations have been made to derive figures for Upper Canada for the later date.

[4] This is a census figure, but it seems inconsistently high by at least 50,000.

Symptoms of the illness that beset French-Canadian society in the nineteenth century appear in many directions.* Selkirk reported in 1804 that "the Canadian peasantry were formerly remarkably sober, but are now much addicted to drink — even the women exceed — formerly it used to be disgraceful in any..."[9] The extent of under-employment is suggested in a description of Boucherville in 1811. About 10 per cent of the population of the town were available as servants at wages of 7s. 6d. per month.[10] Dr. Dunlop declared that the French-Canadians "work for wages not much, if at all, higher than those of a labourer in England."[11] His opinion is supported by a good return of wages, for 1834, which reported that the common labourers of Lower Canada could expect 2s. 3d. per day from May to September, 2s. in October, and 1s. 6d. for the other months.[12] These sums barely exceed contemporary English wage rates, and it is not unlikely that labour was obtainable for less from the immobile sur-pluses of many rural districts. According to the Durham Report, "The circumstances of a new and unsettled country, the operation of the French laws of inheritance, and the absence of any means of accumu-lation, by commerce or manufactures, have produced a remarkable equality of properties and conditions.... The mass of the community exhibited in the New World the characteristics of the peasantry of Europe. Society was dense; and even the wants and poverty which the pressure of population occasions in the Old World, became not to be wholly unknown..." Moreover, "It is impossible to exaggerate the want of education among the habitants."[13]

Not all reports were gloomy. Howison said that, about 1820, the lower classes of Montreal "carry with them an appearance of vigour, contentment, and gayety, very different from the comfortless and desponding looks that characterize the manufacturing population of the large cities of Britain";[14] and Bouchette thought, presumably about 1830, that the food of the habitants was good, though they did not eat nearly as much meat as others.[15] Similarly, a student of habitant living standards, covering the period 1820 to 1850, concluded: "On the whole one gathers that the general type of food consumed was much better than that eaten by the English farmer of the time." But his qualification deserves at least equal attention: "Yet again we must remind ourselves that in the more backward parts of the seigniory (Mille Isles) barley bread, salt pork, and potatoes with a little milk formed the staple diet."[16] There were complaints of widespread distress and famine in the countryside in 1817, and similar reports in 1822.[17] The Durham Report

---

*The "crisis" in the French-Canadian economy in the early years of the nine-teenth century has elicited a lively debate in recent years. See Introduction, pp. xxvi-xxvii. Ed.

said of the habitants that "complaints of distress are constant, and the deterioration of the condition of a great part of the population admitted on all hands..."[18]

The evidence for actual physical deterioration, while not conclusive, is strong. It may be wondered whether physical weakness was not an important reason why the French did not seek the heavy work of canal construction and why employers did not want them (though, occasionally, employers talked differently after a tussle with their Irish labourers). It was claimed of French-Canadian emigrants to the United States: "The incessant work wastes their strength; they die young, or become prematurely infirm. The greater part of those who return are ruined in health."[19] There may have been something more than the penny-pinching labour policies of George Simpson behind a complaint of McLoughlin from Oregon in 1844 about the canoemen sent him: "As you say the Boutes must be trained in this country, but the truth is the men are so miserably small and weak for years past we cannot find men of sufficient physical strength among the recruits to make efficient Boutes to replace the old hands..."[20] Thomas Brassey, who had employed local labour all over the world on his railway projects, recruited a large number of French-Canadians to help with the construction of the Grand Trunk. The result was stated as follows:

> The French Canadians, however, except for very light work, were almost useless. They had not physical strength for anything like heavy work. "They could ballast, but they could not excavate. They could not even ballast as the English navvy does, continuously working at 'filling' for the whole day. The only way they could be worked was by allowing them to fill the wagons, and then ride out with the ballast to the place where the ballast was tipped, giving them an opportunity of resting. Then the empty wagons went back again to be filled; and so, alternately resting during the work, in that way, they did very much more. They could work fast for ten minutes and they were 'done'. This was not through idleness, but physical weakness. They are small men, and they are a class who are not well fed. They live entirely on vegetable food, and they scarcely ever taste meat."[21]

Other evidence of the malaise of French-Canadian society was provided by the unrest in the Assembly of Lower Canada, by the nationalist movement that developed, and by the Rebellion of 1837. If the political leaders of French Canada directed the habitant's discontent into channels that brought him little benefit, that is nevertheless beside the point. The habitant needed help. Some relief was found in emigration, the major palliative in the nineteenth century for the illness of French society. Emigration was important, too, in that it stimulated real efforts to bolster up French society and even to reorganize it. These efforts produced a campaign to have more land opened to settlement under seigneurial tenure, a colonization movement, the abolition of seigneurial tenure, and a movement for the encouragement of manufacturing with the aid of protective tariffs. The effectiveness of these measures depended, of course, upon their relevance to the habitant's situation and difficulties.

The Plight of the Habitant

There is some justice in the view that French Canada's problems stemmed from British domination. Alien seizure of nearly all the places that provided wealth and power not only impoverished the French but discouraged them from attempting to escape their imprisonment in seigneurial agriculture and the professions. Capitalistic practices to which the British were addicted clashed with the institutions and ethics upon which French Canada had been built. It could be held that British governments, somewhat inconsistently, did further damage by supporting the most reactionary features of the earlier regime as an aid to their own control. Some important points can be made, however, on the other side. At no time did the new regime deliberately prevent the French as individuals, or as a whole, from evolving towards more rationalized and workable ways of life: most of the clash of cultures that occurred arose, in fact, out of attempts by the British subjects to promote "progress", and French resistance to it. It was an illusion for French Canada to suppose that it could persist as a seventeenth-century feudal island in a nineteenth-century capitalistic sea. It was a particular and fatal delusion to imagine that seventeenth-century practices, because they had provided a good solution to the problem of how to organize a society in the seventeenth century, could go on giving a good solution indefinitely. By the nineteenth century, they offered no solution at all. Yet the habitant did not want to abandon them, and he escaped from them, when he was forced into it, only with the greatest difficulty. This is the essence of the habitant problem.

The feudal system of New France had been designed to facilitate settlement by requiring the seigneur to provide all the necessary capital, thus leaving "to the settler the free use of all his pecuniary resources at the most critical period".[22] In this aim, the system was extremely successful and compares favourably with the freehold system by which poor men had to struggle desperately to amass the little capitals required for their holdings. The system also took account of the future in that the uncleared lands of the province were held by the seigneurs "in trust for (the habitants) and for their descendents".[23]

The price paid for this facility of settlement was, however, very high. In the seigneurial system the habitant not only had little need to save, he was also hindered in doing so by payments made to seigneur and church, and was generally encouraged in a non-capitalistic and non-accumulative attitude, in which existence was a sufficient justification for existence. In the early years, when there seemed to be plenty of land for all, the attitude assumed by the habitant may not have mattered much, but in the nineteenth century, when French society was in crisis, it closed one door to a solution. It was doubly closed because, by that time, it was much harder for the peasant to save anything, and especially hard for the poor peasant, who needed land the most. These are the men of whom it was fairly said, "that in denying them the power of acquiring land under (feudal) tenure, they are virtually excluded from the market when crown lands are put up for sale."[24] A more general statement of the problem is:

> It is well known on what ruinous terms Canadian settlers will take lands in some situations. Our youth has but two alternatives — to expatriate themselves, or to take up lands at any price.... Thus, there are...almost everywhere in the inhabited part of Lower Canada...a surplus of population ready to overflow any tract of land which is accessible, whatever may be its quality and price. The only requisite is that there shall be nothing to pay down; if the soil is good, they hope to clear themselves in time; if it is bad, and the seller is pressing for payment, they do what they can with the wood and a few exhausting crops, which ruin the soil for ever...[25]

Thus, there arose a class who would do anything to get land except save for it. There was nothing from which to save in their own situations. They might have earned money, as immigrants did, but to do this they would have had to divorce themselves from their society, and they were badly equipped for employment. It is significant that immigrants with a similar background to that of the habitant rarely earned their way to proprietorship either. James Stephen was much less profound than he thought when he ruled that "it is absurd and unmeaning to say, that people will not settle on Lands, merely because they are discharged from feudal services."[26]

Behind the lack of money to pay down was the absence of a system that provided incomes from which surpluses could be squeezed. A freehold system presupposes both substantial income and substantial expense within an exchange economy; the balancing of these, with skill and good fortune, can produce surplus. Such a system also implies surpluses of grain available for sale and, perhaps, export — in capitalistic logic, these are the only justifications of an agriculture in the first place. The habitant had been conditioned in an environment, however, in which it was hard to make surpluses appear and in which surpluses would likely be unsaleable if they did appear. Lunn thought that the agriculture of New France very nearly reached a position by which the market for farm produce and a surplus-producing agriculture might have promoted and justified each other; but this advance was never realized.[27] The habitant turned his back on profit-oriented agriculture and became incapable of responding to price stimuli.[28] His inertia, like that of the Irishman, rested on the improbability of any reward for effort. Again, about 1800, there was promise of an agricultural economy based on exports and profits; but prices became unfavourable, markets closed, and this chance faded away too.[29] The response of the habitant was to fasten even more firmly on an economy of self-sufficiency in which, because sales were small, survival lay in keeping costs smaller. As population thickened, there was progressively less chance of operating in any other way.

A structure oriented against risk, surplus and accumulation is harmful enough, but the French position was made still worse by the system of inheritance in vogue: farms were divided equally between all the direct heirs and, in the name of justice and equality, divided into strips which became narrower and narrower. Already in the 1740s, the administrators of New France had thought it necessary to take drastic steps against this ribbon fragmentation, on which they blamed declining output, at a time when no one could suppose any absolute shortage of land.[30] But subdivision continued, until, according to Munro:

> The habitants were as a class showing signs of restlessness and discontent during the earlier thirties. Under the influence of the French law of succession, their domains had been divided and subdivided until a holding, in the peculiar shape which it retained, would scarcely have sufficed to support a family even had the habitant adopted up-to-date methods of cultivation. This he did not do: his methods were for the most part those of his great-grandfather of the old French epoch. Fertilization of the land was rare; systematic rotation of crops would have been most difficult on the narrow strip of land which he held; and implements showed little or no improvement. If anything, the habitant was at this time worse off than he had been before the conquest; for while his average holding was much smaller, neither his seignior nor his church had in the least relaxed its demands upon him...[31]

It is from this background that the claim that there was a shortage of land in the nineteenth century has to be viewed. The assertion was made very freely and was given particular sanction by the Durham Report, which cited a claim that the population of the seigneuries had more than quadrupled between 1784 and 1826, while the quantity of land under cultivation had increased only by a third.[32] It is surprising to find, then, when the statistics available are examined, that the amount of land per rural household does not appear to have been much different in the nineteenth century than it had been in the eighteenth.[33] Of course the statistics, like the loose claims which they are intended to correct, must be taken with some reserve. In particular, the inclusion as time went on of prosperous farms in the Townships and of extensive but otherwise unattractive holdings of colonists, probably pulled up the averages and thereby obscured the deterioration on the old seigneuries. But the statistics do suggest that shortage of land in an absolute sense was not the basic trouble; and, on the other hand, give a certain support for English puzzlement why, if land was short, habitants did not work their holdings better or take land under freehold tenure. Aversion to freehold has already been touched on. The poor use made of existing lands deserves more attention.

Some reasons for the habitant's bad farming have already been noted: the awkward shape of the farms, the lack of markets to encourage improvement, the lack of capital to finance better methods. There is also the question of the habitant's concentration on wheat as a crop. This emphasis was often condemned, the only question being whether it was primarily a consequence of peasant obstinacy or of the demands made on the habitant by seigneur and church.[34] On the other hand, the habitant's methods have been defended by Professor Fowke. Fowke argued that extensive farming was much better suited to the habitant's situation than zealots of "high farming" allowed; and, therefore, that it was economically sound for him to desire to extend his area of cultivation while the lands presently cultivated were still capable of producing much more. His reason: "The St. Lawrence region did not always produce sufficient cereals for its needs, but it produced cereals with less difficulty than it experienced in the production of livestock."[35]

This is a qualified endorsement of the habitant's methods and efficiency, at best. But what is much more serious, it fails to take account of a capital fact of history: the failure, year after year, of the Lower Canadian wheat crop. That this chronic failure has been so often overlooked is, in retrospect, surprising. There is no lack of evidence about it. It is implicit, for instance, in the statistics of wheat output in Table II, which were recorded without any reference to this matter. It was stated plainly by enough observers. For example, the report of the Legislative Committee on Emigration of 1849 includes the following statement: "As long as the crops continued good, these small farms were sufficient for the wants of their proprietors; but for about ten years past the wheat crop has failed. In no one of the parishes in this part of the country (Quebec district) has a sufficient crop been gathered to feed one half of its population, which increases every year..."[36] Yet, only in the recent work of W.H. Parker, a geographer, has the nature and extent of this phenomenon been set out.[37]

TABLE II
LOWER CANADA: PRESSURE ON THE LAND

Date	Population	Rural Households (Occupied houses excluding Quebec and Montreal)	Farm Lands in Acres Cultivated	Farm Lands in Acres Uncultivated	Farm Lands in Acres Total	Total acres per household	Output wheat (bushels)
1721	24,951		51,788	10,169	61,957		268,565
1739	42,701		156,755				602,875
1765	69,810	7,630			784,450	103	—
1784	113,012	15,000 est.			1,308,180	87	—
1827	471,875	62,900 est.	835,165	1,620,330	2,455,495	39*	2,921,240
1831	553,134	71,322	1,721,844	4,151,519	5,873,363	82	3,407,756
1844	697,084	92,792	2,226,475	3,365,435	5,591,910	60	942,829
		("Occupiers"— 94,819)			Per "occupier"	59	
1851-52	890,261	110,222	3,605,167	4,508,241	8,113,408	74	3,073,943
		("Owners"— 95,813)			Per "owner"	85	
1861	1,111,566	132,684	4,804,235	5,259,614	10,375,418	78	2,654,354
		("Occupiers"— 105,671)			Per "occupier"	98	

Sources: *Census of Canada*, 1870-1, vol. IV, except for land cultivated and output of wheat, 1739, which are taken from Alice J.E. Lunn, "Economic Development in New France, 1713-1760" (Ph.D. thesis, McGill University, 1942), pp. 443-44.

*Acres per household, 1827: The return (not census) for 1827 is shown because it may have been the basis for the claim in the Durham Report and elsewhere that land available to the habitant population had declined disastrously. The population shown for this year is too low, and the farm land shown is fantastically low. The census returns for 1831 demonstrate the fact. However, is would appear that the return of farm lands was too high in 1831, or else too low in 1844. The 1844 figures seem more likely to be at fault.

Parker found that in the 1830s (especially 1835-37) and the 1840s there occurred a catastrophic failure of the wheat crop of Lower Canada. The failure affected in particular the old, rich lands — the seigneuries and the bread basket of the Montreal plain. It was accompanied and obscured by a rapid expansion of wheat production on new lands outside the older (and basically more favourable) regions of cultivation. Parker suggested three aspects of the wheat failure that deserve attention. In the first place, the areas affected by the crop failure were so exactly the areas which actively supported the Rebellion of 1837 in the lower province, that a very close relationship between these phenomena must be supposed. Secondly, though the wheat fly was the immediate "cause" of the crop failures, Parker pointed out that crops on new lands nearby withstood the ravages of the fly with no apparent difficulty; he deduced that the essential cause of the crop failure was exhaustion of the soil. Thirdly, since the agricultural heartland of Lower Canada was forced to turn permanently to other crops than wheat, while newer and poorer lands were incapable of producing wheat very well or in very large quantities, the province was henceforth doomed to be incapable of wheat exports, or even of feeding itself. A fourth conclusion can be drawn: the dense population of the old seigneuries had exceptional reason in the 1830s and 1840s to think that land was short and to seek opportunities elsewhere.

This is bad enough; but to set out the problem fully, it is necessary to notice the total nature of habitant society. It furnishes a classic example of that type of peasant community, most common in Celtic lands, which practised subdivision in the inheritance of land.[38] It is obvious that after division has been several times repeated in such a society, no heir will possess enough land to support him. But this is only part of the difficulty that arises. In communities like this, the prospect of an inheritance (however small) prevents dispersion and encourages rapid population growth. The community contains abudant surplus labour, which is attracted to seasonal employment in an effort to support the farm that cannot support the farmer, but does not seek permanent wage employment. The society does not shake loose the sons made surplus by subdivision. Neither, in its equality of poverty, does it induce them to learn those inventions that are supposed to follow from necessity. Instead, it encourages them to meet their problems by accepting lower and lower living standards, and from the resulting impoverishment of body and mind it inhibits them from seeking out a solution to their problem. The tone of such societies is antagonistic to change and to capitalistic rationality. Their members are incapable of saving themselves — though they might postpone catastrophe by war, or by serving in the British army, or by finding more land under seigneurial tenure. In Europe, societies of this type were finally dissolved by famine. The fate of French-Canadian society was not quite so brutal and simple, but it, too, had to be dissolved through outside pressures and internal ferment.

The Ferment of French Society

The difficulty with efforts of peasant societies like that of French Canada to survive by turning inwards and intensifying peasant values, is that they can be effective only for a time. Eventually, the stresses within the society become unbearable.

The safety valve for French society in the nineteenth century was found in emigration. Emigration is usually dated from about 1837, but it had appeared much earlier. "From the French portion of Lower Canada there has, for a long time, been a large annual emigration of young men to the northern states of the American Union, in which they are highly valued as labourers, and gain good wages, with their savings from which they generally return to their homes in a few months or years."[39] This temporary movement was the typical reaction of overdivided peasant communities, in which heirs receive enough to hinder their permanent departure, but not enough to live on, and in which there is a chronic surplus of unskilled labour. Ireland and many parts of Europe offered parallels for it. The significance of the year 1837 lay in the fact that emigration increased sharply, for both economic and political reasons, and a substantial part of it gave the appearance of permanent removal.[40] There was a notable extension of the districts of Lower Canada affected by emigration from 1841 on, and a relatively large movement from about 1845, related to a variety of unfavourable economic conditions. This emigration was estimated at 4,000 persons per year, consisting two-thirds of habitants and one-third of wage earners. Families went to the cotton mills of New England, but were claimed to return in a few years. Of the young single men, "scarcely two-thirds" returned.[41] Despite the circulation of reports of dreadful experiences of many emigrants in the United States, the outflow increased in volume and took on a character of permanent alienation. The greatest movement, based upon a large-scale penetration of New England's milltowns to replace the Irish, who had so recently replaced Americans, occurred shortly after 1860.[42] It has been estimated that there were 400,000 French-Canadians in the United States in 1873,[43] and that half a million went there between 1860 and 1890.[44] The movement never lost its peasant character of a migratory one; but, more and more, it took on the appearance of that other peasant movement, the final desperate abandonment of the old land. Regarded as a calamity by the leaders of French society, this emigration nevertheless was the means essential to prevent the pressures within French society from reaching an explosive level, and the means also to retard its fundamental reorganization.

Emigration played a great indirect role in the reconstruction of

French society, however, by stimulating questioning and discussion, and by promoting an atmosphere more conducive to change. Those who returned from the United States with "advanced" ideas acquired there, provided a leaven for the society; while local leaders were forced by emigration to wrestle seriously, for the first time, with economic problems. The colonization movement that arose from this self-examination was less helpful than its proponents imagined, and the anti-luxury movement was positively pernicious. But there did emerge a broad nationalist movement which supported on the economic side the extension of settlement in Lower Canada and elsewhere; the modernization of the economy by the abolition of seigneurial tenure and by improved communications; and the growth of manufacturing aided by tariff protection.

The extension of settlement seemed a logical way to meet the pressure on the land in Lower Canada, and, for reasons already noted, there was a widespread desire that the settlement should be made under seigneurial tenure. An approach to the economic conditions of this tenure was made under the colonization movement, promoted vigorously after 1849 and again in the 1870s.[45] Settlement did occur, though probably it should be regarded as the consequence of migrations in progress, rather than of promotion.[46] An important result of it was the rise of vigorous new communities in the north that provided a focus for fresh attacks on the province's problems.[47] Meanwhile in the Eastern Townships, as in New England, French-Canadians showed their capacity to succeed English-speaking farmers by accepting a lower living standard; yet they carried on more efficient practices than those of the seigneuries.[48] Many habitants were provided for in these ways, and yet settlement did not greatly relieve habitant difficulties; for only in a remote sense was land shortage their cause.

A curious campaign by members of the clergy against "luxury" was still less effective or relevant. The clergy opposed emigration to the United States and proposed, in its place, colonization within Lower Canada. Inevitably, this involved hardships for the colonists. Especially was the program undermined when substantial numbers emigrated from the frontier regions marked for colonization, as seems to have been the case in the 1840s.[49] It was more alarming still that a boom in New England manufactures, involving particularly strong demands for labour in 1869 and in 1872-73, could so encourage emigration that a quarter to a third of all the farms in Quebec were abandoned by June 1873 — even though many habitants returned like penitents when the bubble burst.[50] Clerical writers, therefore, put "luxury" as the first cause of emigration, proposing in effect that the habitant should stay home and accept however low a standard of living this implied.[51] The effect of this advice, had it been accepted, would have been to accentuate subsistence farming and withdrawal from exchange economy. In fact, however, the habitant developed new wants in the nineteenth century, providing grounds for clerical complaint, but providing a basis also for a healthier economy.

It seems probable that dependence on home manufactures increased after 1760, when the French were pushed back on the land. However, Bouchette reported that homespun clothing was giving way to goods imported from Britain in about 1830.[52] There was a marked increase in farm indebtedness in the 1840s,[53] due partly to crop failures, but due also to increasing exchange. Clerical spokesmen asserted at this time that home manufactures were being abandoned and that farmers bought "luxury" articles on credit, with the consequence that some fell into hopeless debt and were forced to emigrate.[54] Especially from the 1860s to the 1890s, it was said, contentment with simple home-made things was superseded by a taste for horses and carriages, houses and furnishings (including pianos). The merchants flourished, but the habitants' debts became unbearable, and again they emigrated.[55] The habitants certainly were drawn into exchange economy in these decades, most of all by new means of transportation and by the economic specialization which they supported.[56] Obsolete agricultural methods were replaced by better ones,[57] a benefit which the clergy, more or less unwittingly, would have frustrated. However, the clergy, inconsistently, supported education to improve farming methods and the new system of inheritance that became common about 1850 to protect holdings against further subdivision. The new system required the single heir to a farm to assume obligations towards his brothers and sisters more or less equivalent to their former shares of the land; and it was alleged that this imposed impossible burdens on the heir, and often forced him to abandon his farm.[58] However that might be, the system of single-heir inheritance was of great value in shaking non-inheritors loose from the land and depositing them unequivocally elsewhere.

A product of the new atmosphere sponsored by distress and emigration was the abolition of seigneurial tenure in 1854, hard on the heels of agitation for its extension as an aid to settlement. What the Patriotes of 1838 had in mind in their advocacy of abolition was the repudiation of seigneurial dues and of tithes.[59] In the event, abolition in Lower Canada, as in most other places, imposed slight hardship on the seigneurs and gave little relief to the habitants.[60] What, then, was it supposed to do? By removing the seigneurial class, it eliminated the only check upon the power of the Church within French-Canadian society. But it seems unlikely that abolition was promoted to accomplish this result. A great deal was made of seigneurial obstruction to the use of Lower Canadian water power for manufacturing. In the 1840s, a representative of the clergy denounced seigneurs who "have refused, and still refuse every day to encourage the establishment of profitable works and useful manufactures for the country in order to retain exclusively, without benefit to themselves or the public, the numerous water-powers owned by them, and for which they are offered reasonable prices."[61] But there is little evidence that seigneurs generally were less likely than others to develop their water power, and it is nonsense to suppose that the habitants could have undertaken such development.[62] Cooper found the merit of abolition in the rise of agricultural specialization in Quebec, and in the abandonment of emphasis on grain crops which, he says, was forced by the seigneurs in their hunger for rents.[63] But, aside from the question of who it was, exactly, who had insisted upon the growing of grain, this attributes both greater and quicker effects to abolition than the step entailed. The rather slow shift to livestock that occurred in the province seems to have been a product, first, of the inability of the old lands to grow wheat any longer; secondly, of inability to compete against more efficient wheat producers elsewhere; and, thirdly, of the gradual growth of a market for farm products other than wheat. What can be said for the abolition of seigneurial tenure is that it let the habitant feel the forces of the market more directly than hitherto.

The most plausible explanation for abolition is that it was the work of that important group of French-Canadian leaders of the nineteenth century who accepted capitalistic or English doctrines of economic progress: they desired to transform the peasant into a farmer responsive to the market, and the local economy into one capable of producing surpluses and investing them. While it was not so much the seigneurial system as such, but rather a peasant structure antecedent to it, that "made it difficult, indeed, impossible, for the mass of the French-Canadian population to accumulate capital, to develop industries, to acquire technical skills and to develop attitudes and aspirations conducive to industrial enterprise",[64] abolition helped to overcome the prevailing subsistence economy. Spectacular results were not to be expected, particularly in view of the extreme tenderness shown for the interests of seigneurs. But this change, along with others, made the habitant's behaviour significantly more consistent with economic rationality by the 1870s.

A further major consequence of distress and emigration was a movement to make Lower Canada a manufacturing centre with the aid of protective tariffs. Agitation to this end was not new in the 1840s, but the advantage of the province in respect to abundant cheap labour and abundant cheap water power was given particular emphasis at that time, as was the desirability of protection. Industrialization became a permanent goal of French-Canadian nationalism, as well as of unhyphenated Canadian nationalism. What has to be said of the industrialization movement here is that manufacturing progressed slowly in

Lower Canada; that a great part of what there was, was established and manned by English-speaking residents; and that much of the rest was the work of French-Canadians who had acquired new techniques and new attitudes in the United States.[65] The small size of the local market and the competition of mature foreign producers were hindrances to development. But the ancient structure and attitudes of peasant society were greater ones. The habitants entered the labour market reluctantly. They lacked skill and capital and did not have much will or power to acquire them. Their disabilities were more burdensome for the industries of their home province than for the developed industries of New England. Emigration to New England implied a break with the past and an acceptance of new attitudes. American firms were long established, well equipped with capital and management and technical skill. For the textile mills, the physical weakness of the French-Canadian was of little consequence, while his nimble fingers, his large family, his docility and his willingness to work for small wages, were all points in his favour. What home industries particularly needed — capital and management and technical skill — the habitant could not supply. Even as an unskilled employee, he may have been less willing to accept a new role in his home province. The one thing he could offer was a low rate of wages. But low wages do not necessarily make low labour costs, and the uncertain progress of manufacturing in Quebec raises doubt that they did so there. In any case, cheap unskilled labour can only support a narrow range of industries. Processes that demanded skill arose rather in Ontario or were carried on in Quebec by English-speaking workers. In time, French-Canadians learned them too. But for the most part, French Canada clung to peasant ways and paid the price for them of limited opportunities and low incomes.

French-Canadians and the Labour Market

The paradoxical relation of the French-Canadian population to the labour market in the nineteenth century can now be summarized. A great deal of very cheap unskilled labour was available in particular places at certain seasons. But, except for the small force of traditional wage earners, like the shipyard workers, French-Canadians were not really in the labour market. They were not really out of it, either: they occupied a residual position, for labour could be obtained with some effort. However, the peasant's attitude to the market was negative. He scarcely offered any competition to immigrants for the jobs, many of them labouring jobs, which they took in Lower Canada, and he made little effort to fill other openings, like the annual one for several hundred sailors at Quebec. The chief reason was that the habitant wanted nothing to do with this alien world, except sometimes to earn a few dollars so that he could carry on at home. Eventually and painfully, this peasant economy was transformed. People were shaken loose from peasant society by overpopulation and overdivision and foreclosure. They migrated with the intention to return, and clung to peasant ways in distant cities. In time, they acquired industrial knowledge and discipline and had something to offer besides willingness to work cheaply. Meanwhile, a slow modernization of Quebec agriculture made the transition to industrial life cleaner and easier for later generations. Yet, even in the twentieth century, the French workman carried an awkward burden of traditional restraints.[66] In the nineteenth century, it is more proper to say that he was a potential or marginal member, rather than an actual member, of the labour force.*

The Americans

The second contribution to the settlement of Canada by Europeans was made from the United States after 1760. Some Americans came early as traders; there are supposed to have been about 6,800 Loyalists in Canada in 1785;[67] and during the next thirty years there was a substantial influx of Americans who were relatively indifferent to Canada's form of government, but responsive to its potential farm lands. This inflow represented branches of the westward migration of the American people.[68] To an extent the newcomers settled in the cities and on the old lands of Canada, but their distictive contribution was the peopling of Upper Canada and of the Eastern Townships of Lower Canada. There must have been more than 100,000 of them in Canada by 1815, and they must have provided 80 per cent or more of the English-speaking population at that date.[69] The Americans and their descendants set the tone of Canada's English-speaking regions until about 1840, and they remained influential thereafter, although increasingly indistinguishable from later British arrivals in most parts of the provinces.

Whereas Canada's French population was bound in a pre-capitalist peasant mould, the American immigrants stood at the opposite extreme of full acceptance of economic rationality. That class of Loyalists with aristocratic pretensions who represented a remnant of European feudalism went predominantly to the Maritime provinces. Those who came to Canada carried the political conception of an equalitarian lower-middle-class democracy. In economic affairs, they accepted the market without any evident inhibitions. The bulk of them became proprietors, so that insofar as they entered the labour market at all, it was as employers. But they had no objection in principle to taking wage employment, so completely had money evaluation displaced status evaluation. As McTaggart noted, the Yankees would not work on the Rideau canal, but this was for the plain reason that they could see no money in it.[70]

---

*For a skeptical note on this point, see Introduction, pp. xxvii-xxviii. Ed.

They were very ready to do canal work when they saw a prospect for profit.[71] A little earlier, Americans in Canada seem to have been quite ready to take employment to save for a farm or to support one already possessed.[72] But even the Americans employed as farm labourers displayed markedly bourgeois attitudes. They knew that they were a skilled, scarce kind of farm labourer. They stipulated for good pay, were extremely touchy in respect to any kind of criticism or mistreatment and insisted upon eating at the family table, thus disconcerting some farmers with British notions.[73] It was asserted that "New Englanders will not hire by the year".[74] On the other hand, Americans were ready to work much harder than British immigrants.[75]

The artisans of Canada's English-speaking settlements also were supplied in the early period by the United States. Innkeepers, millers, mechanics and lumbermen, along with farmers, made up the original American immigrations.[76] It was itinerant American pedlars and mechanics who catered to the needs of the pioneer settlements of Upper Canada.[77] As masters and journeymen — the line between was crossed easily — Americans supplied the mechanics' class of the towns of Upper Canada, and established the reputation for political interest and insolence of the mechanics that enraged opponents of democratic institutions.[78] The character of the artisan class was modified after 1837 by a changed political atmosphere and by British immigration. Nevertheless, where there was constant intercourse with the artisans of American cities and migration back and forth, as remained true of Toronto and particularly of Hamilton, a strong American flavour was retained.

But most Americans became proprietors of farms. As the pioneers of agriculture, they were professional and indispensable. As farmers, however, they left so much to be desired that even Fowke's argument that Canada's economic position called for extensive and superficial agriculture, can hardly cover them.[79] But by the criterion that interested them, they were decidedly successful. They quickly amassed cattle and buildings and more or less improved land. It was characteristic of them to build substantial frame houses while others were barely launched upon their log cabin stage.[80] There could be no question that they would produce for the market, foster exchange and endeavour to accumulate. In contrast to the habitants, they responded sharply to changes in the prices of agricultural products.[81] They were more than ready to cultivate new wants and to lower their level of self-sufficiency the moment the market made that possible.[82] When labour was scarce in the 1850s, they shifted readily to mechanized production.[83] Yet, they were perhaps more interested in speculative returns than in pedestrian ones from methodical production, and the appreciation of land values played a central part in their calculations.[84]

People who had developed the attitudes required to deal successfully with the conditions of pioneering could not be expected to excel in social graces. Adverse criticisms of the Americans in Canada rested partly on prejudice, but it was true that in repudiating every piece of cultural equipment that was not essential in the backwoods in preparation for their hard task, they ignored much that lends distinction to societies. They were vulgar; they did not value rank, or learning, or education, or even preachers, but only acquisition.[85] Though active and sagacious, they were litigious and unscrupulous.[86] McTaggart even asserted that Americans showed no passion of love for the opposite sex, as "we" do.[87] If this was so, it represented the most remarkable victory of the economic calculus.

The character of the American influx ensured the rapid economic development of Canada, but it ensured also that Americans would not have a large place in the labour market. This was so, not because the Americans had any inhibitions in respect to the market as the French had, but because the bulk of them became small farmers who were neither employees nor employers. American artisans and the small body of American labourers were keen, capable, but inclined to be individualistic in their bargaining. As employers, Americans usually showed a great deal of capacity. They seem to have got on well with their employees, and they were largely, if not entirely, free from the attitude of outrage cultivated by British-born employers in respect to American wage rates. They accepted the market, paid the wages necessary to attract the labour they required, or abandoned their project if it could not support current wage rates. On the other hand, they were inclined to react with emotional hostility to combination among their employees to a greater extent than employers with different backgrounds. When abundant labour supplies and new market possibilities appeared in Canada, new American immigrants along with old ones hastened to exploit the opportunities and to build a capitalist structure.[88]

Canada's Immigrants: An Overview[89]

Immigration to Canada from Europe set in seriously in 1816. Until the 1850s, when Germans and Norwegians appeared in some numbers, the immigrants were almost exclusively from the British Isles. The great bulk of these immigrants came as individuals or families, without any particular help or direction, though frequently in the company of kinsmen or friends.[90]

By force of geography, most immigrants from overseas entered Canada at Quebec, at which port the British government maintained an Emigration Agent from 1828. It was easy therefore to keep a record of entries, and reasonably accurate statistics of arrivals by way of the St. Lawrence exist from 1816. Some immigrants, perhaps 10 per cent, were not reported.[91] But since a large and variable proportion of the arrivals at Quebec proceeded to the United States, sooner more often than later, accurate immigration statistics would tell little about the numbers of persons added to the Canadian population. In addition, a considerable number of immigrants came to Canada by way of New York. No one counted this traffic across the border, and while attempts have been made to determine its net consequences,[92] the results can only be estimates. The only statistics that afford a measure of the contribution of British immigrants to Canada's population are those of the censuses, and censuses did not show the place of birth of the Canadian population until the 1840s.

The statistics of immigrant arrivals by way of the St. Lawrence are nevertheless useful. They show, in the first place, the extent to which European migration to North America proceeded through Canadian, as against American, ports. From 1819 to 1835, Quebec (and still more, British North America) drew far more British immigrants than did the ports of the United States. From 1835 to 1847, Quebec was behind New York as an immigrant port but retained a respectable competitive position. After 1847, the St. Lawrence could attract only about one-fifth as many British immigrants as American ports did, and its position in respect to total Atlantic migration was more unfavourable still.

Quebec's immigration figures show, secondly, the number of persons exposed each year for hire in Canada. Most immigrants would have been content to accept employment in Canada, had suitable employment been offered. The fact that so many went on to the United States, either directly or after some period of employment, is simply a measure of the limited capacity of the Canadian labour market. But it is important to observe the advantage enjoyed by Canadian employers: a stream of people passing their doors into which they could dip easily for labour when they wanted it. After 1815, population statistics give an inadequate picture of the supply of labour available in Canada; only when the numerous, eminently mobile immigrants are taken into account can the true abundance of supply be appreciated. It may be doubted whether any other country in the world was so favourably placed in this respect. Certainly none which enjoyed a reserve of labour had so convenient a way to rid itself of the surplus as simply to allow the surplus to proceed to the United States.[93]

The statistics tell, finally, what part of the arrivals were English, Irish and Scottish. Since each ethnic group had peculiar traits, this is an important kind of information. The statistics would have been more helpful still if they had distinguished Highland from Lowland Scots and (more important) Ulster from southern Roman Catholic Irish. But some guidance on these points, and on the proportions of peasant-labourers, artisans and substantial farmers, is provided by the comments of Emigration Agents.

TABLE III
NORTH AMERICAN IMMIGRATION: 1815-60

| Date | Emigration from British Isles | | | US immigration from Europe | Immigration at Quebec and Montreal | | | |
	Total	to BNA	to USA		Total	English	Scottish	Irish
1815	2,081	680	1,209		5,000*			
1816	12,510	3,370	9,022		1,250			238
1817	20,634	9,979	10,280		6,796			2,218
1818	27,787	15,136	12,429		8,400			4,599
1819	34,787	23,534	10,674		12,809			5,971
1820	25,729	17,921	6,745		11,239			5,580
1821	18,297	12,995	4,958		8,050			4,041
1822	20,429	16,018	4,137		10,468			8,374
1823	16,550	11,355	5,032		10,258			8,413
1824	14,025	8,774	5,152		6,515			5,168
1825	14,891	8,741	5,551		9,097			
1826	20,900	12,818	7,063		10,731			
1827	28,003	12,648	14,526		16,862			
1828	26,092	12,084	12,817		11,677			
1829	31,198	13,307	15,678		15,945	3,565	2,643	9,614
1830	56,907	30,574	24,887		28,000	6,799	2,450	18,300
1831	83,160	58,067	23,418	13,039	50,254	10,343	5,354	34,133
1832	103,140	66,339	32,872	34,193	51,746			28,204
1833	62,527	28,808	29,109	29,111	21,752	4,112	3,314	12,559
1834	76,222	40,060	33,074	57,510	30,935	5,414	3,710	24,320
1835	44,478	15,573	26,720	41,987	12,527	2,685	1,867	6,850

Year								
1836	75,417	34,226	37,774	70,465	27,728			
1837	72,034	29,884	36,770	71,039	21,901			
1838	33,222	4,577	14,332	34,070	4,992			
1839	62,207	12,658	33,536	64,148	7,439			
1840	90,743	32,293	40,642	80,126	22,234			
1841	118,592	38,164	45,017	76,216	28,086			
1842	128,344	54,123	63,852	99,945	44,374			
1843	57,212	23,518	28,335	49,013	21,727			
1844	70,686	22,924	43,660	74,745	20,142	7,698	2,234	9,993
1845	93,501	31,803	58,538	109,301	25,375	8,833	2,174	14,208
1846	129,851	43,439	82,239	146,315	32,753	9,163	1,645	21,409
1847	258,270	109,680	142,154	229,117	90,150	28,725	3,628	50,360
1848	248,089	31,065	188,233	218,025	27,939	6,034	3,086	16,582
1849	299,498	41,367	219,450	286,501	38,494	8,980	4,984	23,126
1850	280,849	32,961	223,078	308,323				
1851	335,966	42,605	267,357	369,510	41,076	9,677	7,042	22,381
1852	368,764	32,873	244,261	362,484	39,176	9,276	5,477	15,983
1853	329,937	34,522	230,885	361,576	36,669	9,585	4,745	14,417
1854	323,429	43,761	193,065	405,542	53,183	18,175	6,446	16,165
1855	176,807	17,966	103,414	187,729	21,274	6,754	4,859	4,106
1856	176,554	16,378	111,837	186,083	22,439	10,353	2,794	1,688
1857	212,875	21,001	126,905	216,224	32,097	15,471	3,218	2,016
1858	113,972	9,704	59,716	111,354	12,810	6,441	1,424	1,153
1859	120,432	6,689	70,303	110,949	8,778	4,846	793	417
1860	128,469	9,786	87,500	141,209	10,159	6,481	979	376

NOTES TO TABLE III

"Emigration from the British Isles" is from statistics of outgoing passengers from British ports. These statistics appear in S.C. Johnson, *A History of Emigration from the United Kingdom to North America, 1973-1912* (London, 1913), app. I, pp. 344-45, and in W.A. Carrothers, *Emigration from the British Isles with Special Reference to Development of the Overseas Dominions* (London, 1929), app. I, pp. 305-6. These figures understate emigration by 10 per cent or more because they were calculated in terms of adult passengers — i.e., two or three children might be counted as one passenger — and because vessels that could not meet passenger regulations because of overcrowding or other reasons slipped away without registering their passengers. The figures did not take account of incoming passengers, so they are gross rather than net emigration figures.

"U.S. Immigration from Europe" is taken from Brinley Thomas, *Migration and Economic Growth: A Study of Great Britain and the Atlantic Economy* (Cambridge, 1954), app. 4, and derived from *Historical Statistics of the United States, 1789-1945.* The sources from which early American immigration statistics must be computed are not very dependable.

"Immigration at Quebec and Montreal" constitutes the figures provided by A.C. Buchanan, Emigration Agent at Quebec. Figures for particular years often vary slightly in different sources. Generally later and larger figures have been taken on the assumption that they represented a correction of the originals. In the same way, totals of "English", "Scottish" and "Irish" need not add up to the total given because the total may be changed slightly and because there were other small groups of immigrants in the totals. Buchanan's figures count persons, not passenger space, and should ordinarily be pretty accurate, though something like 10 per cent of immigrants evaded count in Quebec. The immigration figures for 1847, in particular, are notoriously too low. The total of English immigrants for that year were largely Irish coming via Liverpool. The totals shown for the years 1816-28 were calculated by Buchanan (who took office in 1828) from the port books. Adams stated that they agreed well with immigrant arrivals shown in the *Quebec Mercury.* The totals of Irish arrivals, 1816-24, were calculated by Adams from the same paper. W.F. Adams, *Ireland and Irish Emigration to the New World from 1815 to the Famine* (New Haven, 1932), appendix.

*Buchanan's estimate of the overseas immigration via the St. Lawrence, 1790-1815 (Q198 *Pt. 1,* p. 59).

Canadian immigration was part of a vast phenomenon, "The Atlantic Migration". Students of this movement have inquired into the goals of the migrants, the means of their transport and the conditions of their homelands. What the migrants sought in America is scarcely in question: it was superior economic opportunities. Means of transport are important because they determined very largely the volume and direction of migration. The technical improvement of ocean and land transport in the nineteenth century facilitated migration; but this was a development that came too late for the migration that made the Canadian labour market, and its predominant effect was to divert immigrants from Canadian to American ports. The development of great staple export trades from North America to Europe was of much greater importance for migration until late in the nineteenth century. The great trades settled into well-defined shipping routes that determined for most migrants their destination in America. Most of the trades involved a bulky cargo to Europe, but bulky return cargoes were rarely to be found. The pressure of overhead costs therefore induced shipowners to cater to emigrants as a return cargo, and to offer very low fares when this was required — as it usually was — to fill their ships.[94] Without attention to the Atlantic traffic in tobacco and cotton, the nature of a great share of Atlantic migration would be unintelligible. But no commodity trade was more momentous for migration than the timber trade of British North America. Its character led to the provision of the most numerous, the cheapest and the roughest passages to America at a crucial time, when multitudes who could barely afford to pay a fare of £2 per adult head wanted to leave Europe, and at crucial places, the ports convenient for these persons. The timber trade does not explain the outpouring from Britain after 1815, but it does explain why so many who wished to go were able to go, and why they shipped for British North American ports.

The question why so many Europeans were willing to go to America is far more intricate. That economic conditions were generally better in America, and that most Europeans so believed, has little direct bearing on this question, since inert humankind seldom abandons a customary environment except under strong or even overwhelming pressures. The most comprehensive approach to an answer was offered by Marcus Lee Hansen. He distinguished three stages in the migration of the nineteenth century, each related to the land-holding system and population pressures of a part of Europe: "Celtic" migration, 1830-60; Teutonic migration, 1860-90; Mediterranean and Slavic migration, 1890-1914. Hansen called his first stage "Celtic" because the regions affected above all others were Ireland, the Highlands of Scotland and the mountains of Wales, "regions where the language and blood were predominantly Celtic and where the land system grew directly out of the agrarian customs of the early tribes".[95] The Upper Rhine and other parts of Europe from which many emigrated before 1860 were held by Hansen also to be Celtic, possibly in blood, certainly in agricultural practices. Hansen does not make the logic of his generalization very precise. Communal systems of holding and cultivating land, excessive subdivision, disruption of ancient economies from the inroads of capitalist concepts in agriculture, rapid population growth facilitated by potato culture, medical advances and the suppression of local wars, all seem to play a part. Ulster, the greatest of all the early sources of emigration, does not fit the "Celtic" pattern closely. But it is a fact that pressures of the type noted by Hansen are all too evident in Celtic lands; it is a fact that regions that followed Celtic agricultural practices developed extreme imbalances which sooner or later provoked wholesale emigration; and finally, it is a fact that early migrations were predominantly from Celtic areas.

The second, Teutonic, migration also found its unity in a system of land-holding, that of undivided farms which forced non-inheriting sons to seek their fortunes elsewhere. The distinction has been pursued by Habakkuk, who argues that regions of subdivision are likely to experience more rapid population growth than others, that they invite seasonal migrant labour and cottage industries, but that they tend to retain their population growth until a crisis drives it forth in a kind of stampede; while regions of undivided land-holdings contribute a smaller but regular and orderly flow of outcasts.[96] It is not difficult to see how the undivided holding can produce potential emigrants, and it is a fact that there was a great outpouring from regions of undivided holdings in England, Germany and Scandinavia, especially between 1860 and 1890. It is not immediately clear, on the other hand, why the Teutonic lands would not have contributed a regular flow of migrants from much earlier times; and it has to be taken into account that a region of undivided holdings, Ulster, did provide a regular stream of emigrants from the seventeenth century. The argument seems to require some refinement. Habakkuk suggests that regions of single-heir inheritance offer the best possibilities for industrial advance, because the mobility of labour facilitates industrial development at the best locations in respect to other factors than labour, and because such regions can produce food surpluses to support industrial cities.[97] On the whole, European industrial development did take place in such Teutonic regions. It would follow that such regions would not require the safety-valve of emigration in some periods because surplus farm population could find a place in industrial centres. On the other hand, because industrial populations were particularly prolific,[98] and because the growth of demand for industrial labour has been notably uneven, such areas might easily provide substantial numbers of emigrants in other periods. It is significant that industrial workers formed a large part of the Teutonic migration of

1860 to 1890, and that this was a period of industrial development in America and of relative stagnation in Great Britain. Ulster may represent a special case, its industrial development being so often impeded by political and economic obstacles, and its labour force so peculiarly placed with one foot in agriculture and the other in manufacturing, that its emigration became a regular phenomenon.[99]

Some other general statements concerning migration can be made. When migrants had a choice of destinations, their decision was likely to rest upon considerations that seem inconsequential in retrospect. The level of ocean fares was particularly vital, even in the period when fares were low by either earlier or later standards. So crucial was this factor that emigrants (especially Irish emigrants) divided themselves into segments whose destinations depended upon the amount of fare they could afford. The poorest Irish went to Lancashire; the next lot to (or via) New Brunswick; those a little better off to Quebec; and the most affluent to New York.[100] The reputation of an emigrant route with respect to mortality — Quebec's reputation was usually bad — also played a part. It follows that emigration could be diverted in a substantial amount from one route to another by propaganda, taxes and subsidies, and it was so diverted from time to time. But it was also true that the volume of emigration to any destination was governed very closely by its level of prosperity and the experience of recent immigrants in finding employment; above all, it was governed by the availability of work for labourers in constructing canals or railways. It was literally true that the great periods of emigration to Canada in the nineteenth century were those during which such labouring jobs were plentiful, and the periods when immigration was low were those during which construction jobs were scarce.[101] In such years as 1819 and 1842 when jobs were easier to find in Canada than in the United States, migration tended to swing in Canada's direction; but the reverse also was true. The powerful pull of the United States in the last half of the nineteenth century mirrored its rapid expansion and Canada's stagnation.

It is a general truth that patterns of social behaviour in European communities, in Celtic ones especially, inhibited movement. But it was equally true that, once emigration had begun, new patterns of behaviour developed which had as much momentum and compulsion as the old. A habit of migration was as difficult to stop as it was to start, and the name "mania" frequently applied to it does not seem inappropriate. Religious and political persecution, though they provoked the emigration of very few people directly, may have been of critical importance in introducing a pattern of migration to old communities. The arrival of immigrants in a new land had a similar cumulative effect, as those behind drew their friends after them. Canada probably benefited from this magnetic effect before 1850, but after that date this advantage was

/ felt by the United States and Australia. In this period,
anada suffered from the cumulative momentum of her own
em_ on to the United States.

Still other forces affected the flow of immigrants to Canada: the early
hostility of Britain's rulers to all emigration; their subsequent efforts to
keep emigration within the Empire; schemes for assisted emigration
and "colonization" and selling land; trade and lending policies; and, ultimately, the whole shape of development of metropolis and hinterland.
However, the purpose of this discussion is not to duplicate the efforts of
those who have studied migration itself, but to mark the relation of
migration to the Canadian labour market. To that end, the nature of the
various ethnic groups of immigrants who came to Canada is next considered.

The English

English emigration to Canada, in contrast to Irish and Scottish emigration, was unspectacular. English arrivals at Quebec numbered 5,000 to
10,000 in most years from 1825 to 1860 — a quarter or less of the total
immigrants in the first half of the century, and about half in the 1850s.
These numbers were especially tiny in proportion to the population of
the country from which the immigrants came, in contrast to the wholesale movements from Scotland and Ireland. The pressures that produced English emigration did not affect whole communities, but rather,
politically impotent and relatively inarticulate segments of society.
Moreover, the evidence indicates that the majority of the Englishmen
who arrived in Canada before 1850 proceeded on to the United States.
Those who remained were neither outstandingly good nor outstandingly bad in adjusting to their new environment.

The striking cultural and material differences in the nineteenth century between England and every other part of the British Isles except
the Scottish Lowlands could hardly fail to produce distinctive kinds of
emigration. There was, for one thing, the custom of inheritance by a
single heir, which applied to an extent to urban occupations as well as to
land, though apparently not to land in all parts of England.[102] Then,
there were the immense structural changes of the Industrial Revolution. They demanded an upheaval of English community life and custom,
and some groups suffered severely. But in England they also provided
great opportunities and great gains, and the prospect of a place of some
sort, eventually, for most of the losers. There was nothing like the
whiplash impact of these changes upon the marginal countries north
and west, where losses were universal, benefits imperceptible. In addition, England had been an immensely richer, more diversified and more
developed country to start with. Its agriculture had been rationalized
long before 1800. Its population, aside from the victims of the Poor Law

system, was inured to mobility and individual adjustment. Its cities furnished the opportunities of a highly-developed commercial structure along with the opportunities for submersion in their slums.

English emigrants, therefore, might be expected to carry a more modern set of skills and attitudes than emigrants from peasant regions. Yet not all the groups ground into emigration by the new order were well-equipped. Younger members of aristocratic families, who brought to the New World the presumption that the world owed them the social and economic status of their ancestors, are sad examples. English pauper immigrants, debilitated by the Speenhamland system, were in a somewhat similar position. They emigrated without enthusiasm and did not have or deserve a very favourable reputation in America. However, it would be difficult to show that they did worse, in the end, than other badly-equipped immigrant groups, the Highland Scots and Catholic Irish. Some thousands were sent to Canada under schemes for relieving the poor rates;[103] but most English labourers, given their own choice, determined that they would fare better elsewhere.

Better equipped, and more important, were English small farmers. Perhaps they were thrust out, ultimately, by population growth, primogeniture and structural changes in the economy, but they emigrated in response to more immediate factors. The first wave, in 1816, reflected the fall in agricultural prices at the end of the war, while costs remained rigid and high. There was a second peak in the early 1830s, when small farmers felt intensely the burden of poor rates, without enjoying the power of the big farmer to profit at the other end from the artificially low wages of parish labourers. A third wave developed from about 1848, when farmers feared that the repeal of the Corn Laws would ruin them. These farmers were valuable settlers, equipped with industry, capital and skill. Generally, they were wise enough to avoid pioneering, for which they were unequipped, and instead took over developed farms which they managed well. Canada got some share of these emigrants, particularly in the 1830s.[104]

The most numerous and important group of English emigrants were industrial workers. Many were those cast out by the new factory system, particularly hand-loom weavers. But it was characteristic even of the expanding parts of the industrial machine to throw off groups of artisans and factory operatives from time to time. It is probable that an important factor in the emigration of some of these elements was the influx of Irish Catholics into Great Britain, which was especially large in the second quarter of the nineteenth century. The Irish invasion did not affect the entrenched crafts very greatly, but it put extreme pressure upon exposed groups: factory workers, hand-loom weavers, dockers and various types of semi-skilled urban labour, as well as on English and Highland Scottish farm labourers. Some feared that Irish migration to

Britain would eventually lower the living standards of the whole British working class to Irish levels. As it was, the Irish contributed directly and indirectly a great part of the widespread distress in England before 1850, as well as fostering rapid industrial expansion by their contribution of unlimited cheap labour.[105] It is arguable, then, that many semi-skilled English workmen, and some skilled, fled to America essentially to recapture the status which the Irish threatened at home. However, the impact of all the unfavourable forces that beset industrial workers depended primarily upon the state of industrial activity in England. Workers emigrated in periods of unemployment or insecurity at home. The course of industrial conflicts, and union support of emigration to relieve the labour market, also played a part.[106]

Skilled and semi-skilled British workmen were distinguished from other immigrant groups by their possession, in their hands and heads, of the most advanced technology of the age. With their high investment in industrial skill and their familiarity with urban life, they were rarely attracted to agriculture in America. Even English weavers seem to have clung to their craft, though Irish and Scottish weavers settled readily on the land and did well there, probably because they had been quasi-farmers at home. The consequence of this preference of the English industrial workers for their accustomed employments was that most of them who came to Canada before 1850 found no adequate use for their skills and went on to the cities of the United States. Even so, the power to fill its limited wants from this large, trained pool of workmen was very convenient for Canada. Then, in the 1850s, Canada experienced an industrial revolution of its own, founded upon railroads. To man the railroads, and the metal and other industries that appeared in connection with them, a very numerous group of English craftsmen appeared and stayed. Several thousands of these workmen were imported for the purpose by railroad companies. This influx gave a new industrial dimension to Canadian urban life in the 1850s, and a British tone to urban craftsmanship tended to replace the earlier American one. In later decades further movements of British artisans occurred in much the same way, as Canada's industrial structure widened and deepened.[107]

Ancient migrations typically resulted in the immigrants (conquerors) occupying the top of the succeeding social structure. But overseas lands settled from Europe tended, after the original seizure of territories, to demand immigrants at the bottom, subordinate labourers for the first-comers. Such immigrants were provided first by the black and white (indentured labour) slave trades; next by the outpouring of peasant, especially Irish Catholic, labourers from northwest Europe; and finally, by another peasant exodus from Mediterranean and Slavic Europe. So effectively has the conception of the immigrant as a social and economic inferior been implanted by this historical experience, that a recent study of British migration to America breathes constant surprise that British immigrants entered American society at high social and economic levels.[108] But, in fact, British migration, and the whole Teutonic migration, involved persons who took positions high on the social and economic scale in Canada and in other countries. Occasionally, Canadians took exception to the unwarranted pretensions of newcomers, such as the would-be gentry around Peterborough and the incompetent managers of the Grand Trunk Railway. But, by and large, the immigrants were accepted at their own high valuation because they had the cultural equipment to support it. The immigrants were neither conquerors or subordinates, but a reinforcement of the existing privileged classes, readily assimilated into them. Racial homogeneity facilitated this assimilation, but the possession of scarce skills was the essential condition of entry.[109] The skilled English artisan, clerk, manager and professional person were members of this favoured group.

It was implicit in the easy entry and complacent attitude of English immigrants that they practised an exclusiveness based upon the status systems of their new, and perhaps more their old, country. This exclusiveness was occupational and social more than racial, its criteria consisting in occupational and social mores. English immigrants practised it easily, perhaps unconsciously, because they refused to contest for employment in which they did not enjoy a substantial advantage, and they were well-enough equipped that they did not have to do so. The consequence was that the English had little to do directly with ethnic conflicts. On the other hand, they showed much less interest or adeptness than the Scots, or even the Irish, in bridging ethnic gaps and working out social compromises. Their enclaves instead provided an element of aloof and somewhat arrogant stability.

English (and similar) artisans also contributed a new sort of stability to the Canadian labour market from the 1850s. Unlike the native and immigrant groups that entered the labour market on a temporary basis, English artisans were committed to lifetime employment at their crafts. Unlike the Irish Catholics and their own fathers of the 1830s, these men made up a mature, disciplined labour force. They did not quarrel with their employers over trifles, or indulge in battles, picturesque but useless, that were really revolts against capitalism and employee status. When conflicts did arise, they did not consider it essential to fight to the last issue, but were constantly on watch for an acceptable settlement. On the other hand, when these artisans were provoked to strike, they conducted their campaign with far more discipline, solidarity and staying-power than peasant-labourers could command.[110] In general, English immigrant craftsmen contributed very greatly to make Canada's labour market a modern one, by providing a regular and dependable supply of skilled labour and by insisting as a counterpart upon a suitable level of wages and conditions.[111]

The Scots

Overseas immigrants to Canada before 1815 were nearly all Scottish Highlanders. The troubled state of outlying Scotland towards the end of the eighteenth century, provision for military settlements, and the inclination of a clannish people to join kinsmen who had preceded them, all worked to make Scottish immigration preponderant. Nevertheless, probably no more than 10,000 came before 1815.[112]

In post-war immigration, the number of Scots, though substantial, always was less than the English and much below the Irish. On the other hand, there is every indication that most Scots stayed in Canada, instead of moving on to the United States as a great part of the English and Irish did. Scotland's effective contribution to Canada, therefore, was greater than immigration statistics would imply, and the number of Scottish-born in Canada was as large as the English-born until after 1852. The size of the Scottish influx, its priority in time, and the Scottish talent for winning the confidence of other ethnic groups which did not trust each other, all assisted to impose a Scottish character to Canadian life that was the more striking because of the slight Scottish influence in the United States.

Two very different groups were included in totals of Scottish immigrants: peasant clansmen from the Isles and Highlands, children of an ancient economic and social structure; and Lowlanders from a region as advanced as any in the world. Highlanders seem to have predominated as immigrants until about 1830, while it was mostly Lowlanders who came in the 1850s and possibly in the preceding two decades as well. However, there is no way to measure the sizes and dates of the respective contributions closely.

Highland emigration, like that of other peasant peoples, grew out of crisis at home. The Highlands and Isles had preserved their clan system, with its communal agriculture, into the eighteenth century. Analogous arrangements had disappeared four centuries earlier in other parts of Great Britain. The logic of the clan system was found in the provision of manpower for war, and it measured success by the number of potential warriors who could be supported by its subsistence agriculture. When the system was broken by English conquest in 1745, capitalist rationality penetrated rapidly to erode traditional institutions and to establish an order for which success depended on a minimum rather than a maximum of population (labour cost). The population was now redundant and had to be cleared or occupied in profitable work. But circumstances conspired to obstruct both these solutions. Population growth, upon which there was no deliberate restraint, proceeded very rapidly. It was facilitated by potato culture, which hid and accentuated the problem. The rural linen industry had helped to support a great part of the population, and under other conditions might have been the salvation of more, but it was put into retreat towards the end of the eighteenth century by cotton, specialization and the factory system. New industries suited to the population, fishing and kelp production, proved to be unstable and incapable of supporting many. Other industries did not appear; as in other such cases, the theoretical advantages of abundant cheap labour were more than offset by the dearth of skill, industrial discipline and capital. From about 1760, there was widespread evidence of overpopulation even in terms of the ancient economy, and the slightest crop failure produced crisis. A final palliative was provided by the Napoleonic Wars, with their opportunities for military employment, the kind envisioned by the clan system. The British armies provided, after their fashion, for an extraordinary number of Highlanders up to 1815. After that, there was nothing but social inertia to slow the renovation of the Highlands.[113]

There really was no solution to the Highland problem except emigration, and the Highlanders, in spite of their distaste for change and removal, had seen that fact much earlier than the mercantilists in the British government. The collapse of the clan system had the merit of removing insititutional obstacles to migration; indeed, some removals to Canada were made in the vain hope of preserving clan life there. The booming Scottish Lowlands seemed to be the obvious place for the Highland surplus to go. The Highlanders could not meet the key demand in the Lowlands for skilled labour, but there were many jobs for labourers in the cities and, seasonally, on the farms. A good many Highlanders went in search of those jobs. But this outlet was obstructed, especially after 1800, by the very large Irish emigration to Scotland.[114] The Irish sought the same types of employment as the Highlanders, could live just as hard as they and worked more efficiently. America, where this pressure could be escaped, was the best destination for Highlanders and, despite their poverty, a surprising number managed to get there. Government assistance to military and (some) civilian settlers, and contributions by the wealthy for reasons of benevolence or land clearing, helped. The influence of those who assisted, patriotic and clan sentiment, and later the pull of the timber trade, took a high proportion of the emigrants to British North America.[115]

The Highlander, despite his hardiness, was badly prepared for life in America. A peasant, he was untutored and undisciplined in the ways of capitalistic communities. His frequent ignorance of the English language put him at a disadvantage with employers. Still more offensive were his unwarranted pride and vanity. Selkirk, noting that a wage differential existed in favour of experienced men, said, " The newcomers however (particularly the Highlanders) are apt to imagine that there should be no difference and to expect the highest rate."[116] Howison gave a savagely unflattering picture of them and declared that Highlanders soon became worse than any others for vanity.[117] Presumably it was these immigrants whom he said acquired "those absurd notions of independence and equality" by the time they reached Kingston, where they greeted those whom they met as equals, instead of uncovering as they had at Montreal, for now they had become "gentelmen".[118] Unhandy and uncooperative, the Highlander was a poor prospect for employment.

However, the Highlander did not compete in the Canadian labour market any more than he could help. Rather, he settled on a farm. His success as a farmer was good by his own standards, bad by other people's. Selkirk said of the Glengarry settlers that the young men became "as good as Americans" with the axe; but they did not get through the work and stuck to the old ways.[119] As for their housing, "These accommodations appear poor to the Amer'ns and English settlers but they are a wonderful advance from the hovels of Glengarry; and the advance in cleanliness seems to keep pace with those of houses..."[120] What the Highlander did was to pitch his agriculture at a level at which efficiency and returns were low; but costs were low too, because he would live on little, and so he survived.[121] Though a marginal farmer, he was reluctant to seek employment. When the Ottawa timber trade expanded, the Highlander did take some part in it, but not the share he might have had if he had been disposed to sacrifice leisure and independence for wages.

Very different were the emigrants from the Scottish Lowlands. Most were small farmers and artisans, and they emigrated at much the same times and for the same reasons as their English counterparts. Relative to the population of Scotland, there were more of them. Emigration was traditional with them, and probably economic pressures were sharper than in England. While Lowland emigration resembled that from England, there were some differences. Scottish farm labourers who emigrated, unlike English parish labourers, tended to be efficient, enterprising men, well prepared to become successful farmers in Canada.[122] Scotland also contributed more merchants of various sorts, and probably more professional men, than other countries. All these Lowland emigrants possessed in a high degree the skills and attitudes suited to a capitalist economy, and they were generally successful by capitalist standards. They were distinguishable from the English by a readier adjustability and a capacity and willingness to make themselves the bridge among the other ethnic strains of the country.

It may be said of both sets of Scots together that most, as small proprietors, were withdrawn from and neutral to the labour market, like most other elements in Canada. Some were substantial employers and generally displayed the traditional ability to deal tactfully with other ethnic elements, especially the French. Artisans were the most notable section of Scottish wage earners, multiplying as industrial employment expanded, especially in the 1850s, and contributing to provide a class of mature, permanent, skilled workers to the Canadian labour market. Scottish immigration also contributed labourers, dependable and capable from the Lowlands, inefficient and mercurial from the Highlands; but both upon a transitory basis. It remained for other ethnic elements to provide a permanent labouring class.

CHAPTER IV

Population Growth
and Migration:
The Irish

The Irish

Irish emigration stands somewhat apart from all other migration. No other country, in modern times, contributed to migration so persistently or in such an enormous proportion to its population. The sheer volume of Irish emigration to America and to Great Britain would make this movement a fact of capital importance. Yet, more significant is the role played by the Irish in their new environment, by reason of their peculiar behaviour patterns, and the consequence for the labour markets of other countries.

Ireland was for centuries an English colony, the one in which the English learned the art of subjecting other peoples, and one in which they practised a remarkable brutality. Irish economic development was continually retarded by destructive wars and by legal disabilities imposed by England. The native Irish, a primitive people, were subjected to social and economic conditions that prevented their developing industrial skill of discipline and robbed them of morale and enterprise. Ireland's population came to consist of two nations: the native Roman Catholic Irish, and a Protestant body of Scottish, English and Welsh origin. As landlords and officials, the Protestants were spread through the island; but the significant element with respect to industry and emigration was the Scottish settlement in Ulster. Ulster, in the hands of these Scots, became by far the most prosperous part of Ireland. The migration that produced the Irish Protestant nation was the earliest British movement overseas, and a remarkably large one. One estimate suggests that in the seventeenth century there were as many as a quarter of a million Protestants in Ireland, as against a million native Irish.[1]

The relations between the two Irish nations were complicated and are easily misstated. There is a tendency in the literature of emigration to treat the two Irelands as one, because they are not distinguished in emigration statistics, because the bulk of Ulster emigrants were not a great deal better off than their southern counterparts, and because Irish of either origin typically took labouring jobs upon their first arrival abroad. But there were differences between the two groups of the greatest importance in skill and knowledge, and above all in morale. Hence, Protestant and Catholic Irish pursued different goals in' America, differed in achievement and played different parts in social and economic development. On the other hand, the difference in outlook of the Irish nations, and the conflict between them, is sometimes overdone. The Protestant Irish rapidly developed a loyalty to Ireland and a hostility to English oppression that matched that of their Catholic neighbours. The Ulster Presbyterians were more chronically rebellious than the Catholics. No Irish rebellion commanded the united support of the factions — the more ardent the one, the more hesitant the other — because the objectives of the two parties were contradictory. The Protestants wanted to win independence from England and to rule Ireland themselves; the Catholics wanted to be rid of all the oppressors. Hence, England could rule by pitting one group against the other. But, on the other hand, they were all Irishmen and retained a certain consideration for each other at home and abroad. In Canada, while Irish Protestants reacted strongly against truculent demonstrations by their Catholic countrymen, they were also ready on occasion to support, relieve and intercede for them. Nor do Canada's Protestant Irish appear ever to have wished to deny their origin, and the phrase "Scotch-Irish" by which their kinsmen in the United States have described themselves, appears to be strictly an American invention in response to the "Native American" movement. As for the tensions that led in the twentieth century to the partition of Ireland, they had not developed into irreconcilable religious and ethnic positions at the time of the great Irish emigrations.

Irish population grew very rapidly in the eighteenth and nineteenth centuries. According to one careful student, there were about two and a half million Irish in 1700 and four million by 1781.[2] The censuses reported six and three-quarter millions of Irish in 1821, seven and three-quarter millions in 1831 and eight and a quarter millions in 1841. Ulster shared in this expansion, but the special circumstances of that province demand separate discussion. The forces that facilitated population growth in the rest of Ireland do not seem to be in dispute. The ancient customs of the country encouraged agricultural subdivision, to which an alien administration added the complete insecurity of annual leases. Until 1815, landlords found grain production profitable and sought to turn their traditionally pastoral island into a land of tillage. To grow grain, by the techniques of the day, required a large labour supply; so landlords encouraged the peasantry to multiply by the ready provision of scraps of land for cottages and potato patches.

The means to support the growing population was, of course, the potato, which had been widely depended on even in the seventeenth century and which became the main and almost exclusive source of food. About an acre of potatoes could support a family (as against ten or more acres of cereals), leaving the rest of the land for grain production for the landlord's profit. The Irish, for their part, were encouraged to marry young and multiply without restraint because there was nothing to stop them — an Irish cottage represented the slightest investment, and potato patches were easily got — and because there was nothing to gain by delay. The peasant could be certain that he would never be better off — for any gain would go to the landlord — and that children represented the one chance for some security in old age.[3] Some of the peasant's misery could have been relieved, it is true, if he had been less inept, indolent and improvident; but tradition bound him to a barbarous subsistence level of life, and the existing structure could not have survived without it. In the same way, the inefficient agricultural practices of Ireland were supported by ignorance and inertia, and aggravated and perpetuated by short leases.

Contemporary England and Scotland provide evidence that a rapidly rising population, even if it outruns local food supplies, need not produce starvation or retrogression. Abundant cheap labour has even been regarded, with slight historical justification, as a great stimulator of economic development. But to most of Ireland, as to the Scottish Highlands, the Industrial Revolution brought only the destruction of existing industries. There had, in fact, been several abortive efforts to foster manufacturing in southern Ireland, and wars and economic disabilities imposed from England were only part of the reason for failure. The fundamental reason seems to have been that, while labour in general was redundant, skilful disciplined labour was extremely scarce. It is instructive that the one part of Ireland in which industry flourished, Ulster, had the highest wages, the highest living standards and the least redundance of labour in Ireland, though labour was cheap by English standards. But Ulster artisans were skilled and industrious, so that labour costs were genuinely low there in the sense that much was produced for each shilling of wages.[4]

The nature of southern Ireland, as of other peasant regions, led its people to ignore signs of strain until monstrous imbalances produced a general collapse. While the population continued to mount, landlords discovered after 1815 that grass would pay better than grain. The large population then appeared as a detriment, and landlords wished to rid themselves of their cottagers. The increasing distress of rural labourers was matched by that of artisans whose trades were being destroyed by outside competition. The logical solution of emigration was resorted to, from about 1818, and the volume of emigration from southern Ireland grew quickly. But the relief was not enough to prevent further population growth, further subdivision of the land, chronic unemployment and distress. In the end, it took the general failure of the potato crop in 1845 and 1846 to induce a gigantic exodus from Ireland, which permitted the emergence of a manageable economy sustained by a system of population control unique in the modern world.[5]

The Ulstermen

English Protestant settlers in Ireland preferred the role of landlord and, except for the Quakers, they contributed only modestly to emigration. Very different were the Scottish Presbyterians of Ulster. They were small farmers or farmer-craftsmen. From the standpoint of agricultural reformers, they were slovenly farmers; partly, no doubt, because they devoted a great part of their time to the spinning and weaving of linen. As pioneer immigrants, these Scots had to face the hostility of the displaced native Irish. As Irishmen, they were confined by the legal prohibitions which England placed on Irish farming and manufacturing. As Presbyterians, they suffered from attempts to enforce conformity to the established church. Nevertheless, besides carrying on their indifferent agriculture, these Scots made Ulster a hive of linen manufacture (with the aid, possibly important, of Huguenot immigrants) and of some lesser manufactures, and a thriving centre of commerce. In the nineteenth century, Belfast became a cotton manufacturing city; and later still, a centre for factory linen manufacture and shipbuilding.[6]

Farms in Ulster were typically held under an "Ulster custom" that provided long leases, lower rents usually than the south and inheritance by a single heir, though subdivision occurred in the development of the linen industry around farmer-craftsmen masters. It is doubtful whether the Ulster custom is a sufficient explanation for the economic superiority of the north.[7] But the necessity for non-inheriting sons to carve out careers for themselves probably promoted manufacturing as well as regular and orderly migration. Emigration was facilitated by the fact that small farmers could realize considerable capital from their improvements and unexpired leases. The periodic falling-in of leases over wide areas appears also to have promoted general emigration movements.[8] But emigration was linked most closely with the fortunes of the linen industry. Bad times for linen weavers were times of large migrations; good times cut the movement to a trickle. It could be said that the high rate of population increase in Ulster was prevented from leading to fragmentation in agriculture by a system of single inheritance; was relieved by the growth of manufactures; and was provided for, ultimately, by an orderly emigration to America.

Ulster emigration in the seventeenth century appears to have been substantial for those times but irregular. In the eighteenth century, the movement became a fairly regular one of possibly 5,000 persons a year, which seems to have provided a greater part of America's immigrants in that period. It has been argued that the early emigration rested upon the desire of Presbyterians (and Quakers) to practise their religion without molestation, and the demand for free religious and political expression may have helped to establish the pattern of movement. But the Ulster people, so recently immigrants to Ireland, could hardly have had any objection to emigration in principle, and it is clear that the great bulk of Ulster emigration was inspired by economic motives.

Until 1820, the typical Ulster emigrant (and the typical American immigrant) was an indentured servant. His destination in America was governed largely by the trade routes between Ireland and America. The trade in flaxseed has been emphasized as promoting emigration, and it took the Irish particularly to Philadelphia.[9] But it seems doubtful whether Ulster depended heavily upon American flaxseed until about 1750: the volume of shipping involved could scarcely have been large, and flax formed only one item in a set of quasi-barter transactions.[10] Emigration as a by-product of the tobacco trade seems to have been much more important.[11] What may be said is that Ireland carried on a lively commerce with America, exchanging linen and some illegal woolens for tobacco, flax, sugar, cotton and other articles; that the trade permitted Ulster emigrants to reach America; and that it took them to ports from Philadelphia southward. But emigration was no mechanical by-product of trade — Ireland had a number of other trade connections that did not draw emigrants.

Opinions concerning the impact of Ulster migration upon America vary enormously. A.C. Buchanan, who knew a great deal about emigration, and about Ulster emigration in particular, estimated that there had been a million and a half emigrants to America up to 1828, of whom at least a million had come from Ireland, and five-sixths of these from Ulster.[12] The American school of "Scotch-Irish" historians similarly argue a large or preponderant Ulster immigration to the United States in the eighteenth century.[13] An opposing authority insists that the average annual Irish migration to the United States before 1774 did not exceed 4,000 persons per year, and notes that the American census of 1790 reported only 44,000 Irish-born in the United States.[14] Another cites, and apparently approves, a calculation that only 6 per cent of the American white population was of "Scotch-Irish" origin in 1790.[15] If these figures were accepted, it would be difficult to explain where the American population came from, or who, aside from negroes, did the heavy work of the country. But it is possible only to offer the general conclusion that Ulster provided a very substantial portion of the people of the United States, especially of its south and west. Perhaps Ulster's greatest contribution qualitatively, was the American frontiersman and pioneer types, without which westward agricultural expansion would have been extremely difficult.[16] Since Ulster settlers had served a long apprenticeship in pioneering, there was every reason why they should fulfil this role.[17]

Ulster made no significant contribution to the peopling of Canada before 1815.[18] But Ulster migration turned from southern to northern ports of the United States after 1783,[19] probably because of the collapse of the north British entrepot trade in tobacco.[20] Then, after 1815, migration shifted still further north, to the St. Lawrence. The convenience and low fares of the timber ships were mainly responsible for this diversion, but it is also important that Ulstermen had developed a new friendliness for the British connection,[21] and that Upper Canada was a mecca for landseekers until about 1835. Immigrants at Quebec after 1815 were about 60 per cent Irish, and these overwhelmingly from Ulster. That is, of the total immigration by way of the St. Lawrence of 335,000 between 1816 and 1835, about 200,000 were Irishmen, perhaps 170,000 of them from Ulster. It is likely that half or more of these immigrants settled in the United States, but Irish immigrants to Canada by way of New York may have offset a third of this loss.

In Canada, as in the United States, the Ulsterman frequently sought employment as a labourer following his arrival. Other British stocks were inclined to regard him as beneath them for this reason, and because his living standard was lower than theirs. On the other hand, the Ulsterman was as determined as any to establish himself as an independent farmer or artisan. He worked to accumulate funds for this purpose and passed rather quickly out of the status of labourer. When the Ulsterman acquired a farm, as most did, he progressed well.[22] Along with these small farmers came many really substantial Irish farmers in the early 1830s, the years when "Edward Everett Hale thought the Protestants (of Ireland) all went to Upper Canada", while New England got the Catholics.[23] The proportion of Ulstermen who became merchants, professional men and artisans seems to have been smaller than for the Lowland Scots and English. A great part of the Ulster population had been trained to a trade, but the trade was weaving in most cases, and weaving had no more future in Canada than elsewhere. In general, the Ulstermen merged readily into the community of English-speaking peoples with which they shared a common outlook and, soon, a common economic status. The Orange Order, most notable of Irish Protestant institutions, was soon joined by large numbers of Canadians of non-Irish origin. Through it, and otherwise, the Irish Protestants had a good deal to do with defeating the Rebellion of 1837, and afterwards they took some share of the Scottish burden of harmonizing the interests of various ethnic and religious groups in Canada.[24]

Ulstermen made up the majority of workmen available for hire in the 1820s and 1830s, and therefore were important for the labour market of that time. However, they passed quickly into the society of small independent farmers. Some were artisans, but Ulstermen were neither numerous nor distinctive in this role. However, Ulster provided the heavy labourers of America until the coming of the Irish Catholics, as well as a vigorous strain of agricultural pioneers.

The Roman Catholic Irish

It has been claimed that the number of native Irish who abandoned their conquered country in the seventeenth and eighteenth centuries, to become mercenaries in all the armies of Europe, reached hundreds of thousands.[25] Some additional thousands were sold by English conquerors as "slaves" to America.[26] From the seventeenth century, if not earlier, some Irish migrated to London and to English ports on the west coast.[27] There was also a movement of migratory Irish labourers in search of seasonal employment in England, which was greatly enlarged when cheap steamship passages across the Irish Sea became available in the 1820s. These Irish often settled down in Great Britain to do the heavy work of the towns, to build canals and roads and to man cotton mills. "Down to 1844-45 Britain was the main destination of the surplus Irish, and in good times they were not unwelcome there."[28]

Of long standing, also, was a regular movement of Irish to Newfoundland in connection with the ancient trade of Waterford with that island and as an incident to the provisioning of West of England vessels.[29] Some part of these Roman Catholic Irish moved on from Newfoundland to Nova Scotia and New England.[30] However, New York attracted more around the beginning of the nineteenth century. Selkirk found, in 1803-4, that "the Roman Catholic Church is very numerous in New York — partly composed of French and Negroes from St. Domingo, but the great majority Irish mostly arrived since 1798 — Dr. O'Brien the priest reckons his parishioners 15000 — viz about ½ of the City — the Irish considerably more than half — a considerable proportion cannot speak English — there are also considerable numbers spread thro' the Country New Jersey &c particularly at the Iron works, and wherever any roads or public works are going on — they are the principal dependence for ordinary labour in New York..."[31]

Yet, allowing all these exceptions, there seems to be adequate evidence for the consensus of scholarly opinion that the native Irish were not an emigrating people until the nineteenth century, that most Irish districts had never been touched by emigration before that time, and that emigration on a substantial scale can be dated from 1818.[32] There were ample economic reasons for Catholic Irish emigration after 1815 and, indeed, before that date. However, pressure was required to break down home attachments and to establish a pattern of movement. A precedent, like the Robinson settlement or a landlord's clearing operation, could set this pattern in a district, and a fairly regular exodus usually followed. Geographically, the custom of emigration spread like a slow epidemic. It affected most of Munster long before Connaught. The better-off and more industrious were more likely to go than others; partly it was poverty that held the Irish back, as it had held the Catholic emigration behind the Protestants. Until the famine, Catholic emigration was substantially less than Protestant in proportion to population. Only with the famine did emigration become a mass flight, affecting all regions and classes.[33]

There can be no doubt, however, that Catholic Irish emigration to America increased steadily in volume after 1815, or that the North American timber trade played a vital role in this movement. Until 1816, passages to America cost about £10. With the rise of timber and relaxation of passenger regulations, the fare to Quebec tumbled in 1817 and 1818 to remain about £2 per head, and often less, from 1820 to 1840. Moreover, Quebec ships took three children for one adult fare, while New York ships did not. Catering to families worsened the crowding aboard ship, but cheapness was vital for poor families, and it was the poorest who most insisted on emigrating as family units. Fares to all ports declined in the 1820s, but American ports could not compete with timber ports for cheapness until the expansion of the cotton trade in the 1830s. As has already been noted, emigrants actually had the choice of three routes, differentiated by price and austerity. The poorest Irish went to the Maritimes and thence to Boston. Quebec drew the next and largest segment. The well-to-do paid twice as much, or more, to go in comfort by way of New York.[34]

Buchanan surmised that a few Catholic Irish had entered Canada before 1815 by way of Newfoundland, but the number must have been insignificant.[35] There were some Catholics, and an increasing proportion of them, among the Irish immigrants at Quebec after 1815; but the ports of origin and Buchanan's comments suggest that they did not exceed 10 per cent of Irish immigrants up to 1830.[36] The names of the labourers employed by the Lachine Canal Commission, 1820-24, to do the type of construction work that always proved attractive to Catholic Irish, indicate that most were Scots and Ulstermen, but that a score or so of genuine native Irish were employed throughout these years.[37] By 1826, however, a substantial number of Irish Catholics seem to have been cared for by the Montreal Emigration Society.[38] Irish Catholics certainly provided the bulk of the thousands of labourers employed on the Rideau Canal, 1827-32, and congregated in this period in Kingston, as well as in Quebec and Montreal.[39] They monopolized the public works' employment of the Canadas from this time and penetrated Toronto in numbers in the early 1830s. However, the fact that Irish Catholics congregated in cities and on canals, whereas other groups disappeared into the back country, gave them more attention than their numbers justified. There may have been about 25,000 of them in Canada in the early 1830s. A further 160,000 Irishmen entered Canada from 1836 to 1842, a larger proportion than hitherto coming from the south. So far as the censuses permit an estimation, there seem to have been about 85,000 "effectively Irish" (Irish Catholics and their Canadian-born offspring) in 1842-44.[40] With the famine migration, the "effectively Irish" seem to have increased to approximately 140,000 in 1848, 200,000 in 1852 and 280,000 in 1861. But by 1871 it does not seem that there could have been more than 260,000 "effectively Irish" in Ontario and Quebec, if so many.[41] A decline is not unlikely, since immigration was slight after 1855, while the Catholic Irish probably migrated readily to the United States.

The pattern of behaviour of Catholic Irish emigrants was surprising and disturbing to those who received them. Since the Irish had been peasants, it might have been supposed that they would settle on land and aspire to small proprietorships like all preceding immigrants to America. In fact, they showed almost no interest in farms. There are a few instances of substantial numbers of Irish Catholics settling on farms, and the settlers were moderately successful; but peculiar compulsions were required to induce settlement. The Robinson settlers in Upper Canada were directed and bribed into settlement.[42] A considerable number of Irish labourers settled along the Chicago Canal; but the reason they did so was that they had been paid for their work in land script, could not obtain cash wages, and considered it better to take land rather than nothing.[43] Irish reluctance to settle on the land has sometimes been put down to the lack of cordiality that Catholics might expect from Protestant predecessors.[44] The hazard was real, but the explanation is not convincing. Catholic peasants who really wanted to be farmers established themselves on the land in America without noticeable difficulty. The Irish did not want to farm, and some other immigrant peasant groups appeared later in America with this same outlook.

Irish Catholics chose the city because of their preference in employment, as British artisans did. But the Irish did not aspire to be merchants or artisans. Their enterprise was confined, in the first generation, to the operation of fourth-rate taverns and boarding houses that catered, primarily, to succeeding Irish immigrants. Few were craftsmen, and these few of an indifferent quality. What the great majority of Irishmen sought was wage employment at labouring jobs in the company of their fellows. Given these conditions, they were willing enough to work in rural places and displayed a partiality for public works' construction in which they could capitalize on their skill with spade and pick. In the cities, they took all the heaviest, most dangerous and most unpleasant jobs, which frequently were the most insecure and badly-paid ones. With their families, they crowded into inferior tenements, made worse by their deficiences in respect to sanitation and handiness.[45]

There was nothing accidental or peculiar to America in this behaviour. Emigrants to England and Scotland, at every period, behaved in exactly the same way. They took all the heavy rough work at whatever wages they could get. They crowded into bad districts, provided a disproportionate share of those requiring public relief, and forced the beginnings of sanitary legislation upon the municipalities in which they lived. They were intemperate and found no objection to public exhibition of the fact. They disputed readily, and resorted to violence easily. All this could be said equally of the Irish in both American and Canadian cities.[46]

Explanations of Irish Catholic behaviour have not been very satisfactory, presumably because those who explained had a very different outlook on life themselves and could not comprehend this behaviour.[47] It has been said that the Irish clung to cities so that they could be near their priests and churches; but this factor can easily be exaggerated and misrepresented. The Irishman was superstitious rather than religious; he ignored the advice of his priests when that did not suit him; and his church was more a national than a religious institution. Nor is there any reason why the Irish could not have enjoyed religious services in rural communities, as others did, if they had wished to live in rural communities. It has also been said, with more cause, that the Irish chose the life they did because they were gregarious. But even this is a superficial statement of the case. Irish love of company was not simply an extreme manifestation of the compulsion that drives even isolates to seek companions sometimes. Rather, the Irishman was a primitive man, half a tribesman still.* In his experience, one man alone could never win success or justice: only the group could do so. The group was the one hope of security: there was no other. But neither is

---

*On this characterization, see Introduction, p. xxviii. Ed.

this a reason for seeking the city. Irish gregariousness had been compatible with rural life in Ireland.

The condition of Ireland suggests much sounder reasons for Irish behaviour abroad. The Irishman eschewed agriculture precisely because he knew a form of it that offered no attractions in his experience. Such preferred employments as Ireland offered were in the cities in strongly unionized fields.[48] Proprietorship, in town or country, no doubt struck the Irishman as a different and desirable status, but he hardly aspired to it himself. It was beyond both his ambition and his wants. It required a capacity for management and a capital, which he had no way to acquire. It implied a standard of life far higher than the one he was used to. Above all, it implied a calculating individualist outlook, a divorce from group life, which was difficult for the Irishman to assume. Another generation, product of a different environment, would be emotionally prepared for proprietorship; yet even they inclined to the small establishment that yielded a living, rather than to the large one that promised a profit. On the other hand, the Irish immigrant enjoyed advantages as a labourer. He was strong, often industrious, willing to do the meanest and hardest work, and excelled all others at digging. He was more reasonable and manageable than the Highlander, more dependable than the English pauper, stronger than the French-Canadian. Above all, he was supremely available.

Dr. Dunlop declared that the Irish Catholic was "far the easiest conciliated of any emigrant", appreciating the relief in Canada from the oppression of his home.[49] There is evidence to support this view.[50] However, when the Irish felt themselves victimized by employers or governmental authorities — and they were victimized frequently — they combined for resistance readily, like most eighteenth-century men, and often struck back violently. Employers and officials were infuriated by this behaviour. They wanted Irish labourers to accept "the law of supply and demand" in economic matters — something no group ever did willingly when it was to its disadvantage — and the regular processes of law in allocating rights, as other groups more or less did. But there was nothing in Irish experience to suggest that law, courts, magistrates and troops had anything to do with assuring justice to the poor: rather, they existed to oppress on behalf of the rich. Justice, so far as there was any, was won by illegal combinations of men arrayed against lawful authority. Irish experience in America served mainly to confirm this view. However, the anarchical attitude of the Irish produced endless distrust.

Still more offensive, probably, was the Irishman's way of living and of handling himself. Earlier residents of America had developed a norm of living enjoyed by many, sought by nearly all. It required fairly substantial housing, made possible by cheap building materials and ubiquitous skills, a plain but plentiful diet and elementary care in handling water supplies and sewage. The standard achieved was well above that of the common man of Europe. The American common man also was a harder, more versatile and, for his environment at least, a more efficient worker than the European. The Irishman, on the other hand, was far below the European standard both in living and working. It is understandable that he should bring his standards to America with him, but it is understandable also that he should earn contempt and dislike by his primitiveness. The following description is of Irish labourers on the Rideau Canal in 1827 and 1828; but it would require very little modification to serve as a description of the labourers' life on other public works, or even in the cities, during the succeeding quarter century:

> The common people of Ireland seem to me to be awkward and unhandy. What they have been used to they can do very well; but when put out of their old track, it is almost impossible to teach them anything.... It is a singular fact, too, with the Irish, that if they can get a *mud-cabin*, they will never think of building one of wood. At By-Town, on the Ottawa, they burrow into the sand hills; smoke is seen to issue out of holes which are opened to answer the purpose of chimneys. Here families contrive to *pig* together worse even than in Ireland; and when any *rows* or such little things are going on, the women are seen to pop their *carroty polls* out of the humble doors, so dirty, sooty, smoke-dried, and ugly, that really one cannot but be disgusted; and do what we will for their benefit, we can obtain no alteration. If you build for them large and comfortable houses, as was done at the place above-mentioned, so that they might become useful labourers on the public works, still they keep as decidedly filthy as before. You cannot get the *low Irish* to wash their faces... you cannot get them to dress decently, although you supply them with ready-made clothes; they will smoke, drink, eat murphies, brawl, box and set the house on fire about their ears, even though you have a sentinel standing over with fixed gun and bayonet to prevent it .... They absolutely die by the dozen, not of hunger, but of disease. They will not provide in summer against the inclemencies of winter. Blankets and stockings they will not purchase.... Surgical aid is not called in by them, until matters get into the last stage. (In summer they drink) *swamp waters*, if there be none nearer their habitations.

The Irish were also accident prone:

> On the public works I was often extremely mortified to observe the poor, ignorant, and careless creatures, running themselves

into places where they either lost their lives, or got themselves so hurt as to become useless ever after. Some of these, for instance, would take jobs of quarrying from contractors, because they thought there were *good wages* for this work, never thinking that they did not understand the business. Of course, many of them were blasted to pieces by their own *shots,* others killed by stones falling on them... it is vain for overseers to warn them of their danger, for they will pay no attention.... Even in their spade and pickaxe work, the poor Irish receive dreadful accidents; as excavating in a *wilderness* is quite a different thing from doing that kind of labour in a cleared country. Thus they have to *pool in*, as the tactics of the art go—that is, dig in beneath the roots of trees, which not infrequently fall down and smother them.[51]

An especially annoying old Irish custom was to pilfer through the countryside for wood, or anything movable, in hard times and particularly in winter.[52] Winter unemployment nearly always reduced Irish labourers in Canada to extreme distress, and it is not surprising that they perpetuated the practice there, nor that gangs of labourers sometimes took provisions forcibly from houses or taverns.[53] But a population for whom property rights were sacred, conscious also of Irish improvidence, found little excuse for these depredations.

In the cities, too, the Irish remained peasants. They crowded into tenements or into shanties which they built on common or unused ground. Their sanitary practices, suited to dispersed cottage life, invited epidemics. They were regularly dependent on charitable agencies for survival. Their penetration of Toronto illustrates the difficulties they raised for city and provincial governments. While the Irish Catholic population of York was small, it appears to have constituted a vigorous community, ready to imbibe democratic ideas from neighbours. It attempted to conduct its church in Presbyterian independence, as some communities in the United States had done before they were borne down by the numbers of their kinsmen.[54] Immigration soon turned Toronto's Irish into religious conformists and social nonconformists. Toronto had barely become Toronto when, in 1834, complaints were raised against squatters erecting "huts and shanties... in the bank and on the beach in front of this City": "haunts of the idle and worthless portion of the inhabitants".[55] William Lyon Mackenzie reported these shanties, and some other poor parts of the city, to "throng with persons unlicensed selling beer, whiskey and other strong liquors, and affording... room for Gambling and Vice in its blackest shapes.... I never saw anything in Europe to exceed the loathsome sights to be met with in Toronto..."[56] A list of the squatters, who were reported to be increasing in numbers daily, makes clear that they were Irish; though it suggests too that most of them were plain poor people, perhaps selling a little unlicensed liquor for a living.[57]

Earlier, in 1831, Sir John Colborne had called attention to the wretched condition of "several" families in Toronto and proposed an enlargement of York Hospital, even at the risk of attracting the destitute from other parts of the province.[58] In the fall of 1831, the hospital reported, "The admissions of this year as compared with the last are more than double owing to the great influx of emigrants.... The prevailing complaint among the Emigrants has been fever of the type denominated Typhus. This disease engendered in poverty and wretchedness has happily been heretofore almost unknown in the Country ..."[59] But the hospital was to be pressed still further by cholera epidemics in 1832 and 1834.[60] The problem of indigence, insofar as it was distinguishable from ill-health, also outgrew the capacity of existing charitable arrangements. At the beginning of 1837, a House of Industry was established in Toronto to overcome, among other things, "the evils of street-begging".[61] Though the House of Industry assisted only those who were not able-bodied, calls upon it were very heavy from the beginning. In the first year of operation, 1837, half those relieved were Roman Catholics, and two-thirds (including Catholics) were Irish.[62] Irish Catholics also accounted for a disproportionate share of convicts in the Provincial Penitentiary.[63]

But the rise of new welfare and regulatory policies induced by the inflow of Irish Catholics, however important, was but a by-product of a far more vital development of the Canadian labour market. Until the Irish came, the market, though fairly well supplied with workmen for permanent and semi-permanent jobs, lacked labour reserves. The immigrant stream that flowed through the country after 1815 made it easy for employers to find craftsmen for more or less permanent positions and also to obtain labourers in season — the annual influx of immigrants was well-timed to meet the peak of harvest work on the farms. But there still was no permanent reserve of readily available workmen. A great part of the immigrants who stayed in Canada dispersed to settle, as potential proprietors, on farms. They often returned to the labour market to support their shaky proprietorships, but they did so at places and times dictated by their own locations and the state of their own crops, rather than by market demand. Habitant labour had the same qualified approach to the market. There were some immigrants who responded more readily to demand, but these were likely to be discouraged by the fitfulness and short duration of calls on their services, and to proceed to the United States in the belief that the labour market was more reliable there.

In Canada, a man with a young family cannot maintain his wife and children in comfort upon the wages of agricultural labour, unless he is in possession of a residence of his own; nor however much the assistance of the labourer may be required at certain seasons, can the farmer afford to give such constant employment, as would enable the labourer to maintain a numerous family throughout the year; — The poorer classes will not, if they can avoid it, resume the conditions of the cottars, or small tenants...[64]

The problem of overcoming this impasse and establishing permanent labour reserves in Canada had two sides. It might be solved, on the one hand, if demand for labour could be sufficiently consolidated so that a body of labourers could be supported in conditions acceptable to them. It might be solved, from another direction, by so narrowing men's opportunities that they would be forced to become "cottars or small tenants" whether they liked it or not. The privileged in Canada leaned heavily to the second solution, arguing their case largely in terms made notorious by Wakefield, whose ideas were very far from being his exclusive property. It has already been noted that an attempt was made in the 1830s to create a cottar class in Upper Canada, available for hire by neighbouring employers, by locating poor immigrants on five-acre lots.[65] The policy of denying free grants of land that was followed fairly consistently after 1831 also owed something to the desire to push poor men into a permanent labour reserve. As late as 1840, a spokesman for privilege still felt required to argue:

> The greatest drawback to the employment of Capital in this Country at present consists in the *high price of wages*, and the *extreme difficulty of procuring the labour* requisite for its profitable employment in *any* pursuit; and more especially in *agricultural* ones. Everything, therefore, that tends to lessen the *quantity of labor in the Market*, will also tend to *exclude capital from it*. But the main cause of the scarcity of hired labor in a new Country is the *Cheapness of Land*, and it seems to follow, as an irresistible conclusion, that the *Free gift of Lands*, must increase that scarcity an hundredfold...[66]

The right course, then, was to make land costly, so as to keep men in the position of labourers, and force down wages. This would avoid a community "composed exclusively of the occupiers of Free Grants — and we can hardly picture to our imagination anything more deplorable than the condition of a community so constituted."[67] But it should not be imagined that this was a policy of sacrificing the poor for the sake of the rich. On the contrary, "the Interests of these two classes, instead of being hostile and opposite, are connected together by so firm and indissoluble a tie, that whatever depresses the Immigrant Capitalist must at the same time injure the labouring Immigrant."[68]

A Wakefieldian program for Canada had to encounter at least two great difficulties. The less formidable of them was that a great number, and probably a majority, of Canada's residents did not want their country overrun by capitalists and proletarians. They were partial, indeed, as William Lyon Mackenzie had been in the 1820s, to the objective of a versatile home-market economy that would free them from their extreme dependence on export markets.[69] But, when they thought of creating manufactures, they probably pictured upstanding free artisans rather than pale and brutalized factory operatives. They, too, believed that "the province should be a community of simple living, hard working, frugal, independent farmers served by honest merchants, craftsmen, small manufacturers, township schools, an honest legislature, and a free press; in short an educated and largely agrarian democracy."[70] Many would have agreed that:

> ... great establishments of manufactures require great numbers of very poor persons to do the work for small wages; that these poor persons are to be found in Europe in large numbers, but that they will not be found in North America until the lands are all taken and cultivated, unless unnatural laws should be framed to prevent the cultivator from being supplied from the cheapest market ... that every one who is able and willing to work may be profitably employed in North America ... (and it is to be hoped that no pauper class will be created to disturb) the happy equality which has hitherto prevailed in this country.[71]

This was not just Reform sentiment, either. R.B. Sullivan, who spoke for the poor settlers of British extraction, was as decidedly conservative as they. But, he found it no evil that in Canada "the only necessary inequality of condition is, between the small and the great, the poor and the wealthy land owner."[72] And there seems to be more than dispassionate logic in his argument, that

> ... the attempt to produce a greater inequality [by restrictions on grants of land], founded upon reasoning applicable to other Countries appears in theory, as it has been found in practice, chimerical ... the attempt to force upon a community like this the universal relation of master and servant, of Capitalist and hired labourer, must be vain; it may be successful in a new Country, from whence the poor man cannot escape; — the trial is in vain here; — but, it may be attended with the worst evil, which, in the ordinary course of events can befall the Colony, namely, the retardment of its progress to the possession of a numerous population.[73]

Sullivan stated in the quotation above, as he stated many times, the second and more serious obstacle to the creation of a reserve of cheap labour in Canada: the natural forces operated against it. If men could not get land in Canada, they would go instead to the United States. Besides, there was not a regular enough demand in the Canadian countryside to support labourers.

Oddly enough, while the case for creating a labour reserve by re-stricting the ownership of land was being debated back and forth, the issue had already been settled. It was settled by the arrival in numbers of Irish Catholic immigrants. Here was a class eager for wage employment and immune to the lure of independent proprietorship. Moreover, their wants were so meagre that they could somehow sur-vive — with a measure of public assistance — upon an assortment of poorly paid jobs of doubtful duration. To be sure, even they could not find enough work and assistance to sustain them in rural areas. But the growing cities proved able to support them in some fashion, and the very coming of the Irish increased the cities' power to do so. As the number of Irish immigrants grew larger and larger in the twenty years after 1830, labourers' wages, which had been pretty stable while immigration was small, came tumbling down.[74] As a consequence, in the United States factory owners abandoned their air of benevolent paternalism and replaced the cultured young ladies of Lowell with cheap Irish immigrants.[75] Canada did not have many factories in the 1840s, but the Irish did find a surprising amount of heavy and un-pleasant work to do in the towns, and they formed labour pools from which factories might be supplied. By 1861, a writer could point to an impressive list of mechanized industries in Toronto which, with the aid of the "cheap labour which in a large city like Toronto can always be commanded", were driving rural craftsmen out of business and overpowering foreign competition as well.[76] The same could have been said of Montreal, Hamilton and perhaps Quebec.[77]

In spite of banker Allan,* men could not be barred from land in Canada. Yet, in spite of Mackenzie and Sullivan, agrarian democracy could not be preserved nor industrial capitalism held back. The Irish, without consideration of any of them, had created the necessary labour reserves to facilitate its growth. In doing so, they had fostered a regu-larity of demand for labour that made it increasingly deserving of a permanent and ample supply, which might include less hardy strains of labouring peoples. It is in this connection that one more thing ought to be said of the Irish. Though they created Canada's reserve of unskil-led labour and made cheap labourers easily available in Canadian cities, it was not altogether Irishmen on whom the industrialists of 1860

---

*William Allan, first President of the Bank of Upper Canada, in 1821, was ap-pointed to the Legislative Council of Upper Canada in 1825, a member of the Executive Council 1836-40, and a "pillar of the Family Compact". Ed.

depended. For, in their way, the Irish advanced rapidly in America. They clung to the cities; but they had the perspicacity to concentrate on the rapidly-growing cities. A second generation retained enough of their heritage that they were still "effectively Irish", but they were no longer peasants. Schooled in city life, they explored new avenues of employment and dispersed among many occupations.* They accepted at the same time an urban version of capitalistic progress, saved to acquire a lot, built a house and put funds in savings banks.[78] Saved from too close a tie with their homeland by the inability of an exhausted Ireland to send out many emigrants after the famine, the Irish in America found new standards, new wants, a new morale and integration into their new environment. The capitalistic labour market, which it had been their dubious role to create, was to be in the last third of the nineteenth century the habitat of French-Canadians, new immigrants and the distressed of many races.

Ethnic Conflict in the Evolution of the Canadian Labour Market

What most clearly distinguishes the Canadian labour market from many others is the way in which its shaping and development were accomplished through the conflicts of ethnic groups. Some nations entered the industrial age as ethnic units. The United States has had sharp ethnic clashes, but has also possessed a dominant group which has been able to impose a high degree of cultural and ideological conformity upon minorities. Canada, in contrast, has remained a land of ethnic diversity. Its main social pattern was determined by its early experience as a bi-ethnic state, containing French-speaking people who were nearly all Catholics, and English-speaking people who were nearly all Protestants, with neither of them able to digest the other. Most of French Canada, and some part of English Canada, have pretended ever since that Canada's population is still of this simple structure. However, immigration produced an ethnically complicated Canadian society early in the nineteenth century, and subsequent immigration has offset cultural assimilation to keep it so.

In the labour market, differences in ethnic background between employers and employed have often accentuated incomprehension and hostility. Division of a work force into two or more ethnic groups of different status has also been common and has encouraged employers to seek bargaining advantage by exploiting the division. The division has sometimes been overcome, when wage earners have been persuaded to sink their ethnic differences in favour of a common allegiance

---

*The idea of intergenerational social mobility is not supported by Katz. See Michael Katz, *The People of Hamilton, Canada West* (Cambridge, Mass.: Harvard University Press, 1975). Ed.

to "the working class"; and the hostility which employers have usually shown to doctrines of "working class unity", and to conceptions of class in any proletarian sense, suggests that they have considered division among their workmen to be of substantial advantage to themselves. The more common experience in Canadian history, however, has been of ethnic groups that stressed their ethnic unity and interest at the expense of their class unity and interest in attempts to win advantages at each others' expense. Superior status groups have done this by cultivating a narrow craft unionism, which frequently displayed energy only in obstructing the organization of other workers, or by demonstrating their immunity to union appeals, counting upon suitable rewards from employers. Groups of inferior status have followed comparable policies. Successive inferior groups of immigrants have accepted inferior working conditions, remained ostentatiously aloof from unions and acted as strike-breakers, partly because the earlier entrenched groups have been disposed to drive off newcomers or to sacrifice their interests. Inferior groups, therefore, have ordinarily offered organized resistance to employers in their early years only where they held a monopoly of employment, as the Irish did in public works' construction. As new groups consolidated their positions and became old groups, they were undermined in turn by new immigrants who depended upon their ethnic communities for whatever support they enjoyed.[79]

Divisions among workers (and employers) may be expressed in terms of race, religion, language, nationality, regional origin, or whatever offers an excuse for dividing. But the amalgams of these factors connoted by ethnic distinctions provided the usual vehicle in Canada to divide the whole and unify the parts. Neither residents nor immigrants, in the early nineteenth century, had been subjected to forces that could produce a deep national consciousness or class consciousness. What most of them had, instead, was an ethnic consciousness of a tribal nature and pre-capitalist origin. The more primitive they were, the more they sought security in the identity with their ethnic community, strength in its unity, and advancement by means of its power to supplant other groups. If the vehicle of the ethnic group had not existed, a substitute would have had to be invented; for the peasantry of Scotland, Ireland and Lower Canada were not conditioned to individualism, and there is little room for individualism in the lives of wage earners. However, the basis of division should not be mistaken for the cause of division. That is found in the contest for economic advantage among groups and classes. Combinations and unions often relied on ethnic loyalties for their cohesion, but it was nevertheless the economic position of wage earners under capitalism that produced the unions, and it was coincidental that ethnic forms were available for them.[80]

Ethnic tensions can be found as early in the labour market as in other spheres of Canadian life. They colour the relations between high-handed English-speaking fur-traders and their canoemen. Strains must have arisen after 1760 between English-speaking artisans who assumed the role of first-class workmen, and French-speaking artisans who were relegated to a second class. But such rivalries were evaded, for the most part, by the device of a customary assignment of certain employments to the French and others to the English. In any case, there was a substantial geographical separation of French and English communities, the labour market was small and most of the population had little to do with it. Where ethnic intermixture in employment nevertheless occurred, it did not have to produce rivalry so long as the supply of labour did not outrun demand. The mixed body of working-men at Kingston appear to have maintained amicable relationships over several decades.[81]

Immigration after 1815 intensified the fears of the French in Lower Canada that the land which they had counted on for the settlement of their future surpluses of population would all be seized by others. It also undermined the position of the old settlers in Upper Canada, where the immigrants ranged themselves on the side of British officialdom, partly from their own conservatism,[82] partly from fear and jealousy of the old settlers.[83] Ethnic rivalries were sharpened to some extent by these developments. But the early immigrants — Scots, English, Ulstermen and the vanguard of Irish Catholics — did not introduce serious strains into the labour market. Most of them moved on quickly to settle as farmers or to pursue richer returns in the United States. Artisans who stayed had found a place of employment. There was room, without displacing anyone, for many thousands of labourers, because some tasks had gone undone for want of applicants and because employment was expanding, particularly in timber and construction. In 1820, in 1827, even in the face of the enormous immigration of 1831, the opinion was expressed that the labour market was not crowded and that there was no visible limit to the opportunities for employment.[84]

It seems clear, however, that the absorptive capacity of Lower Canada was exhausted early in the 1820s. A writer of 1820 declared that Montreal offered openings for craftsmen, but did not hold out the same promise for labourers, and assumed that sensible immigrants would go elsewhere.[85] Very shortly after, it was said that "the hovels" of Quebec and Montreal "contained crowds of British emigrants, who were struggling with those complicated horrors of poverty and disease", too poor and too ignorant to go on to Upper Canada.[86] Many immigrants did find work as "labourers on board Vessels, on the rafts, wharves, and timber yards or are engaged as Servants in Quebec, and lay up a part of their earnings in the Savings Bank... ",[87] and the city held "a profusion of Hibernian porters and knights of the hod..."[88] Efforts were made, with some success, to expand the opportunities of Lower Canada by opening roads in the Townships.[89] But even in good years, the Quebec labour market was flooded long before the peak of the immigrant season which denoted the arrival of the most necessitous Irish.[90] Those immigrants who escaped crowded hospitals often lost their way in poor lodging houses so conducted as to be "exceedingly favorable to the production and diffusion of Typhus Fever".[91] Though 1831 was a prosperous year, "the number of destitute exceeded this year that of any former one", and 5,000 immigrants required the help of the Quebec Emigrant Society.[92] An emigrant guide of 1833 wished "it to be distinctly understood by the labouring classes of society, that there are already sufficient numbers in, and about Quebec, and the lower province..."[93] In 1844, it was remarked that "generally throughout Western Canada these cases [winter unemployment of labourers] prevail to no great extent, and are usually promptly and liberally relieved. At Quebec and Montreal, the first landing points of emigration, and where the winters are much longer and severer, the case is very different..."[94] In 1850, "Quebec will be found to contain loiterers and labourers, who will leave little employment for newcomers. The mechanic, however, may find some work, providing he will accept an engagement at reduced wages."[95] During the economic disturbances of the late 1840s, there was not work enough even for the resident population, and the working classes of Montreal and Quebec contributed substantially to emigraion to the United States.[96] In keeping with this chronic oversupply of labour, wage rates in Lower Canada always were lower than those elsewhere.[97] There was ample room here for contests over employment.

The very optimistic prospectuses for immigrants were based throughout, in fact, upon the opportunities in Upper Canada. It was continually pointed out how superior were the openings and wage rates of that province.[98] It did expand at a remarkable rate and often showed an ability to absorb 20,000 or more immigrants in a year, presumably because a large proportion of them, or of their predecessors, took up farms.[99] But all was not plain sailing here, either. The immigration associated with the construction of the Rideau Canal produced, by 1827, a sharp lowering of labourers' wages in the eastern part of the province and the appearance of a poverty-stricken urban population in Kingston.[100] Widespread distress in 1832 was blamed on the reluctance of residents to hire immigrants for fear they were carriers of cholera.[101] Possibly 10,000 resident labourers were thrown out of employment during the commercial crisis of 1837. The bulk of them moved off, sooner or later, to seek employment in the United

States, and immigration fell to a very low level in 1838 and 1839.[102] In consequence, labour was reported to be short in 1838, and it certainly was so in 1839.[103] Employment expanded rapidly in the 1840s, but so did overseas immigration, and there was an overland immigration of several thousands of Irish labourers who fled to Canada because of the suspension of construction work in the United States.[104] Labour was in surplus in Upper Canada from 1840 to 1843, and its cities held destitute populations reminiscent of the lower province.[105] The immense overseas immigration of 1847 was officially numbered at 90,000 persons, but probably was much larger. Coinciding with economic recession, it was bound to produce overabundance of labour everywhere. The relief burden and social inconveniences imposed by it brought sharp and very widespread demands for limitations on immigration.[106] The labour market recovered by 1850, and the railway boom that followed produced, in spite of a brisk inflow of immigrants, an acute labour shortage by 1854. In that year, Buchanan said, "I have never known such a season as this, and the universal complaint from one end of the country to the other is the impossibility of securing labourers..."[107] But then the Canadian economy lost its buoyancy, and immigration dropped off, in spite of efforts to encourage it by those who imagined that immigration itself was the causal factor in prosperity. There continued to be ample supplies of labour in Upper Canada without outside help.

It is apparent that many of the factors that contributed to the periodic clogging of the Canadian labour market were temporary ones. But there were fundamental weaknesses also. Lower Canada had never possessed a power to expand significantly. Immigrants were more and more likely to be Irish Catholics who flocked to the unskilled labour market rather than the land. The stresses of French-Canadian society forced more and more of its members to seek wage employment from the 1840s onward. These developments were supportable so long as Upper Canada was resilient. But the sharp check felt in Upper Canada after 1835, regarded at the time as a product of temporary factors, in fact marked the permanent end of that soaring buoyancy which had been based upon the attraction of good land. In the view of Hansen and Brebner, "A state of congestion had resulted from the rapid development of land and resources", in contrast to the less cluttered attractions of Iowa and Michigan in the 1830s and other states in succession after them.[108] The trouble was, in short, that "Canada had no Middle West of her own";[109] and this lack is of prime significance, regardless of what exaggerations may have been committed in the name of the frontier thesis. Its effect was that, while expansion continued in Canada, and canals and railways provided an aura of prosperity during their construction and a foundation for deeper development, Canada could not match the new levels of vitality attained by the United States. The commercial extractive economy had reached its limit, and a new economic dimension was required. Agitation for industrialization was general, and the labour supply required for it was available; but market conditions were not very favourable and growth was slow. The labour market remained generally crowded.

In the resulting competition among wage earners for a livelihood, there must have been some rivalry among the old stocks and the senior immigrants. Yet, they did not find it necessary to fall into open conflict for employment. For artisans, the 1840s and 1850s offered a rapid expansion of employment opportunities. The English-speaking, in general, could seek a place in new settlements — in the United States, if not in Canada. The French were hard pressed; but those who became wage earners preferred the light industries of New England to the heavy work at home. Neither they nor the senior English groups seem to have crossed customary lines to take employment from each other.

Most of the "burden of adjustment" seems to have fallen upon the Irish Catholics. They were, in every sense, a marginal group. They stood at the lowest level in skill and income, in the kind of work they were offered and took, and in their standard of living.[110] They were Catholics, like the French, but they were excluded from French society by their failure to share its language and culture. Probably most of the Irish spoke English, but they did not fit into the English-speaking community because they were Catholics and because their behaviour violated many social canons. Others were offended by their heavy calls on relief agencies, their uninhibited traffic in liquor, a similarly uninhibited use of knives and firearms, and strong-arm activities at elections; though politicians often encouraged this last practice on their own behalf. Employers and government officials professed exasperation with Irish unwillingness to bow before "the laws of supply and demand". The Irish were not property owners, and often they were not fixed residents but migratory workers. They were therefore excluded from the body politic by law and custom. The rejection of the Irish by others, and Irish introversion and defiance, tended to intensify each other. Migratory Irish workers were regarded as a third-class, disposable element of the population. "It is notorious that the great majority of the men who have been for some time engaged on public works, become a class of migrating labourers, neither valuable as settlers, nor disposed to fix themselves as such ... "[111] "The over-stock of labourers (are) a floating mass from the States, who, when their services are no longer required, or serviceable to the province will recross the lines."[112]

Yet the Irish felt as much compulsion as others to survive and fought tenaciously for a livelihood. The rise of Orangeism among

English-speaking Protestants is an index of general antipathy; but the
way Orangeism flourished in English-speaking cities suggests a specific
conflict. It seems probable that city workingmen hoped through the
Orange Order to check the influx of Irish Catholics that weakened the
position of all wage earners, and particularly to oppose Irish Catholic
attempts to penetrate the artisan class. In this respect, therefore,
Orangeism is comparable to the "Native American" movement in the
United States.[113]

The major ethnic clash over employment, however, was between
the Irish and the French. Contemporary observers showed a curious
unwillingness to understand this hostility. Buchanan even denied that
it existed.[114] McTaggart said that French-Canadian hatred for the
Irish (and the Yankees) "seems rooted", but did not attempt seriously
to explain it.[115] Another writer said that the French greatly disliked
the Irish, "for what reason I could not find out..."[116] Lord Durham
noted the competition for employment between French and Irish, but
did not "believe that the animosity which exists between the working
classes of the two origins is the necessary result of the collision of
interests." Rather, "The labourers, whom the emigration introduced,
contained a number of very ignorant, turbulent, and demoralized persons, whose conduct and manners alike revolted the well-ordered and
courteous natives of the same class."[117] Such myopia was encouraged
by the fact that the French had not exploited very actively the employments into which the Irish pushed. But the French regarded most
of these tasks as customarily their own, all the same, and they had
increasing need of them as outlets. Yet, until about 1860, the Irish
made the unskilled urban employment of Lower Canada pretty much
their private preserve. Only as they abandoned an economy in decline
were the French able to occupy the place which they thought to be
rightfully theirs.[118]

The French were the one group in Canada that the Irish felt they
could attack with impunity, not only in the labour market, but physically. Superior in physique and aggressiveness, the Irish could and did
terrorize the French when it suited them. It suited them, among other
occasions, when there was acute competition for work in the Ottawa
timber trade, and on some public works in Lower Canada. On the public works, French were employed regularly as carters and woodcut-
ters, but rarely as labourers. They evidently did not like this work, for
they responded poorly to offers of this employment, and contractors
preferred the hardier and more skilful Irish.[119] Still, there are instances
of members of the two groups labouring together in peace. In such
cases, the French seem to have been readier to strike for higher wages
than the Irish; or perhaps the French took this way to end an unpleasant
experience, and the Irish hung back to be rid of their competitors. A

strike of this sort occurred at Beauharnois in 1843, which led to the dismissal of the handful of French involved.[120] There was a larger strike the next year among labourers building the "Cascades Road". It is interesting that in this case the French are supposed to have induced the Irish to join their strike by threats of physical violence.[121] But threats and actual violence were typically levied the other way around. They arose in the course of strikes by Irish labourers, when attempts were made to enroll French-Canadians as strike-breakers. This recourse was tried on the Beauharnois Canal in 1842.[122] A "riot", evidently a fight between the ethnic parties, ensued.[123] The following spring, a similar difficulty arose on the Lachine Canal, where the Irish were conducting a long, bitter struggle against their employer. "A number of Canadians who have been taken on the works since the last strike were, it is said, assailed by a party of Irishmen with stones, &c., and driven off the Canal. One man we hear received a severe blow on the head from a missile, and all were so much intimidated as to render it unlikely they will return."[124]

The way was prepared for strife in the timber trade by the construction of the Rideau Canal, which brought hundreds of Irish labourers to the Ottawa, and its completion, which left them without employment. It was natural that many of the Irish should think of entering the expanding timber industry. They must have been inept woodsmen to start with, but the haphazard method of granting timber limits in vogue produced an opening for their talents:

> These disturbances commenced by persons employing Men to cut Timber [on Crown lands] getting into disputes as to their respective boundaries; and there not being any person to appeal to for settlement of such disputes, they have had recourse to hired bravoes and ruffians of the worst description, to intimidate the more peaceably disposed, this mode of proceeding has been carried on for about 3 years [i.e., since 1832, when most work ended on the Rideau Canal], which has been the means of introducing a number of the worst Characters [Irish] into that part of the Country...[125]

However, there evidently still was not room enough for all the Irish, and that group felt its employment and wage rates to be threatened by the more docile and proficient French, who had hitherto provided the bulk of the employees on the river.

> This year [1835] the disturbances have taken a different shape, these bravoes and ruffians have headed & led on the whole mass of Irish laborers, to drive the Canadian laborers from the lower province off the river, so that they might themselves be enabled to fix a high Standard rate of wages. I am sorry to say, however contrary

this may be to the interests of those who employ men, yet some of the employers have actually headed these disorders, without considering the effect it would eventually have on their own business, but merely looking forward to the immediate gratification of some Jealousy or pique [originating about boundaries] against other employers having Canadian laborers...[126]

The Irish raftsmen called themselves "Shiners".* One of their objects, in which they had considerable success, was to win a closed shop for their quasi-union organization — that is, to induce their employers to hire none but Shiners. For several years, a species of civil war was carried on along the Ottawa, and particularly at Bytown. It involved not only the French and Irish raftsmen, but a "law and order" party that enjoyed the support of Irish Protestant farmers. In the 1840s and 1850s, Bytown's riots were mostly contests between Irish raftsmen and Orangemen†, and mixtures of French and Irish shantymen could be found on the same raft.[127] However, it was considered the wise policy to keep parties employed apart from each other whenever possible.[128]

Ethnic rivalries involving Irish Catholics, like others, declined only slowly. But new conditions allowed the tensions to ease. The Irish won a permanent place in a number of employments, which they tended to make hereditary and customarily their own. The virtual cessation of Irish emigration to Canada, and the continuation of Irish emigration from Canada to the United States, prevented the necessity for the Irish to strike out in new directions. Soon the settled body of Irish workmen, if still suspect, were no longer "strangers" but citizens. On the other hand, they never won the political power in Canada which they exercised in some states of the United States, perhaps because they were too few and scattered to attain a strategic position; and therefore lacked an opportunity to exasperate others from office. After 1855 it was new groups of immigrants who successively entered Canada to produce more, but different, ethnic tensions.[129]

A Note on the Orange Order in Canada

The progress of the Orange Order in Canada affords a useful sidelight upon ethnic and class cleavages and the position of labour. The Order first appeared formally in 1795-96 in the border region of Ulster. However, societies like it, and activities and ceremonies of the sort it adopted,

---

*For a fuller treatment, see Michael Cross, "The Shiners' War", *Canadian Historical Review*, March 1973. Ed.

†For a fuller analysis, see Michael Cross," Stony Monday, 1849: The Rebellion Losses Riots in Bytown", *Ontario History*, Sept. 1971. Ed.

had been carried on in Ireland for a long time before that.[130] Orangeism reflected the lower-middle-class position of the bulk of the Protestant inhabitants of Ireland. They felt continually threatened from two sides: by England and the Irish aristocracy above, and by the mass of Roman Catholic peasantry below. The organizations by which such groups express their resistance are typically ambivalent, sometimes taking on a quasi-socialistic form when the upper millstone is regarded as the more serious pressure, and again, a fascist form when the main attack is from below. Orangeism in Ireland was primarily, but not exclusively, concerned with the threat from below. Pressure on the land, aggravated by the social, political and legal inadequacies of Ireland, provoked the systematic practice of mass violence in the eighteenth century and, still more, in the nineteenth. Violence was ordinarily conducted through the agency of secret societies, and its usual object was to acquire land, or employment on land, by extra-legal means. Those with legal claims to land, against whom the violence was directed, were usually (but not always) Protestants. The attacked sought ways to resist, which included the organization of societies of their own.

A special aggravation, which produced the Orange Order, was the push of land-hungry Catholics from the south of Ireland into Ulster. The landless man would offer any rent for land. Moreover, the southerner actually was able to pay a higher rent than the Ulster Protestant; for though the Protestant was the more efficient farmer, he could not match the Catholic's low and inexpensive standard of living. By tradition, Ulster offered lower rents and greater security of tenure than other parts of Ireland, and a preference for Protestant tenants. Landlords were tempted, however, to reap the harvest of the Catholic influx and to displace Protestants in favour of the newcomers. The very high rate of Protestant emigration from Ulster was, in part, the reflection of this displacement.[131] But the appearance of resistance organizations was another logical consequence. The Orange Order was produced specifically by the physical battles that were a by-product of Catholic immigration. However, it was bound to show hostility to southern penetration of Ulster in general. It was also easy for the unity of Orangeism, and of the things it opposed, to be expressed in religious terms. Once established, the Order acquired a certain amount of independent momentum as a fraternal and benevolent society.

While Orangeism was directed against Catholic competition for land, and its symbolism provoked Irish Catholics by reminding them of the most thorough of the conquests of Ireland, the Order was not simply an anti-Catholic institution, as attacks upon it have commonly assumed. William of Orange symbolized the British Crown, but he symbolized still more the constitutional rights of the subject, including the right on occasion to defy the Crown. When Orangemen talked about the British constitution, which they did constantly, it was usually the rights of the subject that they had in mind. This may explain why, though Orangemen stressed and demonstrated their extreme attachment to the British connection, and were transferring their loyalty in the nineteenth century from Ireland to a loosely-defined community of free Britons, Britain's rulers had an intense dislike for them. The rulers preferred subjects who not only were loyal, but who were docile and asked no questions. They were not very hopeful of finding these desired qualities in any part of the population of Ireland, but thought they detected them in the Roman Catholic population of Canada, which they therefore cultivated. This partiality, and efforts to suppress Orangeism, infuriated Orangemen, conscious of their own superior devotion. This fact, and different economic and political circumstances in Canada from those of Ireland, led Canadian Orangeism into being directed more against groups above the lower middle class than against those below it, and resulted in some remarkable alliances between Orangemen and their reputed enemies.

A conception of Orangeism seems to have been common in Canada from the beginning of the nineteenth century. In 1804, Lord Selkirk had great plans for settling Upper Canada with Roman Catholic Irish from Nova Scotia, New York and Ireland. "I mentioned the prejudice prevailing against Irish Settlers — he [Burke, Roman Catholic priest at Halifax] said Hunter was free of them — frequently expressed his opinion that the intolerance & oppression of the Orange men had occasioned the disorders of the country... "[132] It has been claimed that there were Orangemen among the British troops in Canada during the 1812-1814 war and thereafter, and that Orange parades were held in Toronto in 1818, 1822 and 1824.[133]

The public records suggest that the first Orange Lodge in Canada was established at Perth, in 1823 or 1824, under the leadership of Alexander Matheson, who was then Deputy Sheriff of the Bathurst District.[134] Matheson helped to set up several other lodges in this neighbourhood in the next few years.[135] A circular of 1826, addressed by James FitzGibbon to the "Orangemen of Cavan and Perth", implied that they were the only ones in Canada.[136] The provincial authorities urged the Perth lodge to disband in April 1827, on the ground that "these Associations are discontinued at home and in every part of Canada except this District".[137] However, the secretary replied that Orangeism was flourishing in Ireland under a different name, and that in Canada "there are various lodges in several parts of it with which we ourselves correspond".[138] The Perth lodge bore, in fact, the number "7", and its persistent critic, William Morris, noted another dimension of Orange growth in 1827 when he reported that "several members have lately joined it who have no Connexion with Ireland".[139] By 1827, Orangemen were organized in

Kingston, where a "riot" occurred on July 12 of that year.[140] By 1830, when the Lieutenant-Governor issued a circular to magistrates and clergymen advising them against lending any countenance to Orangeism (the advice met a very mixed reception), the Order seems to have been solidly entrenched from Kingston through Perth and beyond, and in the part of Upper Canada east of this.[141] It was ready for the skilful hand of Ogle R. Gowan, who arrived from Ireland in 1829 and established a newspaper in Brockville. Gowan called a meeting in 1830, attended by representatives from Brockville, Montreal and Perth, which set up a Grand Lodge of British North America. Official recognition of this central authority was received from the parent body in 1832.[142] Gowan became the first Grand Master and was to exercise enormous influence in subsequent decades, usually as Grand Master, but sometimes without the office. It is evident that membership in the Order grew very rapidly. By 1833, it claimed 11,242 members.[143] There are supposed to have been 18,000 members in 1837, 20,000 in 1843 (after some decline about 1840), 40,000 by 1846, about 80,000 in 1856 and perhaps 100,000 in 1861.[144] An ethnic-religious movement of this size could obviously exert great influence. It is useful, therefore, to note conditions that promoted the growth of Orangeism in Canada, and the role that it played.

Though a student of Orangeism remarks that "it was only to be expected that Orange Irishmen, possessing the depth of conviction that their institution inspired, would set up branches of the Orange Order in their new homes in Canada...",[145] the significant fact is that the Order was not at first perpetuated. On the contrary, wherever and so long as conditions reminiscent of Ireland were absent, the Protestant Irish were content to go on without this or any other distinctive organization. In some parts of Canada, this phase lasted for several decades. Nor does there appear to be any mystery concerning the stimulus of its growth. Orangeism arose precisely at the time (1823-24) and the place (Perth and its vicinity) that a large body of Irish Catholic settlers were deposited.[146] It emerged at Kingston exactly at the time that Irish Catholics inundated that town. Its consolidation in the eastern part of the province, especially along the main river routes, reflected the permeation of that area by Irish Catholics (to quite an extent the by-product of canal construction). Its development at Peterborough, Port Hope and Toronto is an index of the numbers and truculence of Irish Catholics.[147]

In keeping with this index, Orangeism scarcely made an appearance in the western part of Upper Canada before 1835.[148] It developed rather slowly in the west in the 1840s and 1850s, attaining its full strength only after 1855, and even then it was nothing like so strong as in the eastern part of the province.[149] In good part, the development of Orange strength in the west was an almost visible reaction to the penetration of that area by gangs of Irish Catholic canal and railway construction workers. Clashes between them and the established population arose on various grounds, and provoked a good deal of hostility. The effect was cumulative: the more construction workers, the more clashes, the more numerous and aggressive the Orange resistance movement; hence, still more clashes, and so on.[150] For the less sophisticated Orangemen — perhaps for a majority at most times — keeping Irish Catholics in their place was the main, or even sole, purpose of the Order. At the lowest level, there was direct competition for employment. The motivation of the small farmer who supported Orangeism is less obvious, but he certainly was concerned to retain control of his social and governmental institutions and to maintain the homogeneous population that made them function smoothly. The Orange Order took particular pains to control as many local offices as it could; and the less an office seeker was entitled to support by his personal merits, the more likely was he to depend on an institutional vehicle and maximize the threat of Popery.[151] Finally, nearly all levels of society were affected by physical clashes between Irish Catholic workmen and others, which were sometimes the product of drink, sometimes of poverty and sometimes of politics.

Yet, though antipathy for Irish Catholics in large numbers was the predominant factor in the rise of Orangeism in Canada, it had little to do with the role the Order played. The leaders of the Perth lodge as early as 1827, and Orange leaders on later occasions, insisted that they wished only to live in friendship with Catholics.[152] The leaders of the movement after 1830 directed their fire consistently against the upper rather than the lower levels of their opposition. Sometimes the objects of their attack were the officials whom British governments had appointed. Mostly, however, they were the entrenched, prosperous, smug "Americans". No love had been lost between these and the new, poorer overseas settlers at any time. As the Reform party became increasingly radical in the 1830s, the Orangemen set themselves solidly against the anti-British Americans who, according to Gowan, "have carried with them to this Province all their republican notions of 'liberty and equality'".[153] The division thus reflected was, in part, a geographical one: Americans were concentrated in the western part of the province and along the front in the east. Overseas settlers were predominant in the eastern back concessions where Orangeism was strongest. The Orange leadership set out with skill and no little spirit of sacrifice to confine and defeat the American threat. That required a working arrangement with others who shared a political loyalty, but little else, with Orangemen: Roman Catholics and government officials. Already, in 1832, Gowan advised the lodges to bow to the government's opposition to Orange processions.[154] In 1836, the Order, upon the request of the Lieutenant-Governor and in view of "the already agitated state of the province",

agreed to dissolve itself altogether — formally, at least.[155] These measures facilitated an alliance with Roman Catholics in defence of the
British connection against the Reformers. "It is estimated that certainly
one third, and probably one half of the Reform defeats [in 1836] were
due to the alliance of the Orangemen under Gowan and the Roman
Catholics."[156] In the Rebellion itself, the Orangemen were zealous,
though undiscriminating, defenders of the government. They provided
an extensive support, which the equally dependable Scottish Catholics
could not give, at a time when neither French-Canadians nor Americans
could be relied on, while the Irish Catholics adopted a policy of neutrality.[157] The solid Orange-Catholic body, dominating the strategic bridge
between Kingston and Montreal, could have doomed a much more formidable rebellion.

The extreme concessions which the Orange leadership was willing to
make were not very popular in the lower ranks, in 1836 or later. Nor did
local High Tories and British governors demonstrate much gratitude;
rather, they continued to try to suppress the Order.[158] Perhaps this was
fortunate for Canada, since it encouraged the Orangemen increasingly
to support moderate reform and responsible government. On the other
hand, Orange distrust of Reformers was heightened in the early 1840s
by violent attacks on the Order, culminating in the discriminatory
"Secret Societies Act" of 1843, which attempted to outlaw it. This, too,
was of profound importance. It consolidated Orange opinion and greatly extended the influence of the Order. It recommended Gowan's policy
of building a moderate centre political bloc. It promoted, therefore, the
disintegration of the Reform movement and the consolidation of a
moderately conservative lower-middle-class interest which provided
the basis of Canada's governments for several decades. The key alliance
of this structure was the one between Orangemen and French Roman
Catholics. The emergence of that alliance may have been facilitated by
the fact that French Canada felt as much antipathy for Irish Catholics as
Orangemen did. However, the Orange leaders desired an accommodation with the Irish Catholic population also, and achieved it, though
it was less easy and stable than the arrangement with the French. The
basic alliance was not an easy one for the Orangemen to maintain either:
large portions of their membership were attracted regularly into the
Clear Grit camp by anti-Catholic appeals. As against this, Orangemen
and French-Canadians found a real bond of unity in demands for protection. Protection was a very popular doctrine, with the additional
advantage that it tended to isolate the Clear Grit element of western
Canada and to neutralize its attractions. Though hardly ever free of
internal division, Orangeism was in the 1860s the mainstay of the government and "the greatest power in the state".[159] It thus underwrote
the coalition that effected Confederation.[160]

The interplay of ethnic and class appeals revealed in the Orange record serves to illustrate the nature of these loyalties. The ethnic and religious alliances achieved are particularly illuminating, for they demonstrate how persistently class interests triumphed over the divisive effects of ethnic and religious differences and contributed to weld diverse elements into a workable community. The capacity of Orangeism to attract members of non-Irish origin emphasizes the same point. One thing remains: the role of Orangeism in the towns and its relation to the Canadian labour market.

Orangeism flourished, above all, in the towns. Kingston and Toronto were its key centres, just as they were the key centres of the province; but the Order was vigorous also in smaller places like London, Hamilton, Brockville and Cornwall.[161] There was room in the towns, of course, for a variety of rivalries. Local town government was a prize, and the concentration of population invited vigorous contests for it. But this does not explain why town organization should be dominated by Orangeism rather than by political parties as such, or by religious denominations or trade unions. The answer seems to be that Orangeism fused these various interests effectively and offered in particular a suitable vehicle by which skilled workmen might advance their interest.

The type of town resident to which Orangeism appealed is indicated from time to time. In Toronto's election riot of 1841, "the (Orange and Green) Combatants seemed all low Irish and directed all their animosity and blows at each other."[162] Though the Orange Order, according to a spokesman, grew rapidly about this time in "respectability" as well as in numbers, enrolling the Mayor of Toronto and several Aldermen,[163] an informant of 1849 emphasized that the leaders of Toronto Orangeism were still persons very low in the social scale.[164] The "political transformation" of William Henry Boulton, "a man popular with the labouring classes" of Toronto, from a High Tory into an Orange spokesman, illustrates both the nature and the strength of Orange suppoprt.[165] After one of the numerous clashes at Bytown between Orangemen and Irish Catholic raftsmen, a correspondent wrote, "I regret to report that the Rioters appear to be of the lower order — Roman Catholics and Irish..."[166] The numerous Orangemen of Kingston who paraded in 1846 were described as "composed generally of labourers and others ..."[167] The Order enjoyed, in fact, the support of a great many "respectable" citizens — its opponents constantly complained of this fact; but it seems clear that the great majority of its urban members were workingmen. Even this does not identify them exactly. The typical urban Orangeman was a skilled worker, a craftsman.* This occupational and

---

*This identification of the Orange Order with craftsmen has been challenged by Gregory Kealey. See Introduction p. xxviii. Ed.

class distinction came out in hearings that followed a "riot" in Kingston on July 12, 1843. The Orange witnesses were carpenters, shoemakers, tailors, carters. Their Irish Roman Catholic opponents, almost without exception, were unskilled labourers.[168] A further aspect of this rivalry among wage earners is suggested in a memorial of February 1844 (part of the active Roman Catholic campaign of that period to have Orangeism outlawed), from two Irish Roman Catholic shoemakers of Kingston. The memorialists were members, they said, of "a private benevolent society", which Orangemen mistakenly believed to be a "ribbon" society. During their society's meeting in the Roman Catholic school, a shot was fired into the room — by Orangemen, they surmised. It seems probable that their surmise was correct, and that these men had incurred particular dislike by entering the ranks of skilled workers, thus raising in some minds the dreadful prospect that the whole skilled labour market might be undermined by an influx of Irish Catholics.[169]

The fact that trade unions were rare phenomena in Canada until the 1850s, and not widely developed even then, may suggest that Canadian workingmen of the early nineteenth century were powerless, terrified or apathetic.* The truth seems to be that workmen advanced their interests with considerable success, by other means. Irish Catholic labourers used their ethnic community to organize their strikes and promote their political interests. So, to an extent, did French workmen. English-speaking Protestant craftsmen, to begin with, were often small proprietors rather than employees. These scarcely had occasion to conduct strikes, but they were great producers of petitions (especially in Kingston) and had considerable success in limiting the competition of convicts and imports. Artisans who worked for wages may often have felt only a slight and temporary difference in status existing between themselves and their employers. On the other hand, they shared with employers an opposition of interest to the dominant commercial forces of the country, which reflected British insistence that Canadians should produce cheap raw materials rather than manufactured goods, and considered labour a commodity to be made as cheap as possible. As the industrial sector of the economy developed, artisans confined permanently to the status of wage earners organized unions, but their differences with their employers tended to be overshadowed for several decades by a common interest in protection. Craftsmen, whether wage earners or proprietors, had an interest in restricting competition in the labour market, particularly in the skilled labour market. The posi-

---

*While unionism may have been rare in central Canada until the 1850s, it developed earlier in Saint John, New Brunswick, from the 1830s. See Eugene Forsey, *The Canadian Labour Movement 1812-1902*, Canadian Historical Association Booklets, no 27 (Ottawa, 1974), pp. 3-4. Ed.

tion and objectives of the artisans being what they were, it may be doubted whether unions would have suited their wants very well, quite aside from the question of legality, which was raised very rarely. Organization along ethnic-religious lines through the Orange Order would appear to have offered a number of advantages over organization through unions. Orangeism was oriented against yielding much room to Irish Catholics — a position that would have been awkward for a union, and a powerful recruiting point at a time when ethnic and religious divisions were uppermost in men's minds. It also stood out against excessive pretensions by the privileged classes, in the general political sort of way that suited craftsmen, and united them before economic development emphasized specific craft interests. Orangeism provided numerous allies from the lower middle class: their outlook was not yet very different from the craftsman's, and their numbers yielded political power. The economic loyalty of the Orange Order was decidedly local and it provided strong support for protection. For all these reasons Orangeism, and the moderate political conservatism which it built, represented the artisan well at a time when capitalism had not advanced enough to subordinate all other divisions to the one between capitalist and proletarian. In that time, the conservatism of the workingman was a fixed point of Canadian politics, and the Orange Order was its typical form of organization. Lord Elgin reported in 1849 that Toronto was the most Tory place in Canada and had no fewer than twenty-five Orange lodges;[170] but he could have found that other urban centres were much the same, for the same reasons.

CHAPTER V

The Transformation of Canada's Economic Structure

The rapid advance of Canada's population in the first half of the nineteenth century made possible the rise of a well-stocked and versatile labour market capable of producing goods and services in variety and volume and, at the same time, the local mass market required to consume them. But, while growth of population was a necessary condition for industrial advance, it was not a sufficient condition. Mere numbers do not assure progress — the persistence of many populous areas of the world in a state of industrial stagnation for centuries would rather suggest the contrary relationship. Neither does the crossing of an ocean by a European population guarantee that an industrial society will arise. Newfoundland and the Caribbean islands are examples of overseas areas that were settled early by Europeans, that have had growing populations, but are.in much the same economic position and stage of development now as centuries ago. Until the nineteenth century, Canada's position, as a producer of one or two staples for overseas metropolitan markets, was parallel to that of Newfoundland and the West Indies; and it was not then obvious that Canada, but not the others, was destined to escape.

Yet, a paramount fact about Canada is that it did develop a national economy of an industrial type in the nineteenth century. The Canada that existed until 1820 needs to be described, and has been very well described, in terms of staple production — a language that is still appropriate to the dependent outposts of the economic world. But this language will not do to describe the Canada of 1870: what is required for that is the terminology of advanced industrial societies. It is true that Canada's transformation was not as rapid, nor as certain and decisive, as that of the United States. It is also true that the Canada of 1870 was a small and rather immature specimen of an industrial country. Nevertheless, Canada's economic integration; its diminished dependence on a single export, or a single market, and on foreign trade in general; the versatility of a labour force shifting from extensive to intensive forms of production; the bustle and variety of activities of its expanding cities: all these distinguish the 1870 economy from its staple-producing predecessor of a half-century before.

The key to the new orientation was development of a home market economy. The Canada of 1820 can barely be said to have had a home market. Its main economic lines were transoceanic, providing for the export of timber and fur and the import of liquor, cloth and hardware. A majority of Canada's European inhabitants had scarcely any connection even with staple trade, but rather carried on a traditional household and local economy which, while it did not amount to complete local self-sufficiency, involved only the most shadowy participation in the Atlantic economy. Local self-sufficiency was encouraged, even necessitated, by the absence of transport facilities capable of drawing the back country into a national or world economy. The St. Lawrence system of waterways was well suited to the gathering up of a few raw materials that were exceptionally exportable, but it was incapable of supporting a market economy for more than a few miles inland from the waterfront. The waterway, in its natural state, also restricted within very narrow limits the range of goods that could be exchanged advantageously up and down the system. While these circumstances persisted, the Canadian provinces did not become a coherent market in their own right, but remained as a series of economic fragments united only by a common dependence upon Great Britain.

An essential step in changing this orientation was the construction of canals, railways and roads, particularly the trunk water and rail lines built in the 1840s and 1850s. All these, except some local roads, had been intended to foster the export of raw materials; but their main effect was quite the opposite: to create an integrated Canadian economy that turned away from the ocean, and to strengthen it with the active participation of the backwoods masses who, until the penetration of the railroad, had had little connection with any market.[1] The new Canadian economy of the 1850s exported wheat and flour rather than fur, more lumber and ships and less timber, and its imports consisted more of capital and less of consumer goods. These changes were the fruit, in part, of the discovery of a second major trading partner — the United States. The really vital development, however, was the trend away from foreign trade to home supply, including the supply of manufactured goods, that the newly-created home market fostered.

There is not much quantitative evidence that can be presented of this change, but the accompanying table showing the occupational distribution of the Canadian people in the second half of the nineteenth century serves, at least, to indicate the direction of development. These data show, for one thing, the preponderant place occupied by agriculture in this period — a preponderance which, in the view of this study, was

TABLE IV

OCCUPATIONAL DISTRIBUTION OF THE CANADIAN POPULATION, 1851-1891

(Data taken or computed from Canadian censuses of 1871, 1881, 1891)

LOWER CANADA

Occupational class	Occupied persons ('000)					Percentage distribution				
	1851	1861	1871	1881	1891	1851	1861	1871	1881	1891
Agricultural	78.4	108.1	160.6	202.0	217.1	38.6	43.3	47.1	46.6	45.5
Commercial*	8.8	19.0	25.5	34.3	50.6	4.3	7.6	7.5	7.9	10.6
Domestic	17.1	19.9	21.2	24.3	73.3	8.4	8.0	6.2	5.6	15.4
Industrial	26.3	44.5	65.6	81.6	93.2	12.9	17.8	19.2	18.8	19.5
Professional	4.8	7.1	15.4	18.4	16.3	2.4	2.9	4.5	4.2	3.4
"Not classified" (including labourers)	67.7	50.9	52.9	72.6	(20.1)†	33.3	20.4	15.5	16.8	(5.5)
Totals	203.2	249.5	341.2	433.3	476.9	99.9	100.0	100.0	99.9	99.9
Rate per 1000 population	228.0	224.2	285.9	318.8	320.4					
Rate per 1000 16 yrs & over	418.9	395.5	513.2	553.5						

*Trade and transportation in 1891.
†"Non-productive" in 1891. Many previously in this class were evidently shifted into the "Domestic" class.

TABLE IV
Occupational Distribution of the Canadian Population, 1851-1891 — continued
UPPER CANADA

Occupational class	Occupied persons ('000)					Percentage distribution				
	1851	1861	1871	1881	1891	1851	1861	1871	1881	1891
Agricultural	86.6	134.3	228.7	304.6	344.8	35.1	39.5	49.4	48.3	46.0
Commercial*	9.3	14.8	29.1	44.9	87.2	3.8	4.3	6.3	7.1	11.6
Domestic	18.0	21.8	26.8	33.4	109.3	7.3	6.4	5.8	5.3	14.6
Industrial	45.0	60.4	93.9	130.2	158.8	18.2	17.8	20.2	20.6	21.2
Professional	6.8	9.4	16.8	23.3	30.1	2.8	2.8	3.6	3.7	4.0
"Not classified" (including labourers)	80.8	99.4	63.2	94.4	(26.4)†	32.8	29.2	14.7	15.0	(2.7)
Totals	246.5	340.2	463.4	630.8	750.3	100.0	100.0	100.0	100.0	100.0
Rate per 1000 population	258.7	243.3	285.8	327.9	354.9					
Rate per 1000 16 yrs & over	471.8	527.3	514.8	549.5						

*Trade and transportation in 1891.
†"Non-productive" in 1891. Many previously in this class were evidently shifted into the "Domestic" class.

The data are taken or computed from the 1871 Census, Table L (vol. V); 1881 Census, Table J (vol. III); and the 1891 Census, Table P (vol. IV).

The statistics shown must be considered in the light not only of the uncertainty that may exist about their accuracy, but of the peculiar occupational classification adopted by the census makers. Their conceptions of economic structure were scarcely adjusted to an industrial capitalist economy. Thus, the category of "Commercial" class included not only those engaged in exchange, but carters, bargemen, "mariners", stevedores, pilots, railway employees and telegraph operators. In the 1891 Census, classification and terminology were improved, and the "Commercial" class received the more accurate name of "Trade and Transportation"; but the shifting of occupations from one group to another that accompanied this reform destroyed the comparability of several classes, possibly including the "Commercial" one. Similarly, the "Domestic" class included not only servants, but barbers, laundresses and hospital attendants. Most unsatisfactory is the "Not classified" class of the 1851-81 censuses, which included not only the voluntarily unemployed, but apprentices, guards and keepers and, notably, labourers. The sharp fluctuations of the totals shown as "Not classified" would appear to arise mostly from variations in demand for construction labourers. The makers of the 1891 census dealt forthrightly with this category. It reappeared as the "Non-productive" class: evidently some aura of gentility still clung to the occupation of being non-productive, and it was still possible then to call it by its right name. The employed persons who had been "Not classified" until 1881 were mostly shifted to the "Domestic" class in 1891, so that at least two classes, and perhaps more, are incomparable as between 1891 and earlier years.

On the other hand, the important "Agricultural", "Industrial" and "Professional" classes apparently covered the occupations that one would expect of them, and, with the possible exception of the "Professional" class, seem to have remained comparable to 1891. The "Industrial" class was conceived to consist of artisans following a trade or craft until 1881, and was given the more modern name of "Manufacturers and mechanical industries" in 1891, but the content of the class does not seem to have changed much. Questions arise, not about the definition, but about the significance of the "Agricultural" statistics. That Canada was becoming more and more an agricultural country up until 1871, and more industrial, commercial and professional at the same time, is consistent with the argument of the text. But the rising rate of participation in the labour force, with which the heavy showing of Agriculture in 1871 and 1881 seems to be associated, deserves attention. A clue seems to be afforded by the 1881 census of occupations (Table XIV, vol. II) which lists 47,223 "Farmers' sons" in Quebec, as against 151,756 male farmers, and 71,642 "Farmers' sons" in Ontario, as against 226,090 male farmers. It seems probable that the "Agricultural" class was swollen out of proportion in some census years by bulges in the birth rate, by which exceptional numbers of young men passed their sixteenth birthday and were counted as part of the agricultural labour force. There were significant bulges in the birth rate (remarkably uniformly as between the two Canadian provinces) in 1878, 1869, 1867, 1864, 1860, 1853 and 1840 (1881 Census, Table 8, vol. II). The size and irregularity of the oversupply of young men on the farms provides reason, not only for the peculiarities of the present statistics, but for the prevalence of Canadian immigration to the United States in the closing decades of the nineteenth century.

The declining weight of the "Domestic" class shown in the table also deserves a brief comment. The decline is what might be expected to accompany the rise of commerce and manufactures. On the other hand, a rise of "service" industries is one of the marks of a tertiary stage of economic development or "economic maturity". Therefore, the distinction must be kept clear between the domestic services that contract in the second or manufacturing stage of an economy, and the commercial services that expand in its third stage.

a necessary condition to shift the weight of the economy away from the gathering of surface products, and to provide the base required for manufactures. They show the considerable expansion of commercial, transportation and professional activities that occurred./They show a marked decline of domestic services and, finally, though this trend is the least emphatic, the correlative expansion of industrial activity, proportionally as well as absolutely. If satisfactory statistics were available for the 1830s and 1840s, it is thought that the same trends would be demonstrated also for the earlier period, and that the growth of industrial occupations would be more pronounced.

Some quantitative evidence is also available to show the growth of towns and cities — a growth so vigorous as to suggest that the urban part of Canada's population was gaining on the rural in the nineteenth century, in spite of the expansion of agriculture.[2] Some of the most striking urban advances occurred in the 1830s and 1840s: Kingston more than tripled its population between 1829 and 1851; Hamilton multiplied nine times between 1835 and 1853; and Toronto, which was to be chief beneficiary of the integrated Canadian economy, increased four and a half times between 1830 and 1851. Some urban centres, dependent on the transatlantic trade in staples, stagnated with growth of the home market. Quebec was only able to double its population between 1830 and 1870, whereas the total population of the Canadas multiplied nearly four times in that period. Kingston remained almost stationary in population for a long time after 1851, while Three Rivers and Sorel, perhaps representative of most centres of the lower province, barely managed to keep pace with the general growth of population between 1851 and 1871. However, there were many places that fared much better. One was Montreal, which had grown more slowly than the total population of the Canadas in the 1830s and 1840s (by about 15 per cent), but which managed to get a firm footing in the home market economy as well as the export economy after 1850. Aided by railways and manufactures, Montreal grew about 20 per cent faster than the total population between 1851 and 1871, almost doubling its population in that period. Toronto, which had enjoyed spectacular growth before 1850, also grew about 20 per cent faster than the total population between 1851 and 1871. The centres that experienced the most striking growth after 1851, however, were the towns and smaller cities of Upper Canada. Nine of these[3] (other than places already mentioned) grew more than 50 per cent faster between 1851 and 1871, on average, than the general population.

It appears that cities, especially in "underdeveloped" countries, may often be mere agglomerations of people, producing only an indistinct and inadequate volume of services.[4] Even these may provide some foundation for a mass market. Canada's cities of the nineteenth century did this but, fortunately, also possessed firm foundations in transport lines, while the places that grew fastest were industrial cities producing flour and beer, shoes and clothing, sewing machines and farm machines, locomotives and sleeping cars.[5] It is, indeed, a new nature and feel of these cities that demonstrates, more clearly than statistics, the appearance of a new level of economy. It is true that cotton mills, the vehicle of industrial capitalism in Great Britain, did poorly in Canada, though some were attempted in the 1840s and 1850s. But railways stimulated the manufacture of iron products, so that in the 1850s Toronto and Montreal and some smaller centres produced locomotives and possessed rolling mills, some of them impressive. In that decade, rubber manufacture and large-scale sugar refining also were inaugurated in Montreal. The Hamilton of the 1850s was an unusually versatile manufacturing centre, turning out foundry products, locomotives, railway cars, sewing machines, ready-made clothing and tobacco.*[6] By 1860, cities generally were supplying the country with shoes, clothing, beer, whisky, furniture, with the aid of power and machinery and at the expense of domestic handicrafts, village artisans and foreign manufacturers. Still the most telling sign of a new atmosphere in Canada was not the volume of physical output itself, but the accompanying emphasis on progress and efficiency, education, science and mechanical invention. The cities with their schools and Mechanics' Institutes, scientific lectures and uplift societies, were the guardians and disseminators of that emphasis.

Canadian farmers of the nineteenth century tended to be suspicious of export markets and conscious of the security of the home market which many of them had always enjoyed and which expanding cities could more fully provide. Nevertheless the cities, by their nature, had to be the most urgent supporters of the national economy. To be sure, they still contained the agents of transatlantic commerce among their populations. But, in the last half of the century, their trade policy was not determined by these but by the far larger numbers who operated the transport systems, effected the exchanges and produced the manufactures required by the national economy. The cities represented an enormous investment in improved real estate and machinery, in streets and docks, in waterworks and gasworks, and in a versatile labour force, largely predicated upon the permanence and growth of the home market. Vulnerable and therefore sensitive, the cities were keenly conscious that the exchange of city manufactures for country produce was the core of their prosperity. They prized the security of

---

*Nevertheless, Katz still refers to Hamilton in the 1850s as a commercial rather than an industrial city. See Michael Katz, *The People of Hamilton, Canada West* (Cambridge, Mass.: Harvard University Press, 1975), p. 7. Ed.

local exchange, not least because of rough lessons in the unreliability of foreign markets, and they were hostile to any who would disrupt it. The cities' own capacity to consume and produce, and the political strength which their concentration and organization permitted, provided powerful weapons to defend their interest.

But, while the vested interests of a national economy are obvious guardians of its survival and expansion, that provides no explanation of how the economy or the interests arose in the first place. It has been pointed out that population increase, similarly, can explain how the transformation of an economy was facilitated, but not how it was effected. Passing reference has been made to new transport media and to agricultural expansion as important elements in Canada's evolution; but their role has not been explored. The great question remains: How was Canada's transition from a staple-producing outpost to an integrated economy brought about? The remainder of this chapter is intended to answer the question so far as the evolution of economic structure is concerned. Changes in attitudes and industrial organization, which are also involved in the answer, will be discussed subsequently.

The Physical Possibility of Canadian Development

The voluminous literature of economic development that has appeared in recent years, with particular reference to the "underdeveloped" countries of the world, appears to be based in many instances upon some questionable assumptions. One is that all countries (and all regions?) are destined to develop economically — that is, to become diversified economies replete with the manufactures, trade, finance and urbanization that characterize "advanced" countries. A second assumption is that the evolution of underdeveloped countries must proceed by much the same stages as those experienced by the countries of Europe and North America. It is perhaps natural that these assumptions should be made by many experts on economic development, who are drawn almost invariably from developed countries, and reflect their common ideas. These assumptions are implicit, too, in the writings of those who, since the days of Comte, Roscher and List, have postulated stages of economic development. Since these writers have had specific developing countries in mind in drawing up their stages, and since the countries now "developed" have followed a reasonably uniform pattern, the stages supposed by various writers usually turn out to be similar to each other. Indeed, a conception of stages of development is desirable, if not essential, when undertaking a discussion of national development such as the present one. The conception of development used as a starting point here is one proposed by Professor J.M. Clark, which has the virtues of extreme simplicity, extreme generality and freedom of mysticism.[7] Clark has proposed that advanced countries have progressed, typically, through three broad stages of development. The first is a stage of extractive industry (including agriculture). In the second stage, emphasis is on manufacturing. In the third stage there is a relative decline in the production of commodities as more and more of the economy becomes devoted to "service" activities. The experience of most countries inhabited by Europeans, including Canada, can be fitted readily into these stages. The comfortable conclusion from this experience that may be conveyed to underdeveloped countries is that all will go well with their development if they will only make themselves, to start with, proficient producers of raw materials.

The view that all countries will proceed neatly from extractive industry to manufactures, and then to an advanced economy of services, is open to at least two serious criticisms, both relevant to Canadian experience. One is that the process by which countries have shifted from raw material production to manufacturing is an obscure one; that the trading partners of raw material producers have usually tried to prevent this shift from taking place; and that, therefore, the change is not effortlessly inevitable, but rather requires unusual circumstances or exceptional national determination, or both. This point, as it bears on Canada, is discussed in subsequent sections. A second criticism is that, in fact, all countries have not had a uniform history of development. On the contrary, physical characteristics and limitations have produced diversity even among areas populated by Europeans. This point is pursued below.

To begin with, some countries and regions have scarcely had a raw material stage. They attracted settlement, failed to develop any satisfactory raw material to export, yet flourished by exploiting commercial opportunities. The Netherlands, New England and Nova Scotia have followed this pattern.[8] At the opposite extreme are territories that never have had, and probably never will have, anything but a raw material stage. In their case, the question is not whether they will go beyond raw materials, but whether they will support staple production permanently. A territory whose known resources can be easily snatched may be discarded after a superficial looting, as the Yukon was. Other areas have an endowment that, though narrow, is permanent and therefore capable of the indefinite support of a population engaged in extractive pursuits. Iceland and Newfoundland, are in this category. Canada's Pre-Cambrian Shield lies somewhere between these types, with resources that are exhaustible, but not readily exhausted. The limited physical capacities of territories of these kinds dooms them always to be dependent, economically and politically, upon better-endowed and better-developed regions.[9]

Between the extremes of the countries that never found an exportable staple, and the countries that must always be raw material producers, are the countries that have managed, or may yet hope, to shift from raw material production to manufactures and maturity. Though these countries vary considerably in their natures, they have a common element: good land capable of supporting a surplus-producing agriculture. This proposition does not rest upon a mystique of Physiocracy (though that is a kind of evidence, too) but upon history and logic.

The historical evidence, though it cannot be conclusive by its nature, establishes a strong *prime facie* case. Every country that has become industrialized possessed the base of a strong agriculture. The countries that have changed most have had the richest farmlands from long ago, and, in the cases of Britain and Japan, which perhaps had the richest farms of all, have come to rely the most on foreign food supplies. Countries that never had a strong agriculture have failed to become important industrially (though it is arguable that Switzerland is an exception). This is not to deny the importance in the history of development of coal, iron, location, wars and institutions. What is most remarkable is that, even when allowance is made for the many other pertinent factors, there remains so high a correlation between agricultural efficiency and industrial advance.*[10]

Beyond the empirical evidence, there is a logical explanation why only countries with a strong agriculture should succeed to manufactures. Agriculture is profoundly different from any of the other primary, easily-undertaken industries that mark the extractive stage of an economy. Agriculture is the activity most able to attract large-scale settlement — and that without any necessary reference to outside markets. It is, similarly, the one primary industry able to support a large population indefinitely, irrespective of the stage of foreign markets. When export markets collapse, mineral economies and timber economies and even fish economies cannot survive upon their own produce: producers are faced with the unpleasant alternatives of starvation or flight. Agricultural economies, on the other hand, can ensure at least that their population will have the elementary means of life; and closing of their export markets, if the inhabitants are not paralyzed by custom or confusion or outside control, may hasten the development of a diversified economy. An agricultural population,

---

*Cf. John Isbister's contention that after 1850 "Quebec's industrial sector was able to develop at a relatively rapid rate without the support of a dynamic agricultural sector." "Agriculture, Balanced Growth, and Social Change in Central Canada since 1850: An Interpretation", *Economic Development and Cultural Change*, July 1977, p. 263. Ed.

characteristically dense until the present century, is an obvious and permanent market for manufactures, in which the local manufacturer enjoys the advantage of cheap food supply and protection against outside producers to the extent of transport costs, at least. However, the agricultural economy that has lost markets or lacks them would seem to be much inferior, as a foundation for manufactures, to the one whose rich soil has permitted the successful expansion of export trade. A population used to maximizing production and surpluses over local needs is likely to exchange, and consume also, at a high level. The attitudes and institutions that flourish in an agricultural economy that is efficient in the capitalistic sense are well suited to produce successful industrial managers and workers. A flourishing export trade invites the transport media that bind the regions they penetrate into a coherent national economy.* The surpluses of a prosperous and market-oriented agriculture are likely to be one important source of the capital required for industrial development. In several respects, then — as a market, as a source of labour and capital, as a coherent economy whose permanence seems certain — a surplus-producing agriculture offers superior conditions for industrialization. When manufactures do become established within agricultural economies, they enhance their own economic justification by providing markets for farm produce that are more diversified, more remunerative and more secure than the export market. There are good reasons, then, why manufactures have flourished on the world's flat plainlands, whereas they have entered only reluctantly, and usually as tentacles of a plainland economy, the areas of impoverished soil, desert, forest and mountain. Among recently settled countries, the United States possessed such rich agricultural resources, and these hardly broken from the seaboard to the Rockies, that only monumental mismanagement could have prevented its rapid industrialization. Canada, on the other hand, has been a marginal case, whose title to intensive development has been kept tenuous by the limitations of its agriculture.

With expansion of European settlement inland from Quebec, Canada has had three chances to acquire a strong agricultural base. These came, successively, with settlement of the Montreal region, of western Upper Canada and of the western prairies. The last of these opportunities needs no extended discussion: prairie settlement gave substantial support for economic diversification after 1900, but the foundations

---

*However, it can be noted that a new transportation system can reduce the natural protection of transport costs and thereby destroy existing small-scale manufacturing in agricultural areas. See J.M. Gilmour, *Spacial Evolution of Manufacturing: Southern Ontario, 1851-1891* (Toronto: University of Toronto Press, 1971). Ed.

for this had already been firmly established with the aid of the earlier agricultural bases. The key question, how the original development of a Canadian national economy was accomplished, concerns the earlier waves of settlement.

As an agricultural base, the Montreal region was limited in area, while its soil and climate were less advantageous than those found in many other parts of the world. Nevertheless, it proved to be more than adequate for the limited population of the eighteenth century. Not only was it the bread basket of the country for more than a century, but it demonstrated a power to make modest surpluses of wheat and other products available whenever there was any prospect of exporting them. It is not true, either, that agriculture was a minor element in the economy of New France, in spite of the contrary impression that preoccupation with the fur trade might produce. It was claimed from time to time during the French regime that too many men were being drawn into the fur trade to the detriment of agriculture, and some very high estimates have been given of the number thus affected. However, it seems doubtful that more than 15 per cent of Canada's labour force was ever involved in the fur trade.[11] Other extractive pursuits, fishing and lumbering, did not take many men, or these for very long. New France was primarily an agricultural colony in which farming was the main occupation and support of the people.[12] Moreover, the colony's agriculture supported in the town and village crafts, the local mills, the ironworks and the shipyards, a respectable level of secondary industry. It was the good farmland around Montreal, finally, that permitted Canada to survive as a distinct entity into the nineteenth century and provided the base that facilitated the development of Upper Canada as a British province rather than as an annex of New York.

However, the inability of Lower Canadian agriculture to support an effective economy of the larger size and intensified tempo required by the dynamic nineteenth century must also be recognized. To begin with, so limited a base could not support a very large expansion of population. Secondly, while it is not unlikely that New France could have developed an efficient surplus-producing agriculture had dependable markets existed to encourage it, it has already been pointed out how continual frustration withered the power and will to produce. The obstacles to efficiency (in a capitalistic sense) imposed by feudal institutions were accentuated by continual subdivision which produced strip farms less and less economic or capable of improvement. Finally, soil exhaustion forced the agricultural heartland of Lower Canada out of wheat production after 1835. If Canada had had no other agricultural base at that time, not only the growth but the survival of its economy would have been doubtful. Rescue was effected, and a really strong foundation for Canadian development laid, by the settlement of Upper Canada. The critical step may be defined even more narrowly, as the push of settlement past the marginal agricultural land in the east of the province to the rich soils around York, and westward from it. Here was a land able to produce large surpluses; or, to turn the matter around, a land in which the cost of production of farm products was exceptionally low. According to the information collected by a Committee of the House of Assembly in 1835, the cost of raising wheat was only about two-thirds as much in the western part of the province as in the east.[13] And this may have understated the western advantage.

It was important, too, that the basic population of the upper province, and especially of its western parts, consisted of American immigrants, and that it was they who established its original institutions and outlook. Whatever their deficiencies, the Americans were never accused of lacking enterprise and efficiency. They were mobile, adaptable and mechanically ingenious. They were not interested in self-sufficiency, as such, and were always ready to produce surpluses and to push exchange to the limits of profitability. Whereas investment, commerce and manufactures tended to be the prerogatives of particular strata of French society, the skill, capital and enterprise which these functions required were widely diffused in American society. American settlers used them from the beginning to establish grist mills and saw mills, stores, stopping places and blacksmith shops, as well as freehold farms oriented to profit-making: to form, in fact, a replica of the highly-developed local economy which they had known in the American colonies. The economic structure thus constructed was one that assumed surplus production and exchange, and provided automatically for intensification and widening of them whenever market conditions permitted.

A third vital fact about western Upper Canada was its remoteness from tidewater and a consequent difficulty and cost of transport that militated against attempts at staple production. The fur trade did not last long after the settlement of Upper Canada, and the province had no great share in it. There was a trade in potash, but this was an ephemeral activity incidental to the clearing of land. The timber trade, when it came, was not very well suited to the small operators typical of the upper province, or to regions other than the Ottawa; and, in any case, the basic economic pattern of most of the province had been established before it appeared. The settlers would have been willing to export wheat, and eventually did so; but there were good reasons — transport costs and the British Corn Laws among them — why this trade was insignificant for a long time. The western part of the province was thus thrown back upon its own resources and made into a relatively self-sufficient or home market economy in spite of itself.

But regions, especially new ones, do not live very conveniently on local exchanges alone, no matter how well endowed in other respects. This was one of the points stressed by R. Hamilton, merchant of Queenston, in an illuminating reply of 1803 to the question whether Canada could feed the whole of the British forces in Canada and the West Indies. Such a market, said Hamilton, would be a great advantage to Canadians, "as offering to them what they most essentially want, a Steady Market for the overplus of their agricultural produce, and the means for paying for many Articles of conveniency, which they can neither raise, nor well dispense with."[14] He noted that western Upper Canada already produced substantial surpluses of wheat and beef, and considered that if there were a steady demand at 3s. 1½d. currency per bushel for wheat and 2½d. average per pound for beef, supply might be indefinitely increased. Pork, though not produced hitherto in any quantity, would be forthcoming also at reasonable prices. But Hamilton then turned to the other aspect of the question — the awkwardness and cost of transportation to seaboard. He suspected that the "Seaport merchants" would be unwilling to promote an export trade in produce from Upper Canada, and that the shipping monopoly, in particular, would yield nothing to assist it. "We have to lament that hitherto, their Capitals generally engaged in perhaps more lucrative, certainly in less bulky Speculation, (they) have been rather unwillingly employed, in encouraging our Agricultural pursuits, and that the Charges on our Produce, have been such as it could hardly bear." The cost of carriage between Montreal and Kingston "has not been decreased for the last Twenty Years — Calculating indeed the fall in the price of Labour, and of Provisions, we are induced to believe that the Charge has augmented."[15]

In actuality, Upper Canada was not required to run the gauntlet of the St. Lawrence for many years — which may be to say that the seaboard merchants overreached themselves and drove the upper province to an independent development. The real secret of Upper Canada's success, and a crucial step as well in the development of a national economy, was the province's discovery of local markets for its produce that were large enough to support it for several decades. The first of these had existed for years before Hamilton wrote the letter quoted above, and was provided by the British taxpayer on behalf of the British military forces. The consumers consisted not, indeed, of all the British forces in the western hemisphere, but of the forces in Upper Canada. Particularly in the western part of the province, around Niagara and York, army agents were pleased to buy local produce rather than face the difficulty and expense of bringing supplies up the St. Lawrence. This military demand typically outran local supply. Despite a certain amount of competition for this business from American farmers, the effect was to establish a price structure for farm produce that, instead of being the Atlantic price minus transportation costs down the St. Lawrence — which an exporting community would have had to accept — was the Atlantic price plus transport costs.[16] Not until about 1830, and later still in some parts, did supply so outstrip demand as to end this happy condition.

To be sure, military garrisons were incapable of consuming the province's surpluses indefinitely, and they did so only till about 1810. But a new market appeared with the influx of immigrant settlers. Long before these persons began to contribute to production, they were buyers of food, of seed and of stock, for which they paid with imported funds or commitments of future income. As long as the inflow of settlers kept increasing, their coming provided a formula for perpetual prosperity. Meanwhile, the War of 1812-14 had made its contribution on behalf of the local market. The war disrupted production, while a good part of the province's population came upon the payroll of the British government by way of the militia. The consequence was an artificial, temporary, but very obvious buoyancy of the market for farm products.

About the same time another home market was opening up with the expansion of the Ottawa Valley timber industry. The demand for food and animal fodder which it entailed was so great that Ottawa Valley farmers were notoriously prosperous far into the nineteenth century, in spite of the poor soils which they farmed.[17] It may be objected that timber camps were not properly a home market: Ottawa Valley farmers were put in the position of being indirectly dependent on the staple timber trade, just as the position of French-Canadian farmers had been rendered ambiguous by an indirect dependence on the fur trade. Nor can it have been entirely accidental that in the Rebellion of 1837 the farmers tributary to the timber regions remained loyal to the British connection which represented staple production, in contrast to the farmers further west whose experience and interest was in an independent economy in which their superior fruitfulness would benefit the local population instead of being drained off to pay shipping costs and to provide cheap food for foreigners.[18] Nevertheless, the market created by the timber trade contributed, along with the military and settler markets, to foster local development. It lifted the whole structure of demand for agricultural products in the province. The highest contours of the price structure were in the timber areas around 1850. Timber production encouraged manufactures of wood directly, and through the demand of the lumbermen and especially of the supporting farmers, stimulated manufacturing in general. When emphasis shifted from timber for Great Britain to lumber production for the local and American markets, this more intensive activity supported a larger number and wider variety of persons, enabled many small and medium-sized operators to flourish and accumulate profits, and unified and strengthened the Canadian economy by providing the timber regions with a clear place and interest in it.

By various local demands, then — of military garrisons, of immigrants and of the timber trade — Upper Canada was protected for nearly half a century from the obligation to find outside markets for its produce, and at the same time discouraged from the attempt by the cost of transport down the St. Lawrence. The inadequacy of that unimproved waterway, indeed, was only part of the problem: settlers who were more than twenty miles from a waterfront were virtually prohibited from exporting by the high cost of wagon transport, and the result was that they produced a volume of output much smaller than they were easily capable of. In spite of this drag on production, the time had eventually to come when, with more and more cleared land coming into use, and immigration checked, local supply would outrun local demand. Such a crisis threatened in the 1830s. The solution offered by exporting down the St. Lawrence had the disadvantages that it required a sharp lowering of the price structure — hitherto supported rather than depressed by transport costs — and that it involved reversal of the principle that prices were highest in the west. If Upper Canada had been new and unformed, the Atlantic price system might nevertheless have been able to impose its directive upon the colony and enroll it as another competing supplier in the Atlantic system. The province, however, was no longer an infant. It had enjoyed several decades of relatively independent and prosperous development; it had acquired a substantial population, a strong agricultural base, thriving towns and cities, a vigorous and even frantic level of political organization and activity. It was not to be expected that this province would be transformed easily into a staple-producing handmaiden of international commerce, even though its first awkward attempt to declare its economic and political independence — the Rebellion of 1837 — seemed to be a complete failure.*[19]

How the province would have fared in a career of independence if new forces had not entered the scene, is difficult to say. As a strictly local economy at the grist and saw mill stage, its prospect would seem to have been an unattractive vegetation. Transformation to a more sophisticated level through construction of railways and factories would have been a difficult and slow process, if it had depended on local capital and incentives. An active and vocal population had plans to speed the transition, however. Roads offered some help, and many

---

*For fuller economic interpretations of the rebellions, see Introduction, pp. xxix-xxx. Ed.

were built. Stimulation of the economy by an increase in the money supply was a fairly popular policy in Canada, as in the contemporary United States: agitation on this side supported Lord Sydenham's* unsuccessful effort to establish a government Bank of Issue in 1841 and a species of free banking in Canada in the 1850s. The remedy most often recommended for the province's ills, however, was protection.[20] Perhaps these devices could have preserved some acceptable level of local prosperity. If they failed, or if the British government prevented use of them, new political crises and some sort of accommodation with the United States might have ensued. As it turned out, however, Canadians did not have to find out what they might have made of their economy by their own efforts, and the policies conceived to deal with the problem remained to take new forms in another context. What altered their situation fundamentally was the successful completion of a canal network in the 1840s and the triumphant sweep of railway lines in the 1850s.

It is a supreme irony of history that the improvements made to the St. Lawrence system of water transport, and the railways that duplicated and supplemented it, were conceived by those who dreamt of binding Canada more firmly into the system of transatlantic exchange, were financed by British governments and investors having the same expectation, yet had the opposite effect of creating a viable Canadian economy. The promoters and investors were not wrong in supposing that the cost of carrying goods from the hinterland to seaboard would be lowered sharply, and exports so far encouraged — when other conditions were favourable in the 1850s, wheat and meat were exported to Britain from Upper Canada, and the accompanying prosperity proved to be highly congenial to Canadians. The planners and investors were no more blind than others to the consequences of two great facts of mid-century: British free trade and New York's enormous success in engrossing the interior trade of North America. They were less perspicacious than they might have been, however, in assessing what canals and railways would do to the Canadian economy. In the first place, their construction involved the expenditure of considerable sums of British money on locally-produced capital goods, and the expenditure of much larger sums, through the medium of armies of construction workers, upon local food and shelter. The canals in a small way, and the railways in a large one, required a smaller but permanent and better paid army of men to operate them. They and their families

---

*Lord Sydenham (Charles Edward Poulett Thomson) was Governor-in-Chief of Canada from 1839 to 1841, following a political career in Britain, and supervised the union of the two Canadas in 1841 as recommended by Lord Durham. He died from a fall from his horse after two years in office. Ed.

filled out the towns and cities, stimulating construction, trade and the local manufacture of consumer goods. Meanwhile, the new transport systems supported local manufacture of rolling stock, while construction had provided a way for some contractors to accumulate the capital required to participate in these manufactures. Canals had promoted manufactures in another way, by making waterpower sites available, at the same time that they widened the market for factory products. The greatest change of all was effected by those railways, like the Northern and Great Western, that penetrated the fertile but hitherto landlocked interior. Masses of subsistence farmers along the lines were suddenly enabled to sell in a wide market and to buy consumer goods and machinery with their earnings, and were encouraged to maximize their outputs. Surpluses thus exposed to the market were sometimes exported — though it was typically lumber consigned to the United States, rather than anything for Britain, that the new lines first carried. But food surpluses brought from remote farms by the railways usually found their market in the growing Canadian cities.[21] Still more important, the purchases of the masses thus introduced to exchange as the normal way of life, while they included cloth and crockery and tea from abroad, were most often of goods produced in Canada. Farm machinery came from the local blacksmith, or from "factories" springing up in town and city, or perhaps from the United States — but not likely from Britain. Clothing and shoes, usually produced hitherto in the home and the village shop but now distributed in a national market, came briefly from a domestic handicraft industry in Lower Canada, but soon from city factories equipped with machines. Manufactures of wood — doors, windows, pails, barrels and, increasingly, furniture — offered an obvious field for local enterprise. But hardware, tobacco and alcoholic manufactures flourished also in the room provided by the railways. The fact was that the British manufacturers' advantages of experience, larger scale and lower wages were more than offset in a wide range of manufactures by lower transport costs of Canadian producers, their acquaintance with the peculiarities of local demand and their greater readiness to cater to them. In the manufacturing fields that developed in the 1850s in Canada, it was American producers who offered the severe competition. In their case, however, it was politically acceptable, and even patriotic, to establish the barrier of a protective tariff. The net effect of Canada's revolution in transport, then, was much less the stimulation of exports, or of imports, than the presentation to Canada's farmers and manufacturers of the coherent home market which they would have had the greatest difficulty in creating for themselves. Meanwhile, immigration provided for the variety of skills necessary to operate a diversified economy, to the extent it had not already done so; while capital, though scarcely abundant, was drawn in several ways from Britain and the United States. Transportation provided the keystone and catalyst of these ingredients of a developing national economy.[22]

The Political Possibility of Canadian Development

Beside the question of the physical possibility of economic development in Canada is the question of its political possibility. That development did become politically possible is an historical fact. It is a fact, however, that needs a political as well as a physical explanation. How was it that the mercantile interest entrenched on the St. Lawrence, backed by British interests which it represented, devoted to staple exchange and indifferent or hostile to diversification, long successful in maintaining Montreal as the leading agency of trade between Europe and interior North America, was forced or persuaded to countenance an independent Canadian development? The answer is that the merchants overreached themselves; that, paradoxically, they lost because they clung so narrowly to the logic of staple exchange; that competing merchants won because they were carried along, in spite of themselves, by slow-moving but massive economies that were *not* devoted to staple production. The answer, again, is that while mercantile aims were simple and narrow, imperial aims were not and frequently provided important support for economic diversification. The answer is also that the transportation facilities constructed in the last desperate attempt to hold the interior for the St. Lawrence system, failed to accomplish that purpose; hence, their incidental function of creating a Canadian economy also became their only function. The answer is, finally that Canadian sentiment in favour of an independent and protected economy, already strong in 1840, received an enormous impetus from Britain's turn to free trade, both because free trade represented a demotion of those staple trades that required preferences, and because Canadians considered that they had been betrayed and abandoned. These points require some elaboration.

Merchants The merchants who operated the world of commercial capitalism and occupied the strategic places in the Canadian economy until about 1850 were engaged in securing differential gains between the prices at which they bought commodities and the prices at which they re-sold them. Any saleable commodity would do, but preference was for staples. Their volume was large and regular, and their marketability was ready and relatively secure. The commodities that provided for the great trades of the eighteenth century — sugar, tobacco, fur and so forth — were few in number, and even such additions as timber and cotton in the nineteenth century did not make the list very long. Nor were the commodities usually indispensable to the relatively self-sufficient Europeans, serving as comforts rather than necessities. However, the staples made up a very high proportion of what overseas commerce there was, and the rest was highly unpredictable, so that merchants concentrated on staples. Though the merchant preferred to limit his risks in this way, his business was perpetual speculation. His gains consisted of other men's losses, seized with the aid of stronger bargaining positions, greater foresight and later intelligence. It may be doubted that the merchant conceived himself as creating "utilities".[23]

The merchant was a capitalist. It was his possession of capital, together with other peoples' lack of it, that allowed him to seize differential advantages in trade. The capital was conceived of as a sum of money beyond the amount required for the owner's maintenance. The more of it the merchant had, and the more expeditiously he could change it into goods and back into (more) money, the more profit he could make. Quick turnover had the additional, and greatly-prized, advantage that it kept the merchant close to liquidity, which was valued because it facilitated early escape or seizure of new opportunities.

The merchant's passion for liquidity was a reason for his concentration on staple goods for which there were relatively dependable markets. It was also a cause of his reluctance to become involved in production. The merchant was particularly loath to sink money into fixed capital. Sums spent in this way diminished the circulating capital with which the merchant conducted his speculations. Moreover, fixed capital might be hard to sell, even at a heavy loss, and the time required to sell it impeded mobility. Merchants sometimes invested in mines, ironworks and shipyards, nevertheless, when they promised to confer an exceptional strategic advantage. The results were often unsatisfactory, however; especially because mercantile emphasis upon immediacy, upon discrete transactions, upon prices, rendered the typical merchant a peculiarly unfit person to handle employees or organize regular production.[24] Where no strategic positions in production were available — and trades based upon immense and ubiquitous supplies of raw materials tended to have this characteristic — the merchant preferred to confine himself to the purchase of finished goods. With the bargaining power afforded by his capital, he might reap large returns from this position, even to the extent of crushing his suppliers out of existence.

The typical merchant was not likely to worry much about that, for his time-horizon was short and his sense of any responsibility towards the area which he exploited of the faintest.[25] These characteristics follow readily enough from the nature of mercantile business; they were no doubt useful, and perhaps essential, for men required to establish first the trade structure of Europe and later the trade structure overseas. They were qualities that tended to set merchants apart as an alien element, however, especially in new countries like Canada where there had been little time for local loyalties to develop. It is notorious that generations of St. Lawrence merchants regarded themselves as temporary residents in Canada, unfortunately exiled from the United Kingdom for as little time as they might require to amass and carry off a fortune. Like the lumberman stripping a woodland which he expects never to see again, the future of the exploited area did not concern them.[26]

It was with this attitude in the background that the empire of the St. Lawrence was constructed as the furthest-extended, the most weakly-based, the most perfectly mercantile of all the seaboard colonies. It approached the mercantile ideal of a colony in which all finished goods were either exported or imported, in which all resources were devoted to the extension and support of overseas trade tentacles, with none frittered away upon local production and exchange. To fight the battle of trade, it was modelled, not upon the standing army, but upon the task force — supremely mobile, effective while resistance was slight, but incapable of facing strong opposition. To the merchant concerned about the next decade (at the longest), the efficiency and economy of the task force was appealing, its ultimate inadequacy of little concern. The St. Lawrence economy succumbed to massive force, in the end, partly because it had been built for maximum success in the age before massive force appeared.

The merchants of the St. Lawrence ought not to be blamed, or credited, too far. The attitude of Montreal merchants was not significantly different from that of other mercantile agents of Britain in New York, Charleston or New Orleans, who likewise accepted scarcely any responsibility for the welfare of the region in which they operated.[27] Rather, agents in many of the American cities had much less success than those in Montreal in preventing the development of a diversified economy, increasingly independent of London. Moreover, the attenuated nature of the St. Lawrence economy, though it owed something to mercantile policy and the exceptionally strong political position which merchants were able to occupy in Canada between 1765 and 1835, owed more to the weakness of Canada's agricultural base before the settlement of Upper Canada. It could be said, also, that St. Lawrence merchants invested to some extent in local enterprises; though their most extensive investments — their seizure of the seigneuries at bargain prices and their looting of the public lands — rather underline their social irresponsibility than prove any other point. But one serious charge can be made against the merchants even within their own code of morality: that they miscalculated in contemptuously dismissing the local economy as being of minor importance — for, as

Fowke has remarked, they depended on it more than they thought.[28] For this neglect they paid a suitable if not an adequate price, for the local economy failed to support them in their hour of need, and then passed beyond them.

Generals and Governors Merchants were not the only, or even the most influential, of imperial agents in Canada. That distinction belonged to military and civil officials, and these, while they supported the economic interests of their principles and therefore supported the mercantile interest in a general way, reflected an imperial interest that was broader and deeper than that of the merchants. The administrators had no particular concern for the welfare of the colony, as such, any more than the merchants: the colony was valuable only as it magnified imperial wealth and power. In practice, however, administrative actions were more likely than mercantile ones to serve the long-term interests of the colony. For one thing, imperial officers had a very much longer time-horizon than merchants and could contemplate investment in developments that might not yield returns for a lifetime. Moreover, their calculations were conducted in a very broad context that emphasized military power and prestige. Until around 1835, administrative thinking about Canada had to do primarily with the problem of survival in war.

A colony's prospect of survival and of useful participation during war was enhanced the more compactly it was settled, and the more its economy was diversified, made self-sufficient and therefore immune to blockade. Since this military criterion stood in flat contradiction to the mercantile ideal of maximum exchange, imperial policy tended to be ambivalent, fluctuating with the imminence of war. French policy supported development of the colony more consistently than British, despite the desire of officials as well as merchants to make money from the fur trade, because a French colony was almost certain to be isolated for long periods in wartime.[29] The French state took extreme steps to populate New France. It fostered agricultural settlement, with arrangements well calculated to make settlement rapid and compact, and to keep the settlers content, obedient and disciplined for war service. It promoted the Quebec shipyards, the St. Maurice ironworks and some other industrial enterprises to make the colony diversified and defensible. It bequeathed to the succeeding British administration a tradition of planning and management of the economy and a structure well started on diversification — foundations from which further development could be launched.

British administrators also worked to strengthen and round the economy, again in preparation for war. During the first few years after the Conquest, the presiding generals operated a military kind of planned economy, intervened vigorously to put markets in order, and used force and conscription to put the shipyards and ironworks back into production.[30] After 1783, encouragement given to settlers from Britain and from the former American colonies, and the care taken to put them in compact blocks at strategic places, paralleled earlier French policies closely. Road building, important for defence, but perhaps more important still for development, was actively pursued. After the War of 1812-14, a very ambitious program was undertaken with development of the elaborate dockyard at Kingston, promotion of the Marmora ironworks that were intended to supply it with iron, and construction of the Rideau Canal. It is true that Marmora was a failure, that the impetus given to lake shipping by the concentration of skills and equipment at the dockyard was indirect, and that the Rideau Canal did not, from a commercial point of view, very much relieve the transportation bottleneck betwen Upper and Lower Canada. On the other hand, the canal's contribution in opening up a large section of Upper Canada; the support which expenditure on all these projects gave to immigration, settlement and town growth; the unmeasurable but scarcely insignificant effect of the skills and processes which they introduced into the colony; above all, the readiness of the authorities to ignore mercantile desires: all deserve to be recognized.[31]

This is not intended to deny that British administrators stood ultimately for Britain's commercial interest, for staple exports from the colony and an open market for British manufacturers in it. They did so during the American Revolution, when Canada was held by naked force in defiance of popular opinion.[32] The independence generally favoured at that time by habitants and artisans implied a more balanced and self-sufficient economic structure, while the consideration that induced most of the merchants to give their tepid support to the British connection was their hope of larger fur exports by way of the St. Lawrence.[33] Again, in 1837, the rebels stood, among other things, for a rounded economy that would provide a local market for agricultural produce; while the official party stood, among other things, for transatlantic trade that might involve very disadvantageous terms for wheat producers. Between times, however, administrators did their part to develop Canada to the point that it could become an independent economy, whether merchants liked that or not.

The Loss of the Empire of the St. Lawrence.

The United States achieved political unity in 1789, if not in 1776, but they were incapable of economic unity for several more decades. Rather the country consisted of a series of discrete economic entities along the Atlantic seaboard, each built around a port city, united in economic matters chiefly by a common dependence upon the British economy for which each of the port cities acted as an agent. The position of these cities, then, paralleled that of the St. Lawrence cities. A realistic economic map of North America around 1790 would have taken small account of political boundaries, but would have emphasized the string of client ports — Montreal, New York, Philadelphia, Baltimore, Charleston (Boston and New England towns standing somewhat apart) — in sharp rivalry with each other for British custom. Each city exercised economic suzerainty over a hinterland, which it sought to extend. Control became shadowy with distance, however, while small ports tried to carve out little hinterlands of their own, so that the boundaries of each commercial empire tended to be vague and fluctuating.[34]

The coastal cities were supported, in the first place, by their immediate hinterlands — the regions so close to the city itself, and so distant from any other port of significance, that they were bound to trade through their local port. Some of these hinterlands were much more extensive and fruitful than others. Philadelphia was especially favoured in this respect, with a great sweep of good land between the sea and the mountains, and it was this circumstance chiefly that made Philadelphia the metropolis of America in the eighteenth century, with a population well ahead of New York's and far above that of any other place.[35] New York, too, had a respectable hinterland close by and a very large potential one up the Hudson. The importance to the city's expansion of the development of upper New York state is suggested by assertions that the spectacular growth of New York's population began about 1783, and that this population doubled between 1786 and 1796.[36] Baltimore, which grew very rapidly in the first half of the nineteenth century, had the aid of a solid agricultural hinterland at its doorstep. New Orleans, when it developed, seemed to command the most sweeping hinterland of all, and might have mounted on it to become the leading emporium of the continent, if it could have arranged for the suppression of canals and railroads. The immediate hinterland of Montreal, on the other hand, was relatively limited, though its Upper Canadian annex bore comparison with the regions commanded by most other cities.

Local hinterlands tended to be taken for granted, however, by the commercial communities that dominated the great ports. These focused their ambitions rather upon that glittering prize, the potential trade of the interior of the continent, which each port hoped to channel through itself to Europe. In the contest for this prize, natural advantages again varied greatly, so that on the Atlantic Seaboard only Montreal and New York commanded natural routes into the interior. But each of the cities insistently used city and state governments, and could sometimes win aid from the American federal government, to advance its interest by the construction of transport routes inland: a competitive endeavour to which North America owes, more than anything else, its characteristic east-west structure. Philadelphia had already benefited in 1758 by the opening of Forbes Road, which linked it to Pittsburgh. The city capitalized upon its exceptionally strong base, by enlisting the money and manpower of Pennsylvania to build other ways across the mountains. The Pennsylvania Canal, a remarkable engineering feat, was completed early in the nineteenth century; and when this work proved incapable of fulfilling its purpose, and the railway age dawned, the Pennsylvania Railroad was promoted.[37] These systems succeeded in holding the trade of the Ohio Valley for Philadelphia for many years. Nevertheless, they failed to win an inland empire for Philadelphia, in spite of the initial power possessed by that city, because the mountains were so high and rough in this region that the routes could not excel in economic competition. In the nineteenth century, therefore, in comparison with other cities, Philadelphia vegetated graciously.

Baltimore was much smaller and also faced a formidable mountain barrier. It had an advantage over all other ports, however, in linear distance from the interior. It also benefited in an exceptional degree, because of its nearness to Washington, from such efforts as the American federal government was allowed to make to unify the states. The most striking of these ventures, and the one that did most for Baltimore, was the Cumberland Road, leading from Washington to the Ohio watershed, and eventually to Illinois.[38] The city's position was also strengthened by construction of the Chesapeake and Ohio Canal, as by the improvements to local waterways that had preceded it. Its greatest achievement, however, was the Baltimore and Ohio Railway, on which Baltimore gambled very early and very heavily — the line which, by its central position and directness to the interior, became the base of North American freight rate structures. Despite these successes, Baltimore never approached a position of engrossing the trade of the interior to the exclusion of other cities. The same could be said of Boston, which had been kept out of competition in the age of waterways by geographic barriers, and which entered a late but powerful railway challenge with the aid of its financial power but never achieved more than a regional supremacy.[39]

A stronger contender than these was New York, already pulling ahead of Philadelphia in population in 1810. Besides its strong local base, it possessed a good strategic position for coastal trade between versatile New England and the bread, tobacco and rice colonies to the south. In respect to the interior, however, its enormous advantage over all other American cities was its command of the Hudson River and the Mohawk Gap through the Alleghenies. This asset enabled

New York to establish its pre-eminence over other cities by means of the Erie Canal, to consolidate that position with the aid of the New York Central and other railways, and to stand unassailable even by 1850 as prime beneficiary of continental expansion. The city's brilliant future was not obvious in 1800, however, or even in 1820, for Montreal seemed to command an even better gap and a far more magnificent water route into the heart of the continent — a route unmatched except by the Mississippi, and superior to the Mississippi in many respects. Montreal was closer to Liverpool than New York by three hundred miles, and the St. Lawrence region disposed of some important staples welcomed in European markets. Montreal's exclusion from the American union involved both advantages and disadvantages for it, but did not seem very important until about 1840, that is, until the United States was forged into a genuine economic unit by railways, by which time New York had already won its continental supremacy in economic competition.[40] How, then, could the city whose mountain gap only led to a difficult portage route to Lake Ontario, triumph over the city that enjoyed a continuous majestic highway to the Upper Lakes? Neither is it apparent, on first sight, how the digging of an insignificant ditch across northern New York state could make the crucial difference.

The Erie Canal[41] was the fruit of masterly political leadership by De Witt Clinton. It represented the first acceptance in America of an engineering fact that a straightforward canal usually produces a better channel for navigation than an improved river bed. The canal was built expeditiously, with the aid of some striking innovations in construction methods. All the same, it was not a very remarkable canal, as canals go; and, decidedly, it did not give New York a cheaper route than Montreal to the interior. Montreal's superiority in this respect was so apparent to New Yorkers that they took care to avoid competition by terminating their canal on Lake Erie rather than Lake Ontario. Even so, when the canal was completed in 1825, the freight rate from Buffalo to New York was $1.00 per hundredweight, whereas the rate to Montreal was 50 cents.[42] Moreover, improvements could be and were being made to the St. Lawrence, which was also to benefit shortly from an imperial preference on exports, regardless of their origin.[43]

New York's victory cannot be attributed, either, to its earlier possession of railroads. The railway was immensely important to New York: its lines to Chicago became the core of the North American transport system and the main muscle by which the whole continent was bent to pay tribute to this port. But New York's pre-eminence over other cities was won before the railways, which merely held and underlined it, and if the case had been otherwise, railways alone could not have brought victory. Railways offer speed, convenience, access to regions remote from water; but they have never afforded cheaper transport than water, and bulky goods for which cheap transport is essential have continued to move by water.[44] It was not until 1839, on the eve of rail communication with the seaboard, that Chicago began to export grain; and the railways had arrived before the grain trade was of any consequence. Yet grain moved by water, in spite of the fact that lake shipping had to expand feverishly to handle it.[45] But Chicago's grain did not go by way of Montreal, in spite of its lower inland freight rates and its shorter ocean route to Britain.[46]

The reason for New York's success was not much of a mystery, though Montrealers acted as if it was, as if their superior inland water route gave them title to the continental trade, and as if some mechanical adjustment which they had so far overlooked would restore their property to its rightful owners. They fussed about the rapids, about the shallowness of Lake St. Peter, about French-Canadian obstruction of the improvements they desired, about imperial preferences: all important in their way, but all beside the main point. The main point was that New York enjoyed a great advantage over Montreal in ocean freight rates — a large enough difference that, once the Erie Canal was opened, it more than made up for Montreal's advantage on inland rates.[47] In addition, New York sailings were more regular and port operating costs were lower. Carriers preferred the volume, regularity and diversity of New York traffic, the better balance of inbound and outbound cargoes; and they preferred these things with increasing urgency as overhead costs became a larger proportion of total shipping costs. Moreover, New York's advantages of economy and convenience were cumulative, as more traffic justified more ships, and more ships justified more traffic.[48]

And what explains the size, balance and dependability of traffic that had given New York its initial advantage of lower ocean rates? Primarily, the volume and diversity of exchanges which a well-developed hinterland occasioned, along with the power to draw on neighbouring communities that were still more versatile, mature and solid. The Canadian provinces, on whose traffic Montreal might reasonably count, contained about 550,000 persons in 1820. New York City itself contained a quarter as many, and there were ten millions in the United States, of whom at least a third must have occasioned some traffic for the port, and that in wide variety. Montreal merchants inclined to the view that local populations did not matter much, that diversification which militated against transatlantic exchanges was pernicious, that it was the great export trades in staples that counted. The local economies of farmers, traders and numerous artisans of many skills which throve in the northern and middle states, but which were rather less developed in Canada and absent from the southern states, indeed possessed a high degree of local self-sufficiency, so that overseas sales and purchases involved perhaps no more than 10 per cent of total receipts and expenditures. On the other hand, the solid prosperity which these economies often enjoyed not only maintained the volume but the regularity and variety of a traffic that provided ports with the kind of commerce that shippers prized. The imbalance of incoming and outgoing traffic on the St. Lawrence on the other hand, and the lack of variety that provides alternative opportunities, diminished the popularity of that route and raised its rates. Local development was more important than the St. Lawrence merchants appreciated, not only in their own businesses, but because it armed New York to defeat them.

The locally-oriented economies typical of the northern states doomed the St. Lawrence route in still another way, by gradual but inexorable settlement of its former trading territories. The frontier of settlement moved slowly, inching its way across the continent, in contrast to the dashing trade parties thrown out from Montreal. But, whereas Montreal's thrusts were flimsy and transitory, the American structure was solid right back to the base port and its advance immensely powerful. American frontier settlements could, and sometimes did, trade through Montreal; but their tendency, even in the absence of tariff barriers, was to trade through agencies familiar to them, which catered to them and to the host of settlers closer in who were like them. A port developed in relation to an expanding area of settlement was not likely to decline very far, though its functions could be transformed as a continental economy developed. But Montreal, geared narrowly to skim the superficial resources of the interior, could lose everything that its merchants considered to be valuable. Its defeat, in fact, was accomplished very quickly. Whereas New York had feared, in planning its Erie Canal, to locate a terminus short of Lake Erie, it was ready to open a branch canal to Lake Ontario in 1828, a branch that drew traffic that had come through the Welland Canal. Waterways and railways integrated the American economy into a meaningful unit by the 1840s, so that, for the first time, American national policy could have been used decisively against Montreal. But fear of Montreal was so quickly replaced by contempt as to permit the insult of American bonding legislation in 1845 and the concession of reciprocity in 1854. Montreal's agreement with New York about the facts of contemporary economic life was expressed in the Annexationist Manifesto of 1849.

Montreal merchants were thus thrown back, unwillingly, upon the support of the home economy which they had scorned. It was a stronger economy than they deserved or than the author of the Annexationist Manifesto appreciated. An efficient agriculture had developed, considerable diversification had already been mounted upon it, and the physical and ideological changes required for the successful operation of a national economy were in progress. The mercantile interest had a last contribution to make: to help to promote the trunk canals and railways by means of which they hoped, vainly, to win back their lost continent, but which served instead to tie the local economy together. Then, the collapse of the empire of the St. Lawrence cleared the way for recognition and promotion of a national alternative. The simultaneous dismissal of Canada from the colonial system, as Britain discovered a superior economic interest in free trade, made it easy for even the most obtuse Canadians to recognize that the old order had ended and that a new orientation was required.[49]

The Popular Demand for a National Economy

Among the remarkable heritages from the nineteenth century is a doctrine widely circulated among succeeding generations that most nineteenth-century men believed in laissez faire and free trade. When this proposition was restricted in its application to the United Kingdom, there was at least the evidence of British free trade to support it.[50] But so persuasive have been the propagandist powers of those who favoured free trade or (a rather different thing) free enterprise and wanted to give them the authority of historical tradition, that scholars have been induced to apply this dogma to other countries, even though they did not necessarily approve of free trade or free enterprise themselves, were in possession of historical evidence that they were not very popular outside the United Kingdom (or, for that matter, in it), and had to deny their own evidence to retain their convictions. Only at mid-twentieth century were American scholars able to dig themselves out of the tradition that nineteenth-century Americans believed in laissez faire and free enterprise,[51] and Canadians appear to have done less well. The "National Policy" of the 1870s of building a Canadian national economy with the aid of protective tariffs is still discussed readily as if there had been no policy of building a national economy with the aid of protection in the 1860s, 1850s and 1840s.[52] Again, with even less reason than Americans, Canadians have declared that their nineteenth-century forebears were proponents and exponents of laissez faire and have treated the obvious evidence to the contrary as "exceptions". Thus, it was stated not long ago in an historical document of some importance that "developmental expenditure, in the pre-Confederation period, was the one permitted exception to the traditional laissez-faire theory of governmental functions..."[53] According to another historian, dealing with Canada's canal building of the 1840s, "The imperial and provincial governments were *laissez faire* governments if ever there were any; yet in the matter of canals they had made an exception, and had gone into business on a large scale."[54] When an historian was required to set out the views that most Canadians actually held in the nineteenth century, he might feel compelled to deprecate them and dissociate himself from them to the point of declaring that historical evidence is unreal:

> The arguments advanced by *Le Nouveau Monde* in favour of protection were certainly not original. They were the stereotyped mercantilist views, set forward with some vigour, but with little attempt to adapt them to conditions in Canada. They may be regarded as a general statement of principles, supported by a large amount of examples drawn from the tariff history of various countries. It was pointed out, for instance, that France was most prosperous under the system of tariffs instituted by Colbert and most powerful when her industries were protected by Napoleon...the triumph of the Northern over the Southern states was traced to the high tariff which had turned the North into a great workshop. Unreal as these arguments appear to-day, in the seventies they were accepted as valid...[55]

The facts are that the tradition familiar to Canadians of the nineteenth century was one of state intervention and regulation; that laissez faire was little honoured in government policy and little admired by the public; that Canadians gave a very qualified approval to free competition in the market — home market or international market; that those few residents educated to the intricacies of British free trade doctrine were likely to argue that Canadian conditions made the doctrine unsuitable for their country; and that the Canadian public favoured protection and other forms of intervention to build a national economy, sporadically in the early nineteenth century, passionately in the 1840s and consistently thereafter.

Most of Canada's residents of the nineteenth century learned their social policy not from Adam Smith, John Stuart Mill or Herbert Spencer, but from French and British social usages and colonial systems. In the eighteenth century, and in the first half of the nineteenth in most parts of the provinces, there was a general acceptance of a class structure derived from, though less clear cut than, European ones, that involved differential rights and duties, and criteria of loyalty and "respectability" rather than of merit by some objective test or success in the marketplace. Until about 1840, the privileged and educated and their supporters were quick to emphasize that Canada was substantially free of the democratic practices of the barbarous Americans and, therefore, of the mob rule and thoughtless measures which they produced. The determination of policy by those born and trained to rule implied the denial of a role not only for the masses but for "natural" laws and forces. What had been learned about policy in the French period was that the king was the father and leader of his people, who arranged the settlement of the country by cajolery and force, provided wives for the settlers, imposed corvées and military service but gave protection against seigneurs and enemies, founded and operated key industries, regulated much and was prepared in principle to regulate all. The Conquest brought a change of masters but little change in policy. For the first few years the new authorities operated the most thoroughly planned and directed economy that Canada has had, and, with a minimum of dependence on market forces, achieved substantial success in restoring production. Soon after, British rulers had their turn at arranging the importation of settlers, providing them with land, equipment and sustenance, and watching paternally over their material and ideological welfare. In the first four decades of the nineteenth century, policy was strongly influenced by military men who believed firmly that defence was more important than opulence. These officers practised a benevolent paternalism in many contexts, but they were perfectly prepared to thwart both the will of the people and the forces of the market on behalf of defence, as in their obstruction of canals and industries on the frontier. It can be said, more sweepingly, that the whole existence and history of Canada has been a denial of laissez faire. Letting economic forces run their course would have involved a much slower rate, and perhaps a different shape, of European settlement. Letting the people determine their own political destiny would have involved the enrolment of the country as part of the United States in 1775, and perhaps at some other times. Letting the military weight in North America reach its own equilibrium would have led similarly to the swallowing of Canada by the United States.

An important ingredient of Canadian attitudes was the very local kind of sustenance economy and market in common necessities that characterized Canada, and other countries, until the middle of the nineteenth century. It was the structure of the eighteenth century, with which the first half of the nineteenth century belongs economically and spiritually; so that it is sharply different from the period after 1850, when a revolution in transport swept up local markets into national and world ones, and produced a "modern" structure and outlook. Before 1850, the cost of transport — especially land transport — was relatively very high. In consequence, every village enjoyed a substantial level of "natural" protection against the producers and merchants of every other village: it was this circumstance that dispersed artisans and their manufacturing places so widely over Europe and northern North America, because differences in labour costs (the ratio was 2:1 approximately as between American and British wage rates) still made less difference than transport costs for most goods. If inhabitants of these protected local economies said that they believed in competition as a check on prices, they were not likely to mean that they wished competitive supplies of goods to be imported from other towns, but rather that they would like to see a larger number of local tailors, shoemakers and blacksmiths competing with each other. If they opposed monopoly and desired its regulation, they might have in mind the Bank of Upper Canada, but more likely, the local baker or grist mill owner. When the cocoons of these localities were ripped asunder by waterways and railways, there was more horror than hope as traditional channels of production and exchange were disrupted by importations of cheaper manufactures from other towns, other regions or other countries, without regard for employment and livelihood. Cheerful doctrines of comparative advantage and division of labour had no more power to quiet this distress than the doctrine of the survival of the fittest consoles the drowning man. The fact is that competition, when it was made a national and international reality by transport media, was thoroughly detested by European societies, and the majority might well have banished it from the earth, if they had known how to do so. Some relief was possible at national frontiers, and tariff protection was increasingly favoured by all except those who thought they could undersell others in "their own" market. Manufacturers hastened to avoid the discipline of competition as much as they could by means of mergers or monopolies. Some wage earners were able to gain a measure of protection through unions.[56] There remained a great number who were no more willing to let nature take its course, but found no means to protect themselves, and were cast naked into the unregulated market. None of this experience raised the reputation of competition, free trade or laissez faire.

A parallel ingredient of Canadian attitudes derived from traditional experience with colonial systems of mercantilist origin. The British colonial system may have operated unfairly against Canada. The colony was required to take its imports from Britain, but among the short list of commodities which Britain might have taken in exchange, wheat was excluded most of the time until the 1840s by the Corn Laws, which contained a colonial preference but seldom let any wheat in. Attempts to substitute the export of hemp from Canada can hardly be regarded as an adequate compensation.[57] Canada was not expected to raise any obstacles against the importation of British manufactures, however. The British government was still making strenuous representations against any Canadian protection in the 1840s, and also opposed measures that, by expanding Canada's money supply, might have encouraged local production indirectly and jeopardized the ready payment of British exporters in sterling.[58] Nevertheless, the colonial system was conceived in theory, and operated in practice, to provide reciprocally secure markets, and it would be difficult to prove that Canada was not a net beneficiary in the system. The timber trade was created in the interest of Britain, not of Canada, and it involved an extraordinarily wasteful exploitation of Canadian forests; but it is also true that for about forty years, Britain gave Canada a secure market for a major export by paying prices much higher than it would have had to pay elsewhere. In the 1840s, favourable preferences on flour, without much restriction upon the origin of the grain used to make it, for a short time made Canadian capitalists still more a favoured group. Not all Canadians admired the colonial system, but objection was nearly always to the balance of advantage, or to the concomitant political control, rather than to intervention in principle. The great majority seem to have regarded the system, even as it stood, as the essential pillar of their existence. Their shock when Britain turned to free trade was that of men whose employment and homestead had disappeared. Their experience, or even the fact that they survived it with the help of a new reciprocal security with the United States, could not persuade them that free trade and laissez faire were desirable.

There is, finally, the point elaborated by historians: that the uncertainties associated with underdeveloped frontiers, the weakness of a small country adrift in the world economy and, above all, the vigour and aggressiveness of the neighbouring United States, have induced Canadian governments to take direct responsibility for economic development, primarily but not exclusively through seeing to the construction of canals and railways.

Suspicion of laissez faire does not lead necessarily to support of a national economy protected by tariffs; until the 1840s, it led mostly to support of the colonial system. Nevertheless, it has been pointed out that there was already in Canada some experience of fairly self-sufficient local economy: the little economies of the village, and a broader local economy in western Upper Canada. Canadians had also an acquaintance with the self-sufficiency of wartime. Then, with a substantial proportion of them drawn from the United States, Canadians had knowledge of and shared in the tradition of those vigorous seaboard colonies that had diversified themselves, partly by deliberate policy, even in the eighteenth century. New England had been aided to self-sufficiency by neglect and by failure to find a staple. As well, diversification was encouraged in most of the American colonies by their chronic shortage of sterling exchange, which provided them with a protective differential of about 10 per cent discount on local currency, in addition to any other protection.[59] Monetary stimulation of local industry became a deliberate policy when several colonies, to counteract depression, resorted to increases of their local money supplies. The marked success of these emissions in creating employment and restoring prosperity informs us that the colonies were already broadly developed.[60] An instructive contrast is provided by the history of New France which, though not without diversification, evidently still had too narrow and torpid an economy to find ready substitutes for imports, so that the attempt to meet the problems of the colony in the 1740s and 1750s by issues of paper money only led to an inflationary chaos.[61] The Americans gained further experience in self-sufficiency in their Revolutionary War. An early fruit of their independence was the promotion of manufactures in the interests of national security and prosperity.[62] New American manufactures not infrequently succumbed to British competition in peacetime. Nevertheless, the conception of an economy making its own employment, prosperity and development secure by local exchanges was familiar to Americans, and to Canadians, at the beginning of the nineteenth century.

It has been pointed out that Papineau, the reform leader in Lower Canada, favoured a protective tariff as a weapon against the colonial office,[63] and there was always some volume of protectionist sentiment in Lower Canada from his time, at least. However, Upper Canada was the heartland of Canadian protection: the early security of its farmers against the need to export has been noted elsewhere, and they consistently sought protection against the American agricultural produce that could weaken or destroy their advantage. They scored occasional victories, too, when tariffs were imposed on American products in 1822 and 1843; though Fowke regards these as tactical manoeuvres in the war between farmers and merchants which merchants always won.[64] Objection to any measure that would make American imports more competitive with Canadian products was also a reason why farmers opposed state expenditure on trunk canals.[65] In the 1820s, William Lyon Mackenzie was inspired by the American, Mathew Carey, to advocate a protected and diversified Canadian economy in which farmers could dispose of their surpluses safely to a local manufacturing population, and draw their supplies from it, without economic or political dependence on Britain.[66] Mackenzie would seem to have represented his constituency more accurately then, and certainly was more popular than in the 1830s, when he was more or less converted to free trade, not by any intellectual merit of the doctrine but because he thought this a means of combatting the colonial office and the progress of class differentiation in Canada.[67] Whatever the influence of Mackenzie, most Canadian farmers were protectionists throughout the nineteenth century, though Fowke has suggested that fervour slackened after 1850.[68] The address of President William Ferguson to the Provincial Agricultural Association in 1858 may serve to illustrate the sentiments circulated in the period. His speech was a plea for political action by farmers as a "class" in support of protection. He called not only for protection of hemp, tobacco and flax (which he thought could produce homespun linen as a good substitute for imported cottons), but also for the protection of industrial products. The province could already be proud, he said, of such industrial products as its fine farm implements and locomotives. With Canada's "inexhaustible water power", nothing but protection was needed to build native industry. The leadership currently enjoyed in the world by Britain and New England was the result of the long period during which they had protected their manufactures. Local manufactures would retain those artisan immigrants who did not wish to farm and who would otherwise be lost to the United States.[69] These views, which seem typical enough, were bound to be opposed vigorously by the mercantile interest, but they provided an obvious basis for alliance between farmers and industrial producers.

The protectionism of industrial producers was the really strategic element in creating a political framework congenial to development in Canada. The industrial population was not outstandingly numerous or powerful, but it — along with smaller interests that sought protection, like lake shippers[70] — could affect the balance between a protectionist agriculture and a mercantile interest devoted to the maximization of external trade. Again, the protection of manufactures could produce what protection of agriculture could not, a sufficiently diversified economy to permit an independent development. Industrial protection occupied a strategic position at mid-century, finally, because new transport facilities had established the possibility of the home market economy which it implied, and Britain's acceptance of free trade whetted public demand for such an economy.

Agitation for the protection of manufactures was desultory until the 1840s,[71] and the question of protection for industry does not appear to have been discussed much in the 1820s and 1830s. This situation cannot be accounted for by lack of an industrial interest; for, while Canada did not have many large industrial establishments, it had a substantial body of handicraft "manufacturers", along with an abundance of grist, saw and carding mills, and tanneries. The reason for silence is, rather, that Canadian manufactures enjoyed substantial protection from American competition throughout these decades: under a bewildering succession of legislation, American manufactures were nearly always either excluded or charged about 30 per cent duty.[72] That there was hardly any protection against British manufactures does not seem to have occasioned complaint. Tact and acceptance of the mutual preferences of the colonial system may be the explanation for this. What the documents suggest, however, is that in the lines of manufacture in which Canadians were interested, they did not fear British competition but greatly feared American competition. This may follow from the peculiarity of British manufactures, that while Britain was far ahead in some lines, she was behind in others, and not much concerned about it. Thus, it was a rather common observation about 1850 that while Britain led the world in heavy machinery, Americans (and Canadians) excelled in light.[73] A similar specialization seems likely to have prevailed earlier.

It was the decade of the 1840s that produced the first wave of protectionism for manufacturers in Canada. There was already an undercurrent of agitation at the beginning of the decade — for specific rather than the *ad valorem* duties that allowed evasion, and against the rising competition of Michigan lumber.[74] However, vigorous agitation was set off by some drastic changes in legislation. In 1842, an imperial act reduced sharply the tariff rates applying on foreign manufactures imported into Canada. There was also an improved preference on Canadian wheat, made still better when Canada imposed a duty of 3 shillings per quarter on wheat imported from the United States. But Canadian farmers, at a peak of protectionist agitation, were far from satisfied with this measure on behalf of the mercantile interests and induced the Legislature in 1843 to impose substantial duties on a wide range of American agricultural products.[75] These measures brought industrial producers into full voice.

Canada's numerous producers of leather and shoes took the lead in this agitation. In 1842 and 1843, those of Montreal and Kingston stressed their sufferings from American competition and did not fail to point out that American producers enjoyed a prohibitive protection in their own market.[76] In 1844 and 1845, the shoemakers of Belleville and Kingston took up the issue, emphasizing this time that farmers now had 30 per cent tariff protection on their products, and that leather producers also should have their traditional 30 per cent defence.[77] Meanwhile, better protection was being demanded also for lumber products and iron products.[78] Petitioners did not point only to the low level of duties on American goods, but also to the frequent evasions through *ad valorem* duties, carelessness and contraband trade.[79] Agitation was sufficiently effective to produce Canadian legislation in 1845 that imposed discriminatory duties on American goods, as well as higher duties on imports from Britain and lower rates on raw materials.[80] The legislation was modified in 1846 under pressure from the British government, but it is significant that discriminatory protection was retained for leather producers, the most vociferous interest.[81]

Up until this time, arguments for protection had been particular and traditional, and supporters of it may have been the same. The shock of free trade in Britain, and the depression of 1847-48 produced a new kind of agitation. It was involved with the whole question of the country's future and culminated in the sweeping proposals of the British American League and the Annexationist Manifesto.[82] British complaints against Canadian protection[83] not only fell upon unsympathetic ears but faced a general and sophisticated argument for protection: Britain's abandonment of the colonial system required a new policy for Canada. "The existence in British North America of the chief elements of Manufacturing Prosperity viz. Cheap Food — Abundance of Water Power — Great Mineral Wealth — extensive lumber Forests — Freedom from direct Taxation — and a healthy and invigorating Climate — point to the encouragement of Domestic Manufactures, as the most certain means of restoring this Provincial Prosperity..." However, as long as the United States had high tariffs, capital would not invest in Canada unless Canada was protected too. Therefore, Canada should have tariffs equal to those of the United States.[84] This position, rounded out by a more concrete conception of the home market, was to be the basis of protectionist sentiment — the dominant sentiment — in succeeding years.

Agitation for protection was usually louder in English Canada than French. However, in spite of this and Lord Elgin's sneer about the "British party", protection was not at all the sole prerogative of the English-speaking, even in the 1840s. But French support for protection had fluctuated, or appeared to fluctuate, in intensity. The root of this apparent instability lies in the fact that the French problem — how to retain and provide for a rapidly-growing population sunk deep in an unprogressive agriculture, with holdings too small and soils exhausted — seemed to Church authorities to be susceptible of several solutions, of which industrialization was not the preferred one.[85] The actual solution of the problem in the nineteenth century was emigration; but emigration was considered to be undesirable, so that agitation for other measures, especially protection, rose and fell with the flow of emigrants. The solution favoured by the Church was "anti-luxury": the acceptance of a lower standard of living so that more people could survive in peasant agriculture.[86] This program was ignored by the habitants, and it could only have postponed collapse and made it spectacular when it came. Colonization was the only solution approved by all elements of French-Canadian nationalism (except, perhaps, the colonists). Settlement of the surplus population on new lands expanded the economy, yet retained the peasant structure desired by the Church, and had therefore been prominent in popular agitation and provincial policy since the 1840s.[87] It did not, however, much reduce the congestion of population on the old lands. Industrialization, the remaining alternative, had been viewed with some suspicion, but was preferred to emigration, and received strong support whenever there seemed to be no other adequate cure for that ill.

It is no accident, therefore, that the clearest expression of French-Canadian protectionism in the 1840s comes in the Report of that Committee of the Assembly charged with discovering "the Causes and Importance" of emigration from Lower Canada to the United States. The Committee and its witnesses, many of them priests, reflected French-Canadian nationalist attitudes, including the uncertainty about method. There was no dispute that emigration had already reached alarming proportions. But the major proposals, referring vaguely to public opinion and education, amounted to an acceptance of the clerical prescription that habitants should be persuaded to stay in Canada at a lower standard of living, rather than seek relief in the United States.[88] However, some of the priests who testified, anxious for the development of manufactures to keep their flocks at home, denounced the seigneurs who, they said, refused to develop the waterpower sites which they owned.[89] And the Committee's Report came out strongly in favour of tariff protection for the benefit of local industry and the promotion of further manufacturing. "Your Committee cannot, no more than Your Honourable House, be blind to the fact that Lower Canada, by its geographical position, its wants, its natural advantages, is destined, as well as the Northern States . . . to become a great manufacturing country."[90]

The home market economy and protection gained support simultaneously from another direction, the considerable economic development of the 1840s. Development was promoted by population growth and the advance of technology in the world, by the particular stimulus of a mining boom north of Lake Huron[91] and, above all, by canals. In the first place, canals tied the market together. Secondly, waterpower sites suitable for mills were a by-product of their construction. In addition, the imported capital spent on construction provided funds, indirectly, to finance manufactures.[92]

The Canadian government was anxious to develop industries at the power sites along its canals to begin with in order to recover part of the costs of the canals in rentals.[93] But if this desire did not arise in a context of general development, it very soon acquired that context.[94] The industries established were predominantly of the older sort, grist and saw and carding mills; but there were also paper mills, a mill for treating copper,[95] the "Thorold Cotton Manufactory", as well as a "formes de souliers" factory, a "métier à tisser", a "métier à filer" and "other machines".[96] What the industrially-minded hoped for is indicated in Samuel Keefer's report of 1847. Keefer was impressed with the possibility of making the Welland Canal area a centre of cotton manufacturing:

> . . . a modification of the duties on various imported articles

would, no doubt, bring cotton and other manufactures along this Canal into successful competition in the western markets, with the fabrics of the Eastern States, because we enjoy the advantage of being situated so much nearer to the farmer. Raw Cotton is now beginning to be forwarded to the New England factories via Buffalo, and if it be found an advantage to them to obtain it by this route, certainly this District ought to benefit by the difference of the double transportation.[97]

Keefer had in mind to end "the draught of our money to the other side". His earnestness is indicated by the fact that it was a Keefer who headed Upper Canada's first cotton mill at Thorold.[98] Lower Canada's first cotton mill, built near Chambly in 1844, seems likely to have been located with reference to the Chambly Canal.[99] And at Montreal, the Lachine Canal was described by W.H. Merritt, Chief Commissioner of the Board of Works, as the most important part of the canal system for manufactures in 1851.[100] It supported, in particular, very large flour mills of advanced design.[101]

Finally, the canals supported development in still another way, through the debt which the province had accumulated in building them. It was stated in 1849 that £3 millions had so far been spent on canals, with another £95,000 still required to complete them; and the outstanding debt of the province in July 1850 was £4,242,263.[102] The level of duties on imports required to carry this burden, taking into account that the canals were far short of being self-supporting and that direct taxation was considered to be impossible to impose, invited and even necessitated that "incidental protection" that was to come into prominence a decade later.

The 1840s had produced a substantial industrial population with a vested interest in protection, inclined to seize the initiative from agricultural protectionists, destined to grow and to strengthen its protectionist sentiment. And it did so until, by 1877, it could topple the citadels of vested mercantile interest, the Boards of Trade of Toronto and Montreal.[103] By 1850, there was also a vigorous public movement for protection that commanded the assent not of the most powerful, but probably of a good majority of citizens. This movement was supplied with a coherent doctrine that appealed, as free trade did, to principles; it could also draw its argument, as free trade could not, from historical examples, the contemporary difficulties of Canada and the contrasting prosperity of the United States. But it was the decade of the 1850s that consolidated these forces to provide what Canada lacked in 1850 in spite of its considerable diversification — a real home market economy. The primary agent of this change was the railway that penetrated the backwoods, wove a fabric of crossing economic lines around the simple line of waterways, yet bound the market together as canals never could. It was upon the basis of this market, and to a great extent with capital supplied directly or indirectly by railway construction, that Canadian manufactures came of age. Also important, however, were the prosperity of the early 1850s and the opening of the American market to Canada's natural products by the Reciprocity Treaty.* In another context, these might have produced the different result that many expected: a willing reversion to the production of food and wood for export, with manufactures left to others. Canadians used them, in fact, as a guard or breathing space while they rounded out their home market economy. Faith in the gospel of development; suspicion of unreliable foreign markets, heightened once more by the decline of prosperity after 1854; the growing fixed investment in factories, homes and cities, and in canals and railways — all hostages to the fortunes of the local economy; the fiscal problems of a government that had committed itself long ago to development: all worked to produce this result.

The 1850s contributed another vital change: the maturing of the labour market. Speakers were inclined, at this time, to put great stress on Canada's resources of water power, in the apparent belief that manufactures swung on this fulcrum alone. It is true that, somewhere among the details of their exposition, they remarked broadly and loosely upon the great untapped reserves of labour in Canada. These reserves did exist. Surpluses of totally unemployed persons, as well as masses of partially employed, appeared in Lower Canada from early in the nineteenth century, at least.[104] It was characteristic of extractive industries to demand active male labour, to the neglect of women, children and older men. Agriculture, freehold as well as feudal, contained an immense potential of labour time that went unused in winter. The more sophisticated, acquainted with classical economic thought, might ground their argument for protection in labour surpluses, rather than water surpluses. The editors of the *Canadian Almanac* of 1859, drawing the conclusion from the depression of 1857 that Canada should seek relief in more home manufacture of wool, cotton and iron, observed, "There is a large amount of waste labour in the country [Upper Canada] which could be cheaply secured if constant employment was given."[105] In fact, as the same *Almanac* pointed out, the more abundant labour surpluses of Lower Canada had long since been directed into a domestic manufacturing industry that still exported a substantial volume of home-made clothing to Upper Canada in 1859.

---

*For a recent reappraisal of the economic impact of the Reciprocity Treaty, see R.E. Ankli, "The Reciprocity Treaty of 1854", *Canadian Journal of Economics*, Feb. 1971; and L.H. Officers and L.B. Smith, "The Canadian-American Reciprocity Treaty of 1855 to 1866", *Journal of Economic History*, no. 4 (1968). Ed.

Nevertheless, this labour reserve was not very suitable for the operation of manufactures, and still less for their initiation. Though plentiful, and cheap in Lower Canada, the labour was not located in the best places, it was not available continuously in most cases, and above all it was unskilled. Even experience in domestic manufactures was a doubtful asset, for, as many realized, machine production was making rapid progress in the United States, and Canada would have to resort to machinery too, if she hoped to compete in manufacturing.[106] What was required along with machines was skilled labour to plan and manage, and unskilled labour that was disciplined to be available, dependable and obedient. In short, as has been argued elsewhere, Canada needed a mature capitalistic labour market, with, on the one hand, adequate supplies of skilled workers and an abundance of cheap unskilled labour, and, on the other hand, the sustained demand for these workers that would retain them in the market.[107] New industries in the 1850s rounded out a sufficient demand in the cities, founded especially on the regular employment offered by railways. But rural Canada could scarcely have met it. Rather, as has also been argued elsewhere, it was a body of British artisans, and a swarm of persistent Irish peasants, who provided that sufficient and always available supply by which, in the 1850s, Canada achieved its mature labour market.[108] Thus, a critical hurdle to industrialization was surmounted; subsequent expansion faced limitations of the market, of capital, of protection, but no serious limitation on account of labour supply. The contribution of surplus farm labour to this solution of the labour problem was not significant, even in Upper Canada. Rather, industrial growth solved problems for agriculture: first, by providing a larger and more secure market for farm produce; and, secondly, by establishing the industrial framework that eventually could, after all, absorb many of those cast off by agriculture.

In 1852, the Boards of Trade of Montreal and Quebec called for the protection of domestic manufactures.[109] In the same year, the Legislature supported protection, as did the finance minister, Francis Hincks. Hincks expressed the view that was becoming common among educated men in Canada: that laissez faire was the proper policy in theory, but that the American tariff forced a different policy in practice on Canada.[110] Agitation for protection continued in spite of — or because of — reciprocity, and in 1855, a Committee of the Legislature, of which W.H. Merritt was chairman, deplored the large imports of cotton, woollen and iron goods from the United States, and the lack of protection for Canadian industry. "Water power with us is abundant. The climate — owing to the long winters in Lower Canada during which the population are unemployed, insures a supply of labor at low prices; and the export of straw hats and bonnets is an evidence at least of the desire for employment among its inhabitants, which might be turned to more profitable account."[111] Duties were raised to 15 per cent on manufactures (20 per cent on leather) in 1856.[112] However, hard times brought more urgent demands for protection, and Isaac Buchanan organized his first "Association for the Promotion of Canadian Industry" in 1858.[113] Tariffs on manufactured goods were raised to 20 per cent (25 per cent on leather and clothing) in 1858, and there were additional increases in 1859. In this way developed Galt's "incidental protection" which Porritt pronounced, not without justification, to be "The First National Policy Tariffs".[114]

"Incidental protection" appears to have been a politically popular and an economically effective policy. John A. Macdonald reported in 1861, "The success of our Policy in this respect [protection] is already shown by the numerous manufactories of every description, which have sprung up in both sections of the Province.... We hear of hundreds of industrious mechanics and artisans combining together to establish woollen and cotton mills, &c."[115]

The dangers associated with the American Civil War, and the increasing evidence of an aggressive American imperialism did nothing directly to decrease protectionist feeling in Canada.[116] However, in preparation for Confederation, there was a reduction of tariffs to 15 per cent in 1866. This provoked new agitation, and the second meeting of Buchanan's "Association".[117] The reduction need not have been taken so seriously: it was required temporarily to persuade the Maritimes into Confederation; but few informed persons can have doubted that protection would be renewed at a suitable time.[118] Macdonald campaigned again in 1872 on a platform of protection, and found it popular; though tariffs were next raised, in fact, by the Liberals in 1874.[119] All this precedes the National Policy of 1879, which is still sometimes proclaimed to be the beginning of deliberate industrialization by the aid of protective tariffs in Canada.

To this brief catalogue of events it seems useful to add a short discussion of two particular details. The first concerns the role in protectionist development of a remarkable man, Isaac Buchanan. This may serve to illuminate the intellectual atmosphere of Canada in the nineteenth century. The second concerns the shifting of economic and political forces in Canada that formed the background of tariff policy. It is intended to clarify the position of labour.

Isaac Buchanan[120] was prominent in Toronto in the 1830s, and then in Hamilton until the 1870s, as a merchant, railway promoter and politician. He was also an indefatigable writer and propagandist. He was widely acquainted with men and books, was close to a contemporary neo-Mercantilist movement in Britain, and to Mathew Carey and

Horace Greeley in the United States. However, he was an independent thinker with a well-integrated body of economic ideas, which he expressed with more lucidity and elegance than most. In Canada in the nineteenth century, he was only surpassed as an economist by John Rae, and as a leader of popular thought about economic matters by, perhaps, R.B. Sullivan.

Buchanan's objective was stability and prosperity: "employment for the people and markets for the produce",[121] which he saw as inseparable complements. He detested free trade, the gold standard and the instability which he believed they produced. He thought that economic harping on the cheapest prices for consumers, while ignoring the welfare of producers, was stupid and immoral: "the distinctive characteristic of the people is that they *labour* — that they are *producers*, and have therefore as their main interest more bidders for their labour, which means *more*, not less, *price* for the commoditiy."[122] Cheap imports were less important than sure markets: the public interest lay in secure exchanges, which were likely also to produce fairer terms of trade. Buchanan approved of mutually free trade among countries, and in 1847 proposed a great imperial Zollverein to which all countries accepting "reciprocal" free trade might be admitted; but he saw no merit in opening national markets to other countries (in particular the United States) which would not open theirs in return. Stability was his goal, and in his time and place it could only be found in a diversified national economy within which farmers and manufacturers supplied each other in mutual security. Diversification was desirable, also, to allow diversification of agriculture, and because "manufacturing employment is better paid than agricultural".[123] To obtain diversification and fair terms of trade, protection was required.

Buchanan was well aware that a national economy could not be made stable if the gold standard was retained. As one who never ceased to denounce the deflationary policy followed in Britain after the Napoleonic wars, who had been in New York for the panic of 1837, and had participated in Sydenham's unsuccessful campaign for a government Bank of Issue in Canada, he distrusted rigid money, and bankers. His proposal for Canada was a managed paper currency — management to be supplied essentially by the Legislature. The amount of issue should be enough to ensure "full employment".[124] If this led to deterioration of the foreign exchange rate, the decline of that rate would provide a self-correcting mechanism, and appropriate additional protection in the meantime.

Buchanan was the Canadian counterpart of Carey and List, and like other protectionist writers, he dealt far more extensively with "economic development" than economists have typically done until very recently. His monetary ideas resemble those of Lord Keynes in many respects. He was more sympathetic to the workingman than any of these writers, and perhaps more clearly convinced than most economists that economic systems were made for man and not the other way round. For nearly forty years he supplied the electors of Hamilton, and the electors of Canada as far as he could manage, with a theory of economic stability that was sophisticated and humane.

In spite of its popularity among farmers and manufacturers, the survival of protection was kept always in question until the 1870s by a powerful alliance of trading and railroad interests.* To preserve their delicate position, manufacturers had to maintain unity among themselves by allowance in the tariff structure for new interests that appeared from time to time.[125] Only with the approach of Confederation did there come a sharp revision of alliances, based, to begin with, on general agreement that a period of low tariffs was necessary to get the Maritimes in.[126] Soon after, the prospect of exploiting the new West appeared. Now protection received for the first time the name of "National Policy" and ceased for the first time to be a national policy. Policy in earlier times had been directed to the production of a secure society in which appropriate proportions of farmers and urban workers would supply each others' needs without dependence on the world market. The National Policy of 1879 was one weapon of a Canadian imperialism. Its purpose was not to make Western Canada also a participant in a self-contained economy; for the West was fully expected to sell its staple product in world markets, as it did, and there appears to have been no regret that the new Canadian structure was thus made more vulnerable and dependent than its predecessor. The task of the National Policy of 1879 was rather to ensure that unprotected Western wheat was carried by protected Canadian railroads and handled by protected Canadian traders, and that the supplies of the West should come from protected manufacturers, in order that as much as possible of the gains to be made would be drawn to old Canada, and not to Britain or the United States. It was this imperial prospect that brought merchants and railroads suddenly around in the 1870s, so that Eastern Canada spoke with an undivided protectionist voice.[127]

This unanimity spelled danger for the labour movement. The new situation promised an expansion of manufactures and richer monopoly positions from which higher wages could be afforded; but it also freed industrialists from any dependence upon the support of labour

---

*The following section outlines Pentland's contribution to the debate on the nature of the Canadian state in its formative years — what has become known as the "national question". For a fuller discussion of the debate, see Introduction, pp. xix-xxii. Ed.

and from any need, therefore, to offer concessions. It is instructive to notice that while manufacturers were protected from foreign competition in the 1880s and 1890s, workingmen were not and made no progress in their efforts to curb the importation of contract labourers and no significant gain in any other direction.[128] When, around 1900, a new boom promised at last to lift the wages and lower the hours that had prevailed through the long depression, manufacturers launched a fierce campaign against unions and made the importation of foreign labour a communal duty.[129]

The period before 1880 is quite different. While there was a real national policy, from 1850 until 1880, both manufacturers and their workmen believed that their livelihood depended on protection and that protection was always in danger from railroads and merchants. In consequence, employers and employees relied on each other for support and showed each other a marked consideration. Manufacturers resisted strikes, but rarely with anything like the arrogance and vindictiveness displayed by railroad officials. Though unions were illegal combinations in law, George Brown was nearly, if not quite, alone in his prosecution of them, and nearly unique, also, in his consistent hostility to their existence. It was with the open support of many employers that the printers defied Brown in the strike of 1872, and were rescued from their predicament by a great upsurge of popular support, and the Trades Union Act rushed through Parliament by the Conservative Party.[130]

What labour gave in return was an accommodating attitude on most occasions to employers and consistent support for protection and the Conservative Party. The wage earners — not least through the Orange Order in Canada West — were a dependable and not insignificant partner in Macdonald's coalitions.* If workmen had not otherwise been drawn to the Conservative Party, they would have been persuaded into it by George Brown. Most provocative of employers, the consistent supporter of everything that would worsen the position of labour and opponent of anything that might raise it, one whose care for the common man only began (as it has been said) at the city limits of Toronto, Brown did more than any other man to cement the alliance of labour with Conservatism.[131]

What the prevailing structure of economic and political forces induced workingmen to think about trade policy and national development may be sufficiently represented by the Secretary's Report to the Canadian Labour Union in 1874, dealing with the Reciprocity

---

*A very full recent treatment of working-class politics in Toronto during this period is included in Gregory Kealey's *Toronto Workers Respond to Industrial Capitalism, 1867-1892* (Toronto: University of Toronto Press, 1980). Ed.

Treaty being currently proposed by the Liberal administration:

> From Reciprocity, pure and simple, with the United Sates, it is my opinion that we have nothing to fear, but [from] where the want of mutuality exists, as shown on the face of the draft treaty, whereby the United States endeavours to secure a more extended market for their manufactures simply at a premium, by securing to their operatives a cheaper living by opening their market for our raw materials. But the evil for this country (and I must call it such) does not end here. The class of manufactures made free to the U.S. is also to be made free to England. In my opinion if there is anything calculated to make this country more dependent than it is now, it is the passage of the proposed treaty. It is high time that this Dominion was laying aside its swaddling clothes and becoming self-sustaining, and we can only become so by a fair and liberal protection. The United States did not build up her iron, woollen and other industrial interests under free trade. Nor did England.... If ever we are to be a self-dependent country, manufacturers must have inducement to make investments amongst us other than a sharp and ruinous competition...[132]

CHAPTER VI

The Transformation of Canadians

Good land, canals and railways, an adequate supply of labour in the market, and protection: all did their work of changing Canada into an integrated economy at the stage of manufactures after 1850. But they could not do it alone. One more important ingredient was required for this transition: a transformation of values and attitudes to provide the "Moral Conditions of Economic Growth".[1] It is true that Canada's European population already manifested in 1800 in a good measure — far more than many populations now undertaking to make industrial people of themselves — the conceptions of precise rights and duties, specialized tasks and exclusive property, those habits of industry, those propensities to economize, maximize and accumulate, the capacity to take account of a long span of future time, that are required to generate capitalistic progress and operate industrial societies. But it is equally true that this population was too slothful, immediate, anarchic and irregular in its work habits, and too easily seduced by non-economic goals and means to goals, to provide very suitable material for a modern society. To make the material suitable required a complex and unknowable educative process, conducted largely in the school of experience with the goad of harsh impersonal penalties for failures.*
Nor was the means to success capable of reduction to a precise dose of new discipline and new ambition that could be injected, once and for all, like a coin in a machine. Success was attained rather by an indistinct and never-completed process of interacting stimulation and response. Human transformation was bound to be partial, and mostly unplanned, because men were remaking themselves without much comprehension or consciousness of it, because deliberate changes sent out other ripples of subtle, unrecognized adjustments to preserve the tension and balance of existence, and because the inanimate machinery of pro-

---

*This remaking of the labour force has become an integral part of the growing literature on the labour or work process. See Introduction, pp. xiii-xiv. Ed.

duction to which man had to fit himself could only itself be transformed bit by bit and year by year. The nature and extent of the changes in the ways men regarded themselves, conducted themselves and dealt with each other, have therefore to be indicated rather than expounded; and the direction of causation suspected rather than proved.

Staple production, like factory production or buffalo production, had particular conditions and demanded its own peculiar labour force. Staple trades were dominated by the traders, who had commercial interests and outlooks and were more likely to think of themselves as middlemen buying cheap and selling dear than as employers directing production. Fixed investment was kept low. When the trades were highly competitive, as they often were, the instability of sales and prices that typified staple markets was accentuated. In fishing, where the settled nature of the producers made it possible, merchants met fluctuations by passing on the low returns (but not necessarily the high returns) to the fisherman. In fur and timber the scale of production required direct employment, and lack of labour reserves forced employers to assume the overhead costs of their labour forces. Rigid costs, together with variable returns, put employers in a vulnerable position for which their main defence was the maintenance of a wage structure that was customarily low. The conditions of staple production — dispersion of the work force, absence of fixed investment and the habits of mobility of employers of the mercantile type — made it extremely difficult for workmen to offer organized resistance to this. Yet, these workmen could assume a sullen malevolence that made a persuasive weapon when one of their canons was violated. The first defence of the employer was that eighteenth-century workmen rarely used their power of combination to deal in any systematic way with wages. Pre-industrial employments went along with pre-industrial workmen, who lacked a discipline appropriate for rebellion as well as the disposition to rebel over stationary wages, though they might be provoked by declining ones. Men in the staple trades were not accustomed to think that their customary standard of living ought to be improved; and, while they took great pride in personal expertness, they inclined to regard their work as a very ordinary employment for which no exceptional wage should be expected. In the same way, agricultural societies take for granted a familiarity with farm practices, and bodies of pickers and shovellers suppose that anyone can pick and shovel. Societies discount the skills that are common among them, and terms like "unskilled" and "common labour" have no further definition than commonness.

The fact is, however, that catching fish, chopping and rafting timber and canoeing after fur each required so intricate a body of skills, knowledge and attitudes that it could hardly be acquired except from childhood. They usually demanded also great physical hardiness and exertion, so that they were only open to adult males in their prime, and working lives tended to be short. These were shortened, in any case, by a high incidence of sudden death; hence, another qualification required for these employments was an impassive (but not casual) attitude towards this prospect. Moreover, the canoeman and (still more) the lumberman were tied to an existence that was "womanless, homeless, voteless".[2] The remote and shifting scene of their employments partly accounts for this; but it seems to have been true also, at least in the case of the lumberman, that wage levels would not have permitted the maintenance of homes and families — that is, the industry developed on the basis of customary wage rates too low to provide for reproduction of the labour force, and survived on a subsidy of manpower drawn from bush farms.* The camaradie of the camps and the annual spree were hardly an adequate compensation for this rootless regimen, which led only to the early discard of a man unsuited for any employment. The position of the fisherman differed, for he worked from a stationary base, might continue to do so into old age and usually had a family. Since, however, his earnings were no more fit to support a family than the lumberman's, the alternative of abject poverty was his lot.[3]

The willingness of the workmen of the staple trades to accept the conditions of their employments without much complaint is not easily explicable in terms of pecuniary calculation. Indeed, although these men were paid money for their work and did many things that industrial men do, their behaviour was largely the product of considerations beyond the measures of economic rationality. They were as disciplined in their way as industrial men, but theirs was a sociable, extroverted discipline distinct from the cold, methodical routine of factories. They invented but had no cult of invention. They were efficient, not from devotion to a religion of capitalism but as a matter of personal pride. They resisted retrogression, but they did not know that man must progress. Their choice of occupation was determined primarily by local custom. The values that kept them to it were familiarity and excitement. They acquired prestige by exhibitions, not of accumulation, but of strength and daring. The pattern of their lives displays an extreme discounting of the future for which the rational foundation, if there was one, must have been that life in any case is short. The direction of men like these was handled best by employers who also felt and understood the pre-industrial values of fellowship,

---

*Sacouman argues that this is still the situation in the Martimes and refers to it as the semi-proletarianism of the domestic mode of primary production. See R.J. Sacouman, "Semi-proletarianism and Rural Underdevelopment in the Maritimes", *Canadian Review of Sociology and Anthropology* 17 (1980): 233-45. Ed.

prowess and tradition. Even late in the nineteenth century, the lumber boss who could break a log jam, knock a man down and lead a song, could expect more enthusiastic production than the flabby competitor whose range of interest was from cover to cover of the account book. Nevertheless, the future belonged to the latter.[4]

Fishing continued to vegetate on the Atlantic, and many lumbermen pursued their occupation westward to the Pacific, but the old staples withered away in Canada, and all these attitudes and skills became obsolescent. The populations that had cradled them had then to rid themselves of old ways and submit to new disciplines. The problem was much larger than this, however: the whole Canadian population was required to abandon the sociable, traditional, leisurely ways that had marked the age of staple production, and to adopt a grimmer discipline, to pursue change instead of stability, as they made an industrial society.

While Canada suffered losses in its economic development from the devaluation of its own obsolete skills, it was assisted by an enormous subsidy of imported skill brought by British artisans and professional men. These were the finished products of the most advanced school of industrial capitalism in the world. The pain of transition had all been suffered, and the cost of training had all been paid for, in another country. Immigrant artisans came prepared to put their abilities to use immediately; and, because the British labour market was usually congested, they came readily and at reasonable rates. Canada was exporting skilled persons, too, especially to the woods of Michigan and the farmland of the Middle West; but she probably realized a net advantage of the exchange and certainly enjoyed a very large advantage as far as the movement of industrial skills is concerned. It was thus made almost as easy for Canada to develop industrially as for the United States to acquire canals and railways. In the ruthless description of R.B. Sullivan: "the greatest part of the money which paid for these improvements ... came from London, and the labourers who made them from Ireland, leaving to America the very easy task of employing other people with other peoples' money."[5] In Canada, too, it was British money that paid for the railways, Irish who built them and British workmen who serviced them — indeed, the British were all too willing to provide as well the doubtful benefit of British railway managers. In the case of Canada's manufactures, capital and direction were supplied mostly by Canadians;* but, again, supplies of skilled

---

*As Naylor has pointed out, a not insignificant part of this capital was provided as government bonuses. See Tom Naylor, *The History of Canadian Business* (Toronto: James Lorimer, 1975), vol.2, chapters 12 and 13. Ed.

workmen were drawn very largely from the stream of British immigrants. Easy acquisition of a body of workmen who were skilled in industrial arts and prepared emotionally for the impersonal discipline and exclusively pecuniary criteria of industrial capitalism, allowed a great saving in time and cost.[6] Nor was this advantage limited to the sum of the economies realized in various particular branches of industry. It was a still greater benefit that ready availability of skill lifted the economy over the hurdles that lie in the way of any country that seeks a new economic direction and a new structure. Economic progress is fraught with dangers that the implements required for a mechanized agriculture, or the iron to make the implements, or houses to shelter the ironworkers, or food to feed them, will not appear at the right time, so that the whole transformation is halted for a while in a round of interdependent stalemates. The great solvent that saved Canada from any serious acquaintance with this problem was an indefinitely expansible supply of diverse skills. Its stalemated situations appeared, but they came from other directions.

A key problem was thus disposed of by the immigration of British artisans who were already seasoned members of an industrial capitalist society. There was still the necessity for the mass of native Canadians, and for a large segment of immigrants — in particular, the Irish who crowded the market for unskilled labour—to be made over. Making modern men involved a general reorientation by which new approaches and outlooks were acquired. It required also a special education of workers and employers in the relationships appropriate to wage employment under the conditions of industrial capitalism.

The New Cosmos

The dissemination of a "modern" outlook on life is apparent in Canada in the 1830s, becomes a powerful and even engulfing movement in the 1840s and is largely an accomplished fact in the 1850s. Its stock in trade was built around concepts like progress, "science" and invention, but it involved as well a new, sentimental humanitarianism whose rationale must be pursued deep into the unconscious. The new precepts were sometimes expressed in religious terms and fostered by religious movements, but the new cosmic view was a secular one, sometimes conveniently described as "the spirit of capitalism".

The campaign of the 1830s and 1840s was, in the first place, that of the privileged classes to convert themselves, with professional and some business groups in the vanguard. Conversion of the lower classes, by example, instruction and force, was not neglected even in the 1830s; but this was not the strategic evolution, and it was one that could not be completed quickly because of the number and stubbornness of the masses. What the privileged had first to emancipate themselves from was a pre-industrial conception of social structure and social goals. For example, it was common even in the 1830s for men to talk as if the world was still made up of "masters" and "servants", a notion derived from a rural society, and a British rural society at that. On the same basis, "respectability" rather than competence was the foundation of social approval and public advancement until about 1840. How this conceptual framework obstructed the future, and was challenged by it, can be illustrated by a controversy about education that happened to take place in Nova Scotia, but may serve, nevertheless, to illuminate a Canadian problem.* In 1817, new ideas had spread sufficiently in Nova Scotia that many were prepared to make better public provision for the education of lower classes. There was a storm of opposition, however, on the grounds that education was unsuited to and undesirable for the lower classes; that it was unnecessary for their manual work and would make them discontented with it; that all they needed to know was how to read the Bible; that, if there was to be education of the lower classes, its purpose should be to attach them to church and state; and so forth.[7] About the same time, the members of the Legislative Council of Lower Canada were demonstrating what they thought of any display of initiative or self-improvement among the working classes. These had sought incorporation of the Friendly and Benevolent Society of Mechanics of Quebec, (later, The Friendly Society of Quebec) a mutual benefit society formed in 1810 that enjoyed the approval of some part of the upper classes and certainly of the Legislature. From 1812 to 1821 the incorporation was consistently obstructed by the Legislative Council, perhaps from the same terror of ideas that gripped the British ruling classes at that time, but not without the advice that friendly societies were a commonplace in Great Britain. In 1812 the Council proposed, among other disabilities, that a Justice of the Peace or Sheriff might attend any meeting of the Society, with power to disperse the meeting if any other subject than the business of the Society was raised.[8] In 1821, the Legislative Council still considered that institutions whereby "a numerous assemblage of persons may be convened, ought not to be made permanent in the first instance, in a country wherein there has been no experience to demonstrate their effects."[9]

New ideas and methods were to appear within a decade of this, however, and to gather support quickly. One index of the change is the expansion of public aid in the education of the masses until there was general provision for free schooling in 1852. Another and a very sensitive index is the progress of temperance movements. These ori-

---

*For the Ontario experience, see Leo Johnson, *History of the County of Ontario*, p. 213 ff. Ed.

ginated among business and professional groups in the cities, the first stratum to subscribe to the goals of efficiency and economy implicit in capitalism. Hartz found that the temperance movement in Pennsylvania consisted largely of the effort of these groups, and of industrial employers in particular, to impose abstinence upon recalcitrant workers and backwoods farmers who had no ready market for their grain.[10] The Canadian temperance movement, to which some discussion has already been devoted, followed much the same pattern. In 1810, the pre-industrial device of using liquor to hold and stimulate workers, and to discount their wages, was still in widespread use. In the 1840s and 1850s, with the advantage of a labour supply that was more than sufficient to meet demand, employers strove to force industry and sobriety on the labour force. Parallel demonstrations of new attitudes were the prison reform movement, which was also discussed earlier, and the movement against slavery anywhere — an importation into Canada from the more advanced British Isles in the 1830s and 1840s, but a native phenomenon in the more capitalistic 1850s and 1860s. The proliferation of literary and scientific societies, a sign of the increased interest of the privileged classes in natural science and in self-improvement, is another aspect of the evolution in progress. However, no developments were more germane to the new age, or more in accord with its spirit, than the rise of Mechanics' Institutes and the progress of inventions. Some detail is therefore provided about them.

Mechanics' Institutes Mechanics' Institutes originated in Great Britain, apparently around Glasgow, shortly after 1800. They spread rapidly, particularly through the efforts of British philanthropist, Dr. George Birkbeck, and a London Institute was established in 1823. The object of the Institutes was to facilitate the self-education of workingmen in technical and scientific matters by making available books, apparatus and lectures that would otherwise be beyond their reach. The Institutes thus presuppose the importance of science and invention and the desirability of wage earners' becoming familiar with new technical processes. Hence, the spread of Mechanics' Institutes provides an index of the diffusion of belief in progress and of industrial ambition.

A Montreal Mechanics' Institute was established in 1828 and was probably the first in Canada.[11] Institutes were established in Quebec and Toronto in 1831, and the Halifax Institutes opened in 1832. As the movement spread, Institutes appeared in the 1840s in Dundas, Peterboro, Niagara and Carleton Place, and in the 1850s, in such places as Goderich, Port Hope, Paris and Owen's Sound.[12] This proliferation is not entirely trustworthy evidence of the spread of belief in science and the dignity of labour, for the possibility of obtaining a government grant for the local Institute also played a part. However, grants seem to have been given sparingly and probably inconsistently. Thus, the Brantford Institute, in compalining that the London Institute had been voted a grant in 1848 whereas it had not, pointed out that it (the Brantford Institute) had existed since 1836, and implied that it had never received help in the interval.[13] Applications for aid, indeed, seem to have been pretty regularly refused. Credit for much honest interest in scientific education must be given, therefore, and the Institutes must surely have enhanced that interest.

The Institutes were everywhere started and dominated by "leading citizens", and it does not appear that genuine mechanics ever had much voice in their operations. Nor do mechanics appear very often to have patronized the Institutes or profited by them.* The Institutes rather provided a means for the local upper classes to conduct a rather dilettantish investigation of the new intellectual atmosphere. But this, in itself, was important, and there were exceptions. It was asserted in 1841 that a "large majority" of the 350 members of the Kingston Institute were "mechanics and apprentices" and that lectures were frequent.[14] When the Montreal Institute was reorganized in 1840, because "increased commercial and industrial activity indicated anew the need for adequate training in the practical sciences", it provided not only day classes in academic and technical subjects for the sons of members, but similar night classes for "the apprentices and workmen employed in the city's various industrial enterprises". It was reported that the night classes (not the day classes) showed great zeal, and they were continued for thirty years.[15]

It may be doubted, nevertheless, that the Institutes are significant primarily for their educational work or for the public libraries that stemmed from them. What their establishment did demonstrate was the outlook and ambition of the business and professional groups that sponsored them. The objects of the sponsors were sometimes made quite explicit. Thus, Joseph Howe, in opening the Halifax Institute in 1832, expected that it would promote the growth of manufactures and permit Nova Scotia to catch up with the United States. He attributed American success in manufacturing, in part, to the prevalence of Institutes there.[16] Similarly, it was said on behalf of the Montreal Institute that "the world was then rapidly being industrialized; important technical inventions were being given practical application. The pioneering work of the Institute at that time thus became a vital factor in promoting the progress and prosperity of Montreal."[17] If the ambitions admitted are somewhat earth-bound, they do bespeak a respect

---

*A recent dissenting view is provided by Bryan Palmer, *A Culture in Conflict* (Montreal: McGill-Queen's University Press, 1979), pp.49-52. Ed.

for science, invention and skilled labour as the means to attain worldly success.

Patents of Invention[18] The flow of applications for patents also provides evidence of the rise of faith in mechanical inventions. Applications came usually from English-speaking craftsmen, and while they often represented ideas imported or borrowed from abroad, a substantial number seem to have been original.

So far as can be judged from the records, the first spate of inventing came between 1831 and 1836. Towards the end of this period, patents were being sought at about the rate of one per month. Applications resumed and became more numerous in the 1840s, until they averaged about two per month. Some of the applications dealt with traditional devices — an improved stove was, understandably, a great favourite. However, most inventors proposed either to improve mechanical devices of fairly recent origin or to apply machinery to work hitherto done by other means. Thus, some planned to improve waterwheels (turbines) or ships' paddles (propellers) or to build better steam engines; while others offered machines for grinding clay, for making bricks or for drilling rock. Most numerous were machines designed for an expanding, commercialized agriculture — this was the age of Fanning Mills and Mechanical Reapers, of the "Metallic Coil Spring Tooth Horse Rake", the "Horse Threshing Machine", and of a new implement, the "Cultivator". Nor did these machines exist only on paper: in this age of individual enterprise, there were more village mechanics turning out machines, with a fine indifference to patent rights, than ever thought of applying for a monopoly.[19]

What the Mechanics' Institutes tell about the privileged classes, the inventions tell about more humble men. Their absence from the Institute lectures did not signify, after all, that they lacked interest in science, but only that they preferred practical experiment in a familiar environment. It was the tinkering craftsman, more than anyone else, who brought machinery inside the range of knowledge of ordinary workers and farmers, and taught loving care of it. The fruit of their labour was a machine-minded society by the 1850s, at least in the English-speaking parts of Canada.

The New Labour Relations

The conditions of production in Canada's industrial sector developed in sharp contrast to those of the staple trades. Manufacturing operations tended to be continuous, not seasonal, and output was directed to the home rather than an export market. Variations in demand were narrowed in that market, while industrial production could often be diversified, so that regularity of operations could be approached from two directions. Investment in fixed capital was a larger proportion of total capital in manufactures and transport than it had been in the staple trades, and tended to become the dominant portion, with the consequence that increasing overhead costs provided employers with an additional motive for stabilizing activity. Moreover, investors of industrial capital were more likely to be real organizers of production and directors of the labour force, more directly and continuously concerned with their ventures, than their mercantile predecessors had been. For all these reasons, industrialization was marked by a great advance towards stability.

At the same time, production became more economical in a real sense, not only because labour and capital were used more continuously, but because the multiplication of employers in the cities raised an aggregate demand for labour that was relatively large and dependable.[20] This allowed surplus workmen to await employment in the market with some confidence; and with their presence, a rationalized labour market could develop. A permanent reserve, even if modest in size, assured employers that they need hire labour only when they wanted it, yet still have their most capricious demands met. Individual employers could thus free themselves of labour's overhead costs, which became the responsibility of the whole market and of the whole society. The net effect was a more economical use of labour, achieved by the consolidation of previously individual reserves. Its price was that the relationships between particular employers and their employees were loosened, made transitory and increasingly impersonal.

The effects of the change to industrial production upon workmen were both beneficial and the reverse. The more continuous and economic use of labour; the more generous combination of capital with it; the reliance on a steadier market: all worked to raise very sharply the annual productivity of labour, so that employers had the capacity to pay higher annual wage bills to workers than in staple production. It was the more likely that larger amounts would, in fact, be paid, in that continuous operations involved a continuous succession of paydays — a circumstance beneficial to workmen on other grounds as well. Again, the employers' heavy fixed investments provided the workmen with a certain guarantee of permanence of employment and a certain leverage in disputes; though, on the other hand, the employers' anxiety to increase their fixed capital furnished the main justification for diverting as much as possible of the returns away from labour. Stable employment and regular incomes also allowed wage earners to lead a more settled life, which was typically a family life.* As a stable resident and family man, the worker took on some characteristics of a citizen. Eventually he might think it proper to ask why the town authorities were always drawn from among the rich, why they legislated in their own interest, and why justice was administered so unequally as between rich and poor.[21]

But not all was for the good of the worker. The same rationalization of production that made possible a higher annual income also made a capitalistic labour market. Total demand for labour was indeed more stable than before, but by this very fact, employers no longer had to make long-term commitments to particular workmen, and individual security was less. There were various other ambiguous novelties. Thus, though the security of particular employments declined, the chance of finding an alternative employer increased. Again, families weakened the workmen's mobility and bargaining power but may have stimulated them to fight harder for better conditions. However, the general effect of the rise of the capitalistic labour market, with its perennial surplus of supply over demand, was to make the bargaining position of the individual workman much weaker than before. But some won exemption from this rule also. Favoured by strategic positions, they built disciplined combinations upon the common interests of their crafts and bargained on more or less equal terms with their employers.

What, then, did employers and workmen have to learn to fit themselves for industrial capitalism, and how did they learn it? The answer is not simple, because there was not a single labour force, but sharply differentiated bodies of skilled and unskilled. The different groups learned different things, and what employers learned about each of them differed also.

The skilled workmen, in spite of the conditions of industrial capitalism, and also because of them, occupied a remarkably strong position. Supply and demand were not likely to get badly out of line in their trades, because they had some resources and were fairly well informed, and surplus men were quick to go elsewhere. Besides, there still remained the option in many trades of setting up a proprietorship. The British craftsmen who formed the core of the skilled labour force in the 1850s were mature men, without illusions, and much superior in this respect to some of their callow employers. The tradition of combi-

---

*The stability of nineteenth-century Canadian society is questioned by Katz, *The People of Hamilton, Canada West,* and by Gagan and Mays. Regarding their township Gagan and Mays say: "From first to last, the landscape of Toronto must be described in terms of a way station on a busy highway." David Gagan and Herbert Mays, "Historical Demography and Canadian Social History: Families and Land in Peel County, Ontario", *Canadian Historical Review,* March 1973, p. 35. Ed.

nation and exclusion ran deep and strong in the trades, allowing these men to form tight and highly disciplined units when they were provoked, which were made more formidable by the strategic position of the craftsmen in expanding industry, their concentration in urban centres and the relative weakness of early industrialists. And yet, the most important advantage of the skilled workmen lay elsewhere, in their indispensability. At the critical moment when industrial development was just getting underway, the artisan was the key man who held the new technology in his hands and brain, and it was nowhere else. In the 1850s, there existed no standardized blueprints for building machines, and each was built by a craftsman to his own design, with specifications that developed as he went along. There was no class of technicians outside the crafts, and few employers, who knew how the work should be done — only the craftsman knew that. The function of the skilled workers was a vital one, and employers acknowledged that in their approval of education for the working classes, their support of technical education, and by a good deal of novel talk about the worth and dignity of labour.[22] Labour in the past had been regarded as a species of draft animal and in the future would be regarded as a type of machinery; but, in the middle of the nineteenth century, skilled labour was an important part of the community. Its position was all the more secure because employers needed political support against the trading interest.

These circumstances offer an explanation for the fact that while employers barked at workmen with the ideological clichés of their age, they did not bite very hard. This was true even while industrialization was only a premonition, in what may have been the first strike in Canada for shorter hours by masons in 1823 against the Lachine Canal Commission. The Commissioners thundered that the custom of the country was to work from 5 o'clock in the morning until 7 o'clock in the evening; that in a country with so short a summer as Canada, every hour of daylight must be used; that to concede a twelve-hour day would make workmen believe that "they could dictate to the Commissioners their own terms, by supposing them at the caprice and mercy of those so employed."[23] Yet, while the workmen did not at this time obtain shorter hours, they received a substantial increase in wages.[24] After a quarter-century during which skilled workers seldom found reason to strike, there were again a great many strikes in the 1850s from about 1852 to 1855. Most employers seem to have recognized that these were occasioned by a violent increase in the cost of living, and to have accepted them gracefully or to have denounced them in very perfunctory terms.[25] Even the railway companies, no ordinary employers in either strength nor attitude, temporized with their workmen on many occasions.[26] George Brown was not alone in appealing to "supply and demand"; but few others seem to have applied the doctrine directly in the employment of skilled men, as giving the employer a unilateral right to determine wages and conditions. Only in respect to the sewing machine, a device that made skill obsolete, was there ready agreement that "supply and demand" must prevail.

Skilled workmen, for their part, do not display any very clear philosophy in such of their statements as are preserved, but their outlook may be surmised to some extent from their actions. They did not believe in laissez faire — rather in livelihood — but they accepted the reality of "supply and demand" to the extent that when labour was plentiful in relation to demand, they made little complaint. Unrest and strikes were occasioned most often by increases in the cost of living. However, the great boom of 1853-55 that raised food prices also made labour very scarce in relation to demand, so that strikes could be undertaken with impunity; and this also must be taken into account. Strikes could also be provoked by wage cuts, long hours and a replacement of men by machines. Workers do not seem to have troubled themselves with the question of whether strikes were legal; and employers and newspaper editors were also very uncertain on this point, saying frequently that men were perfectly free to leave their jobs, though they should not interfere with others' taking them. One reason for not pursuing the point was that in serious strikes there was no need to arrest workers for conspiracy, because they were soon open to arrest for assault and damage to property — matters that commanded wider disapproval and that were more easily prosecuted.[27] What workmen knew, as demonstrated in their actions, was that strikes were won by keeping work suspended; and they were prepared to take extreme measures against strike-breakers. Events justified their opinion, on the whole. On the other hand, workmen were usually prepared to compromise differences with employers. They seem to have had no "ultimate goals", unless a livelihood at customary wages and under familiar conditions should be considered one.

Employers and Unskilled Labour

What employers learned about handling unskilled labour, when it was collected in the armies of labourers that built the canals and railways, is an entirely different matter, and a contradictory one.

One strong trend was an increasing appeal to the doctrine of the free (atomistic) market. From about 1830, employers were clearly conscious of the fact that the old world of scarce labour and paternalistic relationships was disappearing, and they seem to have considered this a good thing. They were ready to operate in the new economy of impersonal relationships in which they could hire labour freely for as short or long a time as it was wanted, at the lowest wages necessary to obtain it; and they declared explicitly that labour was a "commodity" in the market like any other. Employers often added that the (individual) labourer was equally free to leave their employ, if he found that advantageous, but that the employer then had the right to replace him on the best terms he could get.

Considering that there was nearly always a surplus of unskilled labour available — and employers were aware of this, too — the new doctrine implied a cold future for labourers. However, the question of their welfare was rarely raised, and then usually to explain (in connection with strikes) why the wages paid were sufficient and why higher ones could not be afforded. Typical enough is "the opinion of the Board [of Works]... that it [the rate of wages] is already too high, in proportion to the value of agricultural produce — that the price of labor should be allowed to be regulated solely by the ordinary principles of supply and demand, and that any official interference with it would be productive of much mischief."[28] The fact is that Irish labourers, at least, came to be regarded as an alien and entirely disposable stratum of society. The Chairman of the Board of Works, Hamilton Killaly, expressed this view most directly in dismissing the argument that rapid completion of the canals in the 1840s would leave an awkward unemployment problem afterwards. On the contrary, he said, "the overstock of labourers [is] a floating mass from the States, who, when their services are no longer required, or serviceable to the province will recross the lines."[29] Thus was jettisoned the old imperial idea that the state owed some responsibility even to the Irish and ought to encourage them to settle in a British country. The average employer did not attempt the sweeping analyses of Killaly, but he felt much the same way about labourers.

The expectation that unskilled labour would find its own, low price in the market was disappointed from the beginning by the actions of the labourers, who struck time after time for higher wages, demonstrated a remarkable determination and power of organization in their strikes, and not infrequently won their demands. Even when they lost, strikes entailed trouble and expense for employers, who had reason to fear also for their lives. It was never made very clear whether employers and government officials considered that laissez faire encompassed the right of many employees to withdraw their labour at the same time. Sometimes it was said that they had this right, but were foolish to use it. On the other hand, there were rather obscure statements about the actions of the workers being illegal, and strike leaders certainly were considered subject to prosecution, if they could only be identified. What the employing interest did emphatically assert was that employers had the right to hire replacements for strikers, and that strikers ought not to interfere with those (usually declared to be numerous) who wanted to continue at work. Attempts to resume operations during a strike nearly always provoked violence, and the various questions of right dropped into the background as questions of physical attack came to the fore. The final arbiter of most of the disputes was not abstract right but physical force: the power of the massed labourers to do violence against the similar power of the troops that employers were able to call to their assistance.

In this setting, notions about the atomistic market as the governor of labour matters were pushed aside in favour of a system of state intervention. "Labour relations" in the period when canals and railways were built meant troops and mounted police to "overawe" the labourers, government spies to learn their intentions, and priests paid by the government and stationed among the labourers to teach them meekness. Here was a full program of intervention, immediately on behalf of contractors, basically to promote economy for the state by encouraging contractors to make low bids in the expectation that low wages could be paid, and then enforcing the low wages.

The practice of intervention got underway before laissez faire was widely favoured and at the direction of a man who believed in it less than most, Colonel John By. By was a paternalistic employer, who assumed a broad responsibility for the welfare of the men who built the Rideau Canal, especially in respect to the provision of housing and medical care.[30] He was also dictatorial, not at all prepared to negotiate with his employees or countenance strikes by them. There were at least three strikes by the labourers in the spring of 1827 (quite possibly against a cut in wages from their previous level) and a general aura of riot.[31] Then, in June 1827, the first of By's two companies of Sappers and Miners arrived from Britain. These appear to have been raised on the assumption that skilled workers would be scarce on the Rideau, but the men were not needed or used much for actual construction. Nevertheless, By said, they were "most usefully employed; their presence on the ground enables me to check the disorderly conduct of the Labourers."[32] After the troops came there was no more open rebellion on the Rideau. It looks also as if every official and employer in Canada knew about and admired the By method of dealing with labour — in its authoritarian, not its paternalistic, aspects. For decades afterwards, the first instinct of every contractor who faced opposition from his labourers was to send for troops; and every director of a public work dreamed of how, if he only had a few companies at his disposal, he could play the role of a Colonel By.

These tendencies were prominently in view when disputes broke out along the Cornwall Canal in the fall of 1834.[33] The Canal Commissioners belonged to that middle-class element about to assume control of the country under the name of "Responsible Government" — not very honest, without much sense of responsibility towards other people and extremely aggressive. They called at once for troops to be stationed on the works, and thought that they were going to get them. However, they ran into a check from members of an older and more aristocratic generation. John Beverley Robinson's opinion, upheld by the Governor, was that it was open to the contractors to negotiate with their men or to hire others, and that it would be improper to send troops merely to force the labourers to terms; troops should be used only if they were clearly required to preserve the peace.[34] This view infuriated the Commissioners and it was never heard again. A second and more permanent check came from another pre-industrial source: the armed forces themselves. From the 1830s to the 1850s, while troops were regularly used to suppress labourers, commanding officers were protesting just as regularly that this was not a proper function of the army and that police should replace them.[35] A third check upon instruments of repression was their expense.

The real formation time of the Canadian system of intervention was in the 1840s, when the Canadian government strove to complete its vast canal system, and to spend its loan of one and a quarter million pounds, in the shortest possible time. Near the beginning in 1842 at the Cornwall and Beauharnois Canals, the government employed men directly, and another kind of intervention was practised, by which the directing engineers looked after housing and the general welfare of the labourers. Under these conditions, it was testified, the workers were extremely well behaved;[36] nevertheless, this successful experiment was not repeated. Instead, nearly the whole of the construction program was put out to private contractors, in the name of economy. Economies may well have been achieved, for contractors appear to have taken work at very low prices. But they had in mind to recoup themselves by paying as little wages as possible, and, as a result, the succeeding years were marked by chronic unrest and a succession of strikes. One of these might involve five thousand men, whose instinct was direct action; and violence was common against contractors, against strike-breakers, or even by one part of the work force against another. Moreover, people living near the works had an understandable fear of these masses of sullen and violent men, who were inclined when they were hard pressed to take what they wanted without asking or paying. The objects of government attempts at suppression, then, were to keep labourers to their work at low wages and to pacify local residents.

The first intervention came as an aftermath of the strike at Lachine in 1843.[37] A small force of police was recruited and stationed on the Lachine Canal; and trouble being anticipated at the same time at the

Beauharnois Canal, a Stipendiary Magistrate with ten men was also placed there.[38] In addition, the Board of Works asked that a small force of troops be stationed on the Beauharnois Canal,[39] and arranged for "the Special services of a Clergyman" (the Rev. Mr. Falvey) who was located at the same place "to prevent disturbances there".[40] All these steps were undertaken in haste and were apparently considered to be temporary measures; but the police forces became relatively permanent institutions, engaged chiefly in gathering information about the intentions of the labourers,[41] while Falvey, "as he had much influence amongst the Men", was retained for about two years, and perhaps longer.[42]

In 1844, another police force was placed on the works at Prescott;[43] and in April of that year, in defiance of orders to the contrary, 600 labourers from the Lachine Canal struck work, marched into Montreal for the current elections and ensured the election of their favoured candidate (Drummond) by surrounding the polling places.[44] Nevertheless, and in spite of the annoyance of officials by the election maneouvre, there was a relaxation of tension in 1844 until another election occurred in October. Again the labourers entered Montreal to support their candidate; and a force of about 400 of them, of whom some had firearms, clashed with a body of cavalry.[45] No one was hurt, and the twenty-seven labourers arrested were freed by a Grand Jury.[46] However, the events of this election had thoroughly incensed the officers of government, and another step of intervention was taken, based ostensibly on the labourers' possession of firearms.

Some labourers did possess firearms. An engineer had persuaded the labourers at Beauharnois to turn their weapons over to the government in 1843; but these had been returned in 1844, partly because there was no legal basis for keeping them.[47] These arms had sometimes been the instruments of bloodshed, and might be so again. But to take them forcibly from the labourers would be a highly provocative act, not carried out easily. Here is the background of a legislative act of March 1845, "for the better preservation of the Peace, and the prevention of Riots and violent Outrages at and near Public Works while in progress of construction".[48] It provided, first, that firearms must be registered and that possession of unregistered firearms was an offence — but only on designated lines of public works. Thus, a special kind of law was established for labourers, as distinct from any other persons. Secondly, the act authorized "the raising and employment of a mounted police force to be stationed at such parts of these works as may be found necessary . . ."[49] Here is the inspiration of later mounted police forces. Subsequently, some occasional and mostly ineffective attempts were made to seize firearms from labourers. However, the main effect of the act, and probably the effect mainly intended, was to place mounted police forces on the chief public works. Other means of control were not neglected, however; for troops might still be called for, and additional Roman Catholic priests were appointed as "spiritual agents" to look after the main bodies of labourers.[50] Of these unhappy men, only one dared to turn on his paymasters and denounce the shameless exploitation of his charges.[51]

Even more instructive than these developments are the changes that occurred in the 1850s, when the main "public works" were carried out, not by a public agency, but by private railway corporations. The Board of Works had paid the piper for police and priests, and it had called tunes that were only sweet to employers; but it was, at least, a public agent answerable to the government and the Legislature. As far as particular contractors, mayors and Justices of the Peace were concerned, many public officials viewed their demands with a critical eye, for their ruling-class background endowed them with a suspicion of shopkeepers and a degree of sympathy for peasants. When strikes occurred, officials frequently asked what had caused them; they sometimes denounced the contractors for their grasping ways; and they supplied troops with reluctance. There were to be two profound differences in the railway age. First, aided by a certain confusion in the public mind concerning the status of these quasi-public agencies, private railway corporations were to assume the right of the state to keep and direct police forces. Secondly, when the manager of a large private corporation demanded troops, officials learned not to ask whether they were needed, but rushed a force off as soon as possible.

The new order was foreshadowed when the Montreal Mining Company and the Quebec Mining Company, most important of the firms produced by the mining boom of 1846, asked for the support of troops at their Lake Huron properties in November 1849. It was claimed that employees and property were in danger of attack by armed bodies of Indians and half-breeds, led by a few white men. Officials reacted to this demand in much the same way as they treated the plea of an excited Justice of the Peace. In such cases, it was the custom to say that troops could not be sent unless the danger was real and immediate, and not merely apprehended. The evidence in the public record of danger on Lake Huron is no more than shadowy hearsay. It was usually said also, though it could not be said in this case, that troops could not be used until the resources of the local civil authorities were exhausted. Finally, it was common enough to say that troops could not be spared from other duties. In the case of the mining companies' demand, the commander of the forces considered the logistic problems associated with sending troops into the northern wilderness just as winter was setting in, and flatly refused. But officials were now to learn something about the power of private companies that represented substantial wealth. The companies went directly to the Governor General. They succeeded in so alarming him that he overruled his subordinates and ordered that troops be sent. Sent they were, to pass a very uncomfortable winter at Sault Ste. Marie.[52]

The next step came at the beginning of 1851, when construction of the Great Western Railway was underway in the vicinity of Hamilton. Representatives of that city and its environs petitioned the government to station troops in Hamilton for their protection. They pointed out that a great number of labourers were congregated there and that they had already conducted two strikes for higher wages, during which large bodies of them had marched about armed with clubs, and they asserted that the civil power was incapable of coping with such groups.[53] The government's replies made two points: first, that the use of police should precede the use of troops, and that Hamilton ought to equip itself with a police force; secondly, that troops could be sent only if a suitable building for them to occupy was provided.[54] Hitherto, local governments usually had dropped their requests at this point, for they were typically appalled at the thought of paying for even one policeman. Hamilton, however, was raising a special police force of 27 men within two months; and within four months, a suitable building had been erected and was occupied by a military force of 80 men.[55] All this had been accomplished not only expeditiously, but smoothly and cheerfully. It was plain that a new force was at work — the Great Western Railway — and it seems evident, though there is no document to prove it, that it was the Railway which had undertaken to pay most or all of the costs of policemen and barracks. And, since taxation without representation is inequitable, the committee that governed the new police force included not only the Mayor of Hamilton and the wardens of local counties, but "a Member of the Rail-Way Board" as well.[56] The Railway's influence is indicated, too, by the alacrity with which the government had sent troops, though there was no evidence of any challenge to the civil power.[57]

What was done at Hamilton was legalized and generalized in two pieces of legislation of August 1851. These, if not drawn by a railway official, were drawn with an eye to his wants. The first act[58] renewed for three and a half years the act of 1845, which had created "The Mounted Police Force". There were important modifications, however. The provisions for a special kind of law and for special policing which had hitherto been applicable to canal lines, could now be extended to the line of any railway under construction by an incorporated company. And, just as the Board of Works had been chargeable with the costs of police in the original act, these costs might now be recovered from a railway company. A complementary act[59] was passed at the same time (for five years) providing for the creation of a force of up to 500 military pensioners to be used as local police. These men could be called out, in any desired number, by local mayors and wardens. Thus, manpower was provided to make the other act effective. It was now possible for a railway company, if it could get the cooperation of a municipal officer and if it would agree to bear the expense, to place a quasi-private and quasi-military police force at any point along its line of construction.

Some events of 1853 may illustrate how this legislation could be used. Construction of the Great Western Railway was then being carried on actively near London. However, there had been one or more strikes and some unsuccessful attempts to arrest some of the labourers for "riot" — meaning, perhaps, that they had threatened or dealt roughly with strike-breakers. C.J. Brydges, Managing Director of the Great Western (and later, manager of the Grand Trunk), next applied to the Court of Quarter Sessions of Middlesex and Elgin to call out a police force. The Court, in view of the Railway's willingness to bear expenses, was easily persuaded. It applied to the government for a warrant to call out the force, appending Brydges's statement as a sufficient reason.[60] The provincial government issued the warrant and sent warrants to several other mayors and wardens at the same time.[61]

This was in August 1853. The labourers appear to have been quiet when the warrant arrived, and nothing was done immediately. However, the power bestowed by the warrant was used in January 1854, when the Mayor of London called out a force of pensioners "to aid in the preservation of the peace".[62] The dispatch of the officer in command of this force[63] makes it very clear that no disturbance had occurred and that none seemed to be impending; all that was in progress was a peaceful strike a few miles outside London, which the workmen seem to have abandoned when the troops appeared among them. The officer thought that he should then withdraw, particularly because he was several miles outside the area in which the government's warrant gave him legal authority to act. The Mayor considered, however, that it would be best to have troops remain awhile, to prevent "any further interruption".[64] In keeping with this view, ten men were left on the works for several days, in a building provided by the contractor. Not less significant is the procedure by which the detachment was withdrawn. It was called back after about a week, when the Law Agent and the Chief Engineer of the Great Western Railway decided that its presence was no longer necessary. However, also at the recommendation of these officials, some of the men were retained as special constables under the orders of the Mayor of London.[65]

Thus did a doctrine that wages ought to be determined in the market at the lowest rates for which men could be got, lead on to a regime of intervention, by which the state sought by force to suppress any resistance to the unilateral determination of wages by employers. The additional peculiarity of the 1850s was that a new type of employer had appeared in the railway corporation, that was more powerful than the state and able to shape the state to meet its wants.[66] In keeping with this shift in power, railways assumed some of the state's prerogative to dispose of the force. Interventions by the Canadian state on behalf of employers in more recent years should not be regarded as novelties, but as fruits of what employers and officials learned about "labour problems" in the middle of the nineteenth century.

There is the question, finally, of the unskilled Irish labourers, and what they learned. They learned that work was irregular and uncertain, and rare in the cold winter. They learned that hours were from five in the morning until seven in the evening, or longer. They learned that wages were usually 50 cents a day, often 40 cents a day, rarely 60 cents a day; that they were paid only after a long period of waiting and often discounted by store pay. They learned that the provision of housing on the public works usually consisted of letting the labourers erect such shelters as they could on waste or swamp land. They learned that many hundreds of their kind died of disease, especially on the Rideau and the Welland Canals. They learned that contractors, officials and most other Canadians only cared to make them work cheaply, and concerned themselves not at all about their welfare.

They learned that their only protection lay in cohesion among their own people. They might have learned, but they already knew, how to conduct effective strikes that depended upon able leadership and disciplined obedience of the rank and file. Scarcely ever did the labourers expose their leaders to the knowledge and arrest of the authorities, and rarely did their ranks break. It is impossible to say how much they won and lost by their strikes, but they won their demands a good many times, and the dread in which they were held may have prevented many impositions. As for their habit of violence, it afforded a number of short-run advantages. It disposed pretty effectively of the danger of strike-breaking. It saved many from arrest on charges that were valid only from an employer viewpoint. It had some point in negotiations: it was difficult for an employer to say "No" when he was surrounded by a number of armed ruffians who promised to beat his brains out. In Canada, however, unlike Ireland, the law was not altogether regarded as an alien oppression, and the settlement of differences by physical contests was not widely approved. In the long run, by provoking repressive measures, violence may have done more harm than good to the labourers. What might be said for them is that they learned to be increasingly judicious in their use of it. In the 1830s and even in the 1840s, violence appears often as an unconsidered and indiscriminate response to unfair treatment. By the 1850s, the labourers had learned to wait, and they used force selectively, when it would do the most good and the least harm.

Another thing the Irish labourers learned was to act less like tribesmen and more like a nationality or class. They had the bad habit of carrying on civil war among themselves, usually between the Cork (Munster) and Connaught parties, which may each have included some other elements. There was a basis for division, not unlike that between skilled and unskilled labour, in the fact that southern Irish were more advanced, more disciplined and better armed, while Connaught produced the least-taught wild men of the west. There was a basis, too, in the fact that often there was not enough work for both parties: then the stronger party sometimes drove the others away in order to monopolize the jobs.[67] All the same, this division was a poor preparation for presenting a united front to contractors, and a poor advertisement also for the cause of Ireland, to which all the labourers appear to have been passionately devoted. The authorities had no liking for this internecine strife either; for besides the interruptions to work which it occasioned, it produced the most unpredictable and uncontrollable of civil disorders. The labourers seem gradually to have overcome their division by what must have been an admirable statesmanship, and not without what looks like a brilliant contribution to unity, in 1844, by the Rev. McDonough, who was the government's "moral agent" on the Welland Canal.[68] At the same time, the Irish were gradually assuming a place in the political structure of Canada, little as their contribution to election struggles was welcomed by some.

The labourers were also learning, and helping to form to an extent, the rules of the game in a capitalistic labour market. Contests about wages were not new to them, and what they had to learn did not relate to principle but to the methods of effective combination in Canada. Store pay, delayed payment of wages and the exactions of moneylenders were not very new either; but revolt by the labourers led to a more prompt payment of wages and avoidance of company stores along the public works.[69] One thing that was obscure to the Irish, and obscure to anyone else with a different viewpoint from that of the capitalistic market, was why it was permissible within some limits to argue about the rate of wages, the hours and some of the conditions of employment, but never about the quantity of work to be offered, which was a matter completely within the discretion of the employer.

Sometimes they tried to bargain for the quantity of work to be provided — the right quantity being enough to employ all the available
labourers.[70] Their approach did not rest only on a pre-industrial
conception of livelihood, though it probably owed something to that.
But the fact is that in large-scale canal and railway construction by
hand methods, the amount of work that might be carried on at any
one time was almost indefinitely expansible. More than that, as the
labourers knew very well, the quantity of canal work offered had
frequently been expanded or contracted for reasons unknown to
economic calculation. For example, the bankrupt province of Upper
Canada had kept work going on the Cornwall Canal in 1837 to minimize distress and the danger of disturbances.[71] In 1844, when the
labourers wanted to take part in elections and officials wanted to
prevent them, the work available had been deliberately expanded at
election time, in order to take up all the available labourers and leave
no excuse of unemployment for their entry into Montreal.[72] Then,
when the labourers participated anyway, the Board of Works suspended work for a time as a punishment.[73] Again, the Board sometimes
closed down a work when the men struck, rather than consider their
demands.[74] In view of this flexibility, it should not be surprising that
labourers occasionally tried to stipulate for a sufficient amount of
employment. Their attempts met with no success and seem to have
been abandoned by the 1850s. Evidently the labourers had learned
that in the capitalistic market, for whatever reason, common labourers
had no voice in determining the amount of work to be offered them.

The Irish peasants came to Canada as strong and willing labourers,
but men profoundly untutored in the ways of an industrial economy.
They brought many troubles on themselves by their own ignorance
and lack of self-discipline, and many more were thrust upon them as
the predatory and unscrupulous sought to exploit them. But they
contributed much: they did the heavy work, and built the canals and
railways, and made the well-supplied market in common labour that
supported industrial capitalism. They taught much: that there was
not, after all, an atomistic labour market; that beyond a certain point
of exploitation, labourers would combine and revolt; that it was sometimes necessary to negotiate terms rather than dictate them. They
learned much: that the rules of capitalism allow some discussion of
wages, but none of employment; that unity, to be very effective, had
to encompass all labourers; that life in a capitalistic society demanded
a more calculating, more informed and more disciplined behaviour than
they had been used to. The labourers of the 1850s were pretty rough
still, but they were far more sophisticated than their counterparts of
twenty years before, and were close to becoming citizens of the modern

world. The next generation proliferated into the diverse employments offered by an industrial society, became indistinguishable from anyone else and made another contribution — to the conformity that seems to be the price of economic progress.

Notes

Chapter I

1 J.M. LeMoine, "Slavery at Quebec", *Canadian Antiquarian and Numismatic Journal*, April 1872, p. 158.

2 Allana G. Reid, "The Development and Importance of the Town of Quebec, 1606-1760." (Ph.D. thesis, McGill University, 1950), pp. 129-30, ascribes the agitation to merchants. However, in the *Report of the Canadian Archives, 1899*, p. 86, the statement is "that the farmers have made up their minds to bring in negroes to work their farms".

3 Alice J.E. Lunn, "Economic Development in New France, 1713-1760" (Ph.D. thesis, McGill University, 1942), p. 12.

4 J.N. Fauteux, *Essai sur l'industrie au Canada*, 2 vols. (Quebec, 1927), vol. II, p. 477. (Fauteux's reference to the *Canadian Archives Report, 1904*, p. 28, should be to app. K. in that volume.)

5 Reid, *op. cit.*, p. 130.

6 Lunn, *op. cit.*, p. 12.

7 "L'Esclavage" in *Dictionnaire général du Canada*, ed. R.P.L. Le Jeune (Ottawa, 1931), vol. I, p. 599.

8 Lunn, *op. cit.*, p. 12.

9 *Ibid.*, p. 14.

10 *Ibid.*, p. 13.

11 *Canadian Archives Report, 1899*, p. 142. Reid, *op. cit.*, p. 130, states that this ordinance covered negro as well as Indian slavery.

12 Lunn, *op. cit.*, p. 192.

13 *Ibid.*, p. 13.

14 L.H. LaFontaine and M.J. Viger, "De l'esclavage en Canada". *Mémoires et documents relatifs à l'histoire du Canada publiés par la Société historique de Montréal* (Montreal, 1859) pp. 8-9.

15 Chief Justice William Smith will serve as an instructive, though scarcely typical, example. Writing from Quebec to his wife in New York in 1786, he discussed his numerous hired servants, who altogether would cost him £100 sterling per year, and then, "the lower servants who, I think, must be negroes from New York, as cheapest and least likely to find difficulties.... If you bring blacks from New York with you, let them be such as you can depend upon. Our table will always want four attendants of decent appearance." (Quoted in J.M. LeMoine, *Picturesque Quebec* [Montreal, 1882], p. 388.)

   The marketing of slaves and indentured servants at Quebec and Montreal, 1764-83, is shown in H.A. Innis, *Select Documents in Canadian Economic History, 1479-1783* (Toronto, 1929), pp. 466-68.

16 William Canniff, *History of the Settlement of Upper Canada* (Toronto, 1869), p. 575.

17 *U.C. Land Petitions A-Al: 1792-1840* (P.A.C.), Affidavit of Elisha Anderson, April 14, 1796: concerns runaway slaves in Upper Canada taken in charge by Indians; William Dummer Powell, "First Days in Upper Canada" (MS, P.A.C.): concerns slaves seized by Indians from the Leforce family and sold at Detroit about 1780.

18 *Census of Canada*, 1871, vol. IV, p. 74.

19 Canniff, *op. cit.*, p. 570.

20 *Ibid.*, p. 573. Scadding posed the nice question of the legality of the enslavement of an Indian girl at Niagara in 1802 (H. Scadding, *Toronto of Old: Collections and Recollections* [Toronto, 1873], p. 295).

21 There is a precise statement of the law in Robert Gourlay, *Statistical Account of Upper Canada...*, 3 vols. (London, 1822), vol. I, p. 240. The law also prohibited indentures of over nine years duration. Evidently it was already the fashion in Upper Canada, as elsewhere, to replace slave status with an indenture for life or for some shorter period (Thomas Conant, *Upper Canada Sketches* [Toronto, 1898], p. 128). A very odd case appears in *Upper Canada Sundries* (P.A.C.), Oct. 10, 1819, the petition of Eve Fry. Eve was the old (76 years) slave of one Robinson, of Kingston. She had received her freedom for nine years "slavery". She petitioned on behalf of her female child who was held by Robinson as a slave "though born free".

22 Canniff, *op. cit.*, p. 576.

23 There is a review of these events in L.A. Paquet, "L'Esclavage au Canada", *Transactions of the Royal Society of Canada*, section I, 1913, pp. 139-49, but the clearest record is in the documents and commentary collected in La-Fontaine and Viger, *"De l'esclavage en Canada"*. Scadding (*op. cit.*, p. 294) thought that the Lower Canadian courts acted in emulation of Mansfield in England, who had held slavery to be incompatible with English air. However, so far as the Lower Canadian courts pretended to have any legal basis for their actions, they referred to an imperial act of 1797 which removed slaves from the list of chattels which could be seized for debt in British possessions in America (LaFontaine and Viger, *op. cit.*, p. 51). It is significant also that the Chief Justice who presided over a key case (Osgoode) of 1800 had served previously in Upper Canada (*ibid.*, pp. 62-63).

24 Ida Greaves, *The Negro in Canada* (Orillia, n.d. [1930]), held (ch. 1) that the Canadian climate was no more detrimental to negroes than whites. The facts seem to be on her side, but it is important as well that the question whether negroes could survive in Canada was the single one raised by authorities in France when they considered negro importation, both in 1689 and 1721. L.J. Greene, *The Negro in Colonial New England* (New York, 1942), pp. 224-25, states that the severe climate of New England really was supported with difficulty by negroes; though this handicap did not prevent wide use of them there.

25 The employment of slaves as craftsmen in the British Caribbean is noted in L.J. Ragatz, *The Fall of the Planter Class in the British Caribbean, 1763-1833* (New York and London, 1928), p. 25.

26 Greene, *The Negro in Colonial New England*, pp. 100-123.

27 The figure of "one-third" was put forward too regularly to indicate exactness. It is found, for example, in I. Finch, *Travels in the United States and Canada* (London, 1833), p. 236. Finch describes a most suggestive experiment in South Carolina, about 1833, by which it was hoped to overcome the inattention of the slaves. The slaves were given individual plots and huts and told that upon completing a prearranged quota of general field work, they might devote whatever time was left to their own plots. It is striking that this case reproduces the problem, and to a large extent, the answer, that led to the rise of serfdom in the declining years of the Roman Empire. Somewhat analogous to this was a "task system" that found some use in Rhode Island (Green, *op. cit.*, p. 106).

28 Slaves, the generally available labour supply of the American South, were used for all sorts of tasks in the eighteenth century. They raised the first coal mined in the United States, near Richmond, 1750 (McAlister Coleman, *Men and Coal* [New York, 1943], p. 4). It was they who dug a canal to extend the navigation of the James River (Finch, *op. cit.*, p. 263). They were used extensively in southern ironworks (Carl Bridenbaugh, *The Colonial Craftsman* [New York and London, 1950], pp. 16-18). Indeed, Bridenbaugh indicates that in the eighteenth century the dearth of skilled free labour led to many slaves being trained in some degree as artisans, and that the tendency then was for these slave artisans to drive their teachers out of the South. But he makes equally plain how dependent the South often was upon the availability of genuinely skilled (free) artisans and how anxious its leading men were to have more of them; and he suggests, what I think is true, that as free labour became more plentiful in the nineteenth century, and even allowing that the South never had a large share of it, the trend in the southern states was against slave artisans and in favour of free ones (*ibid.*, ch. 1 especially pp. 15, 29-30).

29 Louis Hartz, *Economic Policy and Democratic Thought: Pennsylvania, 1776-1860* (Cambridge, Mass., 1948). The unsuitability of slavery is noted, p. 181.

30 Earl of Selkirk, *Selkirk Papers* (P.A.C.), vol. 75, p. 214. It is noted also that New Yorkers thought slaves to be the worst servants.

31 It has been stated that four negro families were living in the Western District in 1817 (H.A. Tanser, *The Settlement of Negroes in Kent County, Ontario* [Chatham, 1939], pp. 33-34). However, Gourlay's *Statistical Account...*, which is very detailed and has a very full coverage of the Western District, fails to mention negroes (in 1817). Thomas Rolph, *Emigration and Colonization...* (London, 1844), p. 313, refers to a petition of the magistrates of the Western District of 1840 which refers to the negro immigration as occurring during the five preceding years.

32 A colony near Amherstburgh, dating perhaps from the early 1830s, was described by the whites as small and orderly. A much larger one, the "Black Settlement", formed "some years" before 1841, was "well known to the Coloured people in the U.S. as they have been arriving in considerable numbers lately (June 1841) directly to this place" (*P.S.O., C.W.*, 1841-3, no. 534). Another colony was established by two missionaries about the beginning of 1842 (*P.S.O., C.W.*, 1843-4, no. 6214). The Buxton Settlement, established later, is described in Kent Historical Society, *Papers and Addresses*, vol. IV, 1919, pp. 40-44, by W.N. Sexsmith. Tanser, *Kent County Negroes*, contains considerable information concerning the Buxton Settlement, the Dawn Settlement and negro life in Chatham. In the London District, the Wilberforce Settlement was established in the 1830s; but "though much has been said about (it), it has never flourished, and is now inferior to several other settlements" (Rolph, *op. cit.*, p. 311). Rolph's authority locates a second settlement in the London District in 1843. The petition of about thirty negroes of St. Catharines who want to buy land for a colony is found in *Upper Canada Sundries*, Dec. 20, 1828.

33 Rolph, *op. cit.*, p. 311.

34 Scobie and Balfour, *Canadian Almanac*, 1850. But the returns upon which, presumably, these figures were based, showed only 1,000 negroes in the Western District and 300 in the London District in 1841 (*P.S.O., C.W.*, 1841-3, nos. 19 and 534).

35 *Canadian Almanac*, 1853.

36 W.L. Mackenzie estimated the negro population of Upper Canada at 12,510, and that of Lower Canada at 5,100 in 1840 (Mackenzie's *Gazette*, Rochester, June 1, 1840, in *P.S.O.*, *C.W.*, 1840-1, no. 1480). Rolph, who should have had a very good basis for estimation, thought there were nearly 20,000 negroes in Canada in 1843 (Rolph, *op. cit.*, p. 310). Landon's estimate of 30,000 to 40,000 negroes in Canada might be appropriate for the 1850s (Fred Landon, "Social Conditions among the Negroes in Upper Canada before 1865", Ontario Historical Society, *Papers and Proceedings*, XXII [1925], p. 144).

37 In 1840, two Western District juries "...(and among them some very respectable men in point of property)...refused to convict in the face of the clearest testimony, when parties were indicted for a Riot and pulling down a small building belonging to Coloured people in Amherstburgh..." *P.S.O.*, *C.W.*, 1840, no. 1948.

38 *P.S.O.*, *C.W.*, 1840-1, no. 1948; 1841-3, no. 1680 (doc. missing — the Nelson Hackett case); 1843-4, no. 5949.

39 *P.S.O.*, *C.W.*, 1843-4, nos. 5601 and 5949.

40 *P.S.O.*, *C.W.*, 1846-7, no. 15696 (doc. missing).

41 *P.S.O.*, *C.W.*, 1841-3, no. 4958.

42 *P.S.O.*, *C.W.*, 1843-4, nos. 6529 and 6557 (docs. missing).

43 *P.S.O.*, *C.W.*, 1841-3, no. 4958.

44 *P.S.O.*, *C.W.*, 1856, no. 596.

45 Landon, *op. cit.*, pp. 147-48.

46 Rolph's account would indicate more severe discrimination, if anything, in 1843. From the "strong and unconquerable aversion on the part of the white inhabitants", he says, the negroes "are excluded from the public schools; they are appointed to no public situations; they have great difficulty in obtaining land..." (Rolph, *Migration and Colonization*, p. 310). Writers from the region — i.e., Landon, Tanser — are satisfied to establish that negroes preferred Canada to the United States, but seem anxious to minimize the disabilities imposed on them.

47 *P.S.O.*, *C.W.*, 1853, no. 913. It was stated in this document that a substantial part of the negroes entering Canada were not escaped slaves, but old or infirm persons driven off by their masters because they were no longer useful.

48 Landon, *op. cit.*, p. 155.

49 *P.S.O.*, *C.W.*, 1840-1, no. 1948; 1843-4, no. 5949.

50 R.W. Camm, "History of the Great Western Railway of Canada", (M.A. thesis, University of Western Ontario, 1947), pp. 95-96. The source cited is the Chatham (tri-weekly) *Planet*, Oct. 8, 1858.

51 Rolph, *op. cit.*, p. 311.

52 *P.S.O.*, *C.W.*, 1841-3, no. 4958. Rolph's correspondent put it that "Prejudice in this country, as in the States, obtains rather among the ignorant and vicious than among the intelligent and respectable" (Rolph, *op. cit.*, p. 312).

53 Landon, *op. cit.*, p. 150, citing the "Anti-Slavery Reporter", 1852.

54 Oscar Handlin, *Boston's Immigrants, 1790-1865* (Cambridge, Mass., 1941) demonstrates particularly well the weak position of Irish workmen in the

United States that underlay their hostility to negroes and abolition. See, for example, pp. 67-68.

55 J.G. Rayback, "The American Workingman and the Antislavery Crusade", *Journal of Economic History*, November 1943, p. 152.

56 The rise of tobacco culture is noted in R.L. Jones, *History of Agriculture in Ontario, 1613-1880* (Toronto, 1946), pp. 40-41. There was a rather high proportion of slaves at Detroit under the French regime (Lunn, *op. cit.,* p. 47) and the same appears to have been true in the early years of British rule.

57 The difference in wages stands out in wage surveys — see Gourlay, *Statistical Account*, vol. I, pp. 610-11. Amusing evidence of the labour scarcity in the Western District is found in *Upper Canada Sundries*, Jan. 5, 1828 (North to Hillier). It had been discovered that the officers of the detachment of troops at Amherstburgh were using their men for family servants. Their excuse was that it was next to impossible to obtain servants of any other kind in the area.

Local labour shortage led to another unusual expedient, the employment of Indians. Farmers employed them for harvest work in the years before 1817 at least (Gourlay, *op. cit.,* p. 297). The military authorities had found it desirable to hire Indians to repair a schooner and to build a house and forge at Amherstburgh in 1797. (*C105*, P.A.C., p. 141).

58 Rolph, *op. cit.,* p. 316.

59 *P.S.O., C.W.,* 1841-3, no. 534; 1850, no. 1222.

60 Rolph, *op. cit.,* p. 312.

61 Rolph's great project to transport a large body of Canadian negroes to Trinidad fell through, though this does not seem to have been from lack of support by the negroes (Rolph, *op. cit.,* pp. 309-19; also *P.S.O., C.W.,* 1840-1, nos. 3350, 3351). But Rolph spoke familiarly of negroes who before 1843 had gone to Jamaica and to Trinidad (*loc. cit.*). There was a substantial movement of negroes to Haiti in 1861 and at some earlier period (Camm, *The Great Western Railway*, p. 96). Landon's statement (*op. cit.,* p. 158) that negroes showed no disposition to leave Canada before 1860 is clearly wrong.

62 Landon, *op. cit.,* p. 158.

63 M.L. Hansen, *The Atlantic Migration, 1607-1860...* (Cambridge, Mass., 1940), pp. 44-52, 98, 105. R.G. Lounsbury, *The British Fishery at Newfoundland, 1634-1763* (New Haven, 1934), pp. 259-60, describes the very large indenture traffic from Newfoundland to New England. The practice of sentencing various kinds of rebels, especially Irish ones, to this form of slavery, contributed greatly to the rise of the indenture system (Eric Williams, *Capitalism and Slavery* [Chapel Hill, 1944], pp. 13-15; George O'Brien, *The Economic History of Ireland in the Seventeenth Century*, [Dublin and London, 1919], p. 214).

64 Georges Langlois, *Histoire de la population canadienne-française* (Montreal, 1934), p. 60; Reid, *op. cit.,* ch. 2; Lunn, *op. cit.,* pp. 1-12.

65 The flexibility of the indenture form, and the fact that persons in America as well as Europe might feel compelled to indent themselves, is illustrated by the indenture of one Margret Sellinger (or Zelner) "to learn house keeping" for eight years and four months. This indenture was undertaken in 1769 in "Albany county on the Mohack river", apparently as the means

by which a widow provided for her daughter. The indenture is in the *Mitchell Collection* (P.A.C.), vol. III, part II.

66 Contrary to what might be imagined, white slaves received worse treatment than black ones because they represented a less permanent capital investment. Williams, *op. cit.*, p. 17, points this out with particular reference to the Caribbean. In the same way, Greene noted that New Englanders made little distinction between their black and white slaves, but such distinction as there was favoured the negroes (*The Negro in Colonial New England*, pp. 128, 135, 231-32).

67 A.E. Smith, "Indentured Servants: New Light on Some of America's 'First' Families", *Journal of Economic History*, May 1942, pp. 40-52.

68 Though this seems clear from various sources, including Lord Selkirk's writings, it is also true that Selkirk thought that nearly all the Irish (i.e., wage earners) in New York in 1804 were indentured (*Papers*, vol. 75, p. 300).

69 Hansen, *op. cit.*, p. 105.

70 Selkirk, *op. cit.*, p. 105. However, in 1805 Selkirk himself was still using indentured labour and meeting precisely the same sort of trouble (p. 330). The unruliness of indentured servants was also discussed by John Howison, *Sketches of Upper Canada* (3rd ed., Edinburgh and London, 1825), p. 274.

71 John R. Commons and associates, *History of Labour in the United States,* 2 vols. (New York, 1918), vol. I, p. 413, citing McMaster, *History of the People of the United States*, vol. IV, p. 81.

72 In the 1830s and 1840s especially, unplaced immigrants are reported far more often than disappointed employers. However, the assiduous "emigration" agents could not always discover where labour was wanted. An employer in an out-of-the-way port, like Oakville (William Chisholm), found it wise to recruit at the port of entry (Hazel C. Mathews, *Oakville and the Sixteen: The History of an Ontario Port* [Toronto, 1953], p. 27).

73 A great many statements to the contrary were made, arguing that labour could not be plentiful in Canada since wage rates had not been beaten down to the European level. While distress in Canada did not equal that in Britain, there were often large numbers of men in the former country unable to find jobs; and their failure had very little to do — if anything — with wage rates. Those who expected otherwise were misled by the oversimplified model of the labour market propagated by classical economists. They also took insufficient account of the floor placed under Canadian wage rates by the nearness of the United States market.

74 *Upper Canada Sundries*, July 13, 1829 (Radcliffe to Colborne); July 18, 1834 (Memorandum on the immigration of indentured children by W.H. Dunn); October 17, 1834 (Orrocks to P.S. concerning children successfully placed).

75 Edward Gibbon Wakefield, *A View of the Art of Colonization*, Collier ed. (Oxford, 1914). Wakefield's ideas were not so new to the world in 1830 as he claimed. Notions much like his seem to have been almost common property much earlier — they appear, for instance, in Robert Gourlay's work. With respect to the reversal of the policy of easy land grants in Canada after 1815, Henry Golbourn was anxious in 1817 to confine land holding to large farmers with capital, with other immigrants sent out as

their employees (*C.O. 43*, vol. 56, Golbourn to Harrison, Dec. 6, 1817). There were, of course, often important differences also. Golbourn was primarily concerned to limit the settlement of "idle persons who having neither means of maintenance nor energy sufficient to make use of the Land offered to them are a burthen and a disgrace to [the Colony]." Gourlay was a single-taxer, while Wakefield emphatically was not.

76 Wakefield, *op. cit.*, especially pp.327-44; also pp.481-82.

77 *Statistical Account*, vol. III, pp. ccclxxxvi-ccclxxxix.

78 Canada, *Journals of the Assembly*, 1848, app. Q. Also Sir Arthur G. Doughty, ed., *The Elgin-Grey Papers, 1846-52*, 4 vols. (Ottawa, 1937), vol. III, pp. 1159-62. A very similar plan was proposed in 1847 by the Toronto and Lake Huron Railroad (Northern Railway) (*Executive Council Papers, 1847*, no. 392 [P.A.C.]).

79 Wakefield, *op. cit.*, passim.

80 Doughty, *Elgin-Grey Papers*, vol. I, p.11; see also vol. III, pp.1080, 1100-113.

81 *Ibid.*, vol. I, pp.146-48. R.B. Sullivan's objections to this scheme appear on pp. 203-5. Cf. R.S. Longley, *Sir Francis Hincks* (Toronto, 1843), p. 181 ff.

82 Doughty, *Elgin-Grey Papers*, vol. IV, pp.1436-57 (Sullivan's objections to this plan).

83 There is an undercurrent in nearly all British discussion of colonization in this period of what might be called misapplied classical political economy. With respect to public works, the implicit chain of argument seems to have been something like this: Public works are capital. Capital is congealed labour. If labourers (redundant in Britain) are taken to Canada, their labour can provide the public works. But it is also known that labourers will work hard to get land. Therefore give them land, a useless asset or free good to the authorities, in exchange for their labour. Thus, public works can be provided out of land. A useful analysis of British thinking about colonies at this time appears in Brinley Thomas, *Migration and Economic Growth* (Cambridge, 1954), ch. I.

84 Doughty, *Elgin-Grey Papers*, vol. IV, pp.1427-36 (Memorandum on Immigration and Public Works) and pp. 1436-57 (Tulloch's Plan). The gist of what the Canadians said was that it was not labour but capital that was short in Canada; and that if capital were provided, any deficiency of the labour force would be very rapidly made up by voluntary immigration (i.e., if jobs were available). In terms of the British argument cited above, the Canadians were declaring that capital was something distinct from labour in a sense not made clear in classical value theory. The British had overlooked that while labour in general makes capital in general, Canada could not be expected always to have the particular kinds of labour necessary to make the particular kinds of capital goods wanted. Nor do they seem to have appreciated that capital must afford labourers their subsistence while they are producing capital goods; and classical political economy certainly gave no warrant for that omission.

The Canadians also took strong exception to the plans to pay for labour with land; or more exactly they argued against Wakefield ideas in general. Land had to be cheap in Canada to compete with the United States, and it was desirable to have it settled by any means.

85 *Ibid.*, vol. I, p. 207.

86 *Ibid.*, vol. IV, p. 1457.

87 On the other hand, practitioners of unusual trades could show up in a New France that could not offer them a market — a marble-polisher, an artificial flower maker, a gilder, a perfumer. See E.Z. Massicette, "Les Métiers rares d'autrefois", *Recherches historiques*, vol. 36 (Oct. 1930), pp. 609-13.

88 In 1732, when a rope-making industry was being established in New France, two rope-makers came of their own accord, along with eight others brought under contract (Lunn, *op. cit.*, p. 78). But this was a most unusual occurrence.

89 The references to skilled labour brought under such contracts are very numerous (Lunn, *Economic Development in New France*; Fauteux, *Essai sur l'industrie au Canada*). Repeated calls for labour were made in part because young colonists showed a decided reluctance to learn most trades. Lunn ascribes this to the attraction of alternative opportunities (p. 260), but the contemporary view, that the basic trouble was the distaste of the Canadien for regular work of any kind (*ibid.*, p. 259) — i.e., the attitudes of a pre-capitalistic society — may be closer to the mark.

Another difficulty was that some workers returned to France at the expiry of their short contracts, but it does not seem true that "for the most part, these men came reluctantly, remained a few years at excessively high wages, and then joyfully returned to France" (*ibid.*, p. 12). It may have been their initial intention to return to France, but usually they did not. Moreover, the secret by which Canada held them is revealed in at least the case of the iron workers at Three Rivers. There, workers had scarcely arrived when they are found, typically, marrying a Canadienne (B. Sulte, *Mélanges historiques* [Montreal, 1920], vol. 6, "Les Forges Saint-Maurice", pp. 63-80).

90 This was Thomas Brassey's view. He sent 2,000 workers to Australia, without attaching any condition, content in the knowledge that there was no sizable employer but himself. (Sir Arthur Helps, *Life and Labour of Mr. Brassey, 1805-1870*, 3rd ed. [London, 1872], p. 239.)

91 Toronto (daily) *Leader*, March 30 and May 26, 1854. The extreme labour shortage of 1854 gave a particular justification to large-scale subsidized immigration. But according to the press accounts, the shortage of shipping in 1854 prevented more than 1,250 workers being sent before May 5. The ones sent were likely skilled workers, and those delayed, the labourers whom the Brassey firm had planned to send out that spring in sailing vessels.

92 *Report of the Directors of the Great Western Railway to the Shareholders, upon The Report made by the Commission appointed to enquire into certain accidents upon the Great Western Railway* (n.p., April 1855) (uncatalogued pamphlet, P.A.C.), p. 19.

93 L.G. Reynolds, *The British Immigrant: His Social and Economic Adjustment in Canada* (Toronto, 1935), p. 125.

94 This seems clearly the case in respect to the Grand Trunk and to the Canadian Pacific. An American case in which the problem of labour supply is still more obviously paramount is found in Daniel Creamer, "Recruiting Laborers for Amoskeag Mills", *Journal of Economic History*, May 1941, pp. 42-48.

95 Harold A. Logan, *The History of Trade-Union Organization in Canada* (Chicago, 1928), pp. 38, 67-71. The Alien Labor Act passed in 1898 was purely a retaliatory measure against the United States and had no bearing upon the Canadian contract labour problem.

96 Wakefield, *Art of Colonization*, pp. 174-78, 322-29. Wakefield thought that classical slavery also rested upon labour scarcity (p. 326), but the directly contrary interpretation is offered in J.H. Clapham and E. Power, eds., *The Cambridge Economic History of Europe*, vol. I, *The Agrarian Life of the Middle Ages* (Cambridge, 1941), pp. 234-43.

97 Wakefield, *op. cit.*, p. 175. Wakefield inserted a fourth class of "voluntary" slaves, the Irish Roman Catholics, in reference to the menial work and wretched conditions accepted by these Irish in the colonies. But it will be argued in this study that the status of the Irish in the nineteenth century was as far as might be from the protected one of a slave, and that the presence of the Irish was of the greatest importance to the rise of industrial capitalism.

98 *Ibid.*, p. 177.

99 *Ibid.*, p. 329.

100 Reid, *op. cit.*, p. 42.

101 Lunn, *op. cit.*, p. 6.

102 *Ibid.*, pp. 5-11; Reid, *op. cit.*, p. 42-46. Peter Kalm reported in 1749 that no restriction was placed upon the *faux sauniers* in Canada except that they were forbidden to leave the colony. Peter Kalm, *Travels in North America*, trans. J.R. Forster, 3 vols. (Warrington and London, 1770-71), vol. III, p. 307.

103 Cf. S.C. Clark, *The Social Development of Canada* (Toronto, 1942), pp. 256-60.

104 Mayor W.L. Mackenzie of Toronto was aroused over the inadequate segregation of the sexes in prisons, among other things. Having committed two alleged prostitutes to the Toronto Gaol in November 1834, Mackenzie was shocked to find them at large in the debtor portion of the prison. (*U.C. Sundries*, Nov. 23, 1834 [Mackenzie to Rowan]. This is reproduced in Clark, *Social Development of Canada*, p. 251.) The Sheriff argued in reply that the only part of the prison in which women could be segregated was unheated, and the women had to be allowed out sometimes to get warm (*U.C. Sundries*, Nov. 27, 1834 [Jarvis to Rowan]). Another case concerns two female prisoners who arrived at the penitentiary in 1839 and were found to be pregnant. The penitentiary authorities complained bitterly of the carelessness of the Toronto gaolers. These, in turn, denied that the women could have become pregnant in their gaol, and suggested that they might have become so during the boat trip to Kingston (*U.C. Sundries*, Oct. 12, 1839 [Nickalls to Harrison] and Oct. 17, 1839 [Jarvis to Harrison]). Very shortly afterwards, a young girl sentenced to serve a week in Toronto Gaol for having broken her contract under the Master and Servant's Act, was in fact released immediately because there was no place for her in the gaol except with the "abandoned females". The discussion suggests that this was a usual course in such cases (*U.C. Sundries*, Nov. 1, 1839 [O'Hara to Harrison], enclosures).

105 The need for a special institution to care for the insane, and the particular problems raised by the makeshift expedient of keeping them in gaols, provoked much discussion in the 1830s and 1840s in Upper Canada. Cf. *U.C. Sundries*, May 14, 1831 (Baby to Mudge); Feb. 16, 1839 (Petition); *P.S.O.*, *C.W.*, 1841-3, no. 3021; 1846-7, nos. 15559, 17428; and Clark, *Social Development of Canada*, pp. 234-35. A "temporary" asylum was finally provided in 1841 when a new gaol was built at Toronto and the old one turned over exclusively to the care of the insane (*P.S.O.*, *C.W.*, 1840-1, nos. 257, 397, 2895); but, presumably from lack of space, many insane persons continued

to be left in local gaols. Meanwhile provision had been made for a perma-nent asylum (*U.C. Sundries*, Sept. 20, 1839 [appointment of commissioners]), but this did not open until the end of 1847 (*P.S.O., C.W.*, 1847-8, no. 18607). A report of 1848, indicating that the new institution left much to be desired, is reproduced in Clark, *op. cit.*, pp. 235-36. It may be that the temporary asylum, though it could not care for all applicants, did a better job with the inmates it had (*P.S.O., C.W.*, 1846-7, no. 16616; Canada, *Journals of the Assembly*, 1841, appendix L.L.; 1842, appendix U). That accommodation for the insane was still grossly insufficient, especially in Canada East, in 1858, is stated in *Executive Council Papers, Put By*, no. 602, Report of the Ordnance Land Agent, May 26, 1858.

106 On the condition of gaols, see *U.C. Sundries*, June 9, 1834 (Mackenzie to Rowan); June 17, 1834 (Jarvis to Rowan); July 26, 1834 (Powell to Rowan); *P.S.O., C.W.*, 1841-3, nos. 3021, 4282; and Clark, *op. cit.*, pp. 256-58. The conditions were bad, but statements from imprisoned debtors (*U.C. Sundries*, April 28, 1835 [Petition of Debtors in Amherst Gaol]; June 28, 1838 [Gray to Arthur]; Clark, *op. cit.* p. 258) should be taken with some reservation. There is something in the position taken by contemporaries that these men did not merit a great amount of sympathy because they could win release at any time by taking an oath that they were worth no more than £5.

107 For the construction of the penitentiary, the failure of the Assembly to provide any money to run it, and its eventual opening, see *U.C. Sundries*, July 10 and July 22, 1834 (Macaulay to Rowan): May 4, May 7 and June 1, 1835 (Macaulay to Rowan).

108 *U.C. Sundries*, Oct. 5, 1836 (Macaulay to Joseph); Oct. 12, 1839 (Nickalls to Harrison).

109 Canada, *Journals of the Assembly*, 1842, appendix H; *P.S.O., C.W.*, 1846-7, no. 16866.

110 *U.C. Sundries*, June 9, 1834 (Mackenzie to Rowan).

111 *U.C. Sundries*, Oct. 2, 1835 (Macaulay to Rowan); Dec. 23, 1836 (Macaulay to Rowan); Oct. 12, 1839 (Nickalls to Harrison).

112 *U.C. Sundries*, Dec. 23, 1836 (Macaulay to Rowan); April 11, 1839 (Grant to Macaulay); June 10, 1839 (Nickalls to Macaulay); Oct. 12, 1839 (Nickalls to Harrison, enclosed Annual Report).

113 *U.C. Sundries*, April 11, 1839 (Grant to Macaulay); June 10, 1839 (Nickalls to Macaulay). In the early days, convicts had evidently been employed as personal servants of Warden Smith. The Legislature forbade this in 1837-38. The warden yielded his privileges with reluctance (*U.C. Sundries*, Nov. 26, 1839 [Smith to P.S.]).

114 *U.C. Sundries*, Oct. 12, 1839 (Nickalls to Harrison, enclosure).

115 Upper Canada, *Journals of the Assembly*, 1839, Appendix vol. II, part I, p. 203.

116 *Loc. cit.; U.C. Sundries*, Oct. 12, 1839 (Nickalls to Harrison, enclosure).

117 Commons and associates, *History of Labour in the United States*, vol. I, p. 492.

118 *P.S.O., C.W.*, 1841-3, no. 3226.

119 *P.S.O., C.W.*, 1844-5, no. 9127.

120 *P.S.O., C.W.*, 1844-5, no. 9382.

121 *Executive Council Papers*, Dispatch from Grey, Nov. 2, 1846.

122 *P.S.O., C.W.*, n.d. (June) 1850, Petition of the Inhabitants of Kingston.

123 *U.C. Sundries*, Dec. 23, 1836 (Macaulay to Rowan).

[124] Upper Canada, *Journals of the Assembly*, 1839, Appendix vol II, part I, pp. 236-55.

[125] *U.C. Sundries*, Oct. 12, 1839 (Nickalls to Harrison, enclosure).

[126] Canada, *Journals of the Assembly*, 1842, app. H.

[127] *U.C. Sundries*, June 10, 1839 (Nickalls to Macaulay); Oct. 12, 1839, (Nickalls to Harrison, enclosure).

[128] Canada, *Journals of the Assembly*, 1842, app. H. It is stated here that the labour of the convicts will be required for construction for years to come.

[129] *P.S.O., C.W.*, 1846-7, no. 16866.

[130] *U.C. Sundries*, Dec. 20, 1836 (Hinton, Richmond Temperance Society, to P.S.); Clark, *op. cit.*, pp. 253-54.

[131] *P.S.O., C.W.*, 1841-3, nos. 2385, 4117, 4694, 5256. The Legislature tried to make spirits too expensive for the labouring classes by imposing a severe tax in 1845 (*P.S.O., C.W.*, 1844-5, no. 9521) but, as might have been expected, enforcement proved to be impossible (*P.S.O., C.W.*, 1847-8, no. 19492). Some were then ready to advocate complete prohibition (*P.S.O., C.W.* 1849, no 34).

[132] *P.S.O., C.W.*, Aug.-Sept. 1852; also 1853, no. 246, and Jan.-Feb. 1856.

[133] A claim of this sort is made in a great many of the documents cited above. In addition see Canada, *Journals of the Assembly*, 1849, app. ZZZ (Report of Select Committee on Intemperance). This committee was very anxious to demonstrate that crime was a result of intemperance. It analyzed the population of Montreal Gaol in great detail, and proved, at least, that the inmates were almost wholly from the lowest economic and social levels of society. The Keeper of the Gaol said that the intemperance of his charges was a consequence of their unhappy lives, not a cause of them; but the Committee ignored this opinion.

This Report is valuable in showing the power which the temperance movement had gained in a short time; in making clear that to a great extent temperance was something to be imposed on other parties — i.e., the labouring classes; and in bringing out the capitalist puritanism that inspired the movement. Thus, an educational system was advocated (the savings on police would pay for it) since: "Temperance, therefore, the parent of economy, is closely allied to knowledge. Labour, too, is the source of wealth — it produces capital ....Riches, then, which are proverbially the reward of industry, are incompatible with intemperance."

[134] *U.C. Sundries*, Dec. 23, 1836 (Macaulay to Joseph); Oct. 12, 1839 (Nickalls to Harrison, enclosure).

[135] *P.S.O., C.W.*, 1846-7, no. 15130.

[136] *P.S.O., C.W.*, 1846-7, no. 15666.

[137] *P.S.O., C.W.*, 1846-7, no. 15726. Convict meals were obtained by contract "at a very low rate". Before 1839 they had cost 8d. per day per convict, but rose to 12d. in 1839, which was attributed to the increase in the cost of living. The daily ration is set out for 1839. (Upper Canada, *Journals of the Assembly*, App. vol. II, part I, pp. 203, 215; *U.C. Sundries*, Oct. 12, 1839 [Nickalls to Harrison, enclosure].)

[138] *P.S.O., C.W.*, 1846-7, nos. 16868, 18564.

[139] *P.S.O., C.W.*, 1849, no. 646. This petition is from the inhabitants of Perth, March 1849. Many others appear in the record about the same time, usually employing the same words.

140 Clark, *op. cit.*, pp. 261-62.

141 Donald Creighton, *John A. Macdonald; The Young Politician*, (Toronto, 1952), pp.158-59.

142 Toronto *Globe*, March 7, 1850.

143 *P.S.O., C.W.*, n.d. (June) 1850.

144 Toronto (daily) *Leader*, April 13, 1854, reproducing an item for the Kingston *Commercial Intelligencer*. A consequence of the attacks was to force the penitentiary to advertise the convicts it planned to contract out. A newspaper advertisement of April 15, 1853, reads:

CONVICT LABOR — The undersigned (Warden, P.P.) is prepared to Hire Out the Labor of the following Gangs of Convicts in the Provincial Penitentiary, viz, CABINET MAKERS from 1st Feb., 1855. BLACKSMITHS, from 1st July, 1855. TAILORS and other Tradesmen, at any time that may be agreed upon. TENDERS for same will be received at the Office of the Penitentiary from responsible parties...

However, the authorities could, and evidently did, select tenders other than the highest offer (*P.S.O., C.W.*, 1853, no. 1490 and enclosure [the newspaper advertisement]).

145 Toronto (semi-weekly) *Leader*, Sept. 16, 1853.

146 *P.S.O., C.W.*, 1856, no. 659 and enclosures.

147 Toronto *Mail*, Sept. 24, 1873.

148 Toronto *Mail*, Sept. 26, 1873.

149 Ottawa *Times*, Aug. 8, 1874; Toronto *Globe*, Aug. 5, 1875; Toronto *Globe*, Aug. 4, 1876; Toronto *Globe*, Aug. 10, 1877.

150 Logan, *The History of Trade-Union Organization in Canada*, pp. 63-64, 80, 189-90.

151 "Minutes of the Toronto Trades Assembly" (1871-1878) (typescript), reports of meetings Oct. 4, Oct. 18, Nov. 15, 1872. (There are copies of the "Minutes of the Toronto Trades Assembly" in a number of libraries. The copy consulted by the author is in the Library of the Department of Labour, Ottawa.)

152 *Ibid.*, Meetings of Dec. 10, 1872, Jan. 17 and Feb. 7, 1873.

153 *Ibid.*, Meeting of Jan. 16, 1878.

154 Dr. Coats stated that the Ontario government repealed the Master and Servant Act (R.H. Coats, "The Labour Movement in Canada", *Canada and Its Provinces* [Toronto, 1914], vol. IX, p. 299). So far as I can make out, the law was rather amended so that a breach of contract on the part of an employee was no longer a criminal offence, but a civil offence, as a breach by the employer always had been.

Chapter II

1 In law, the wage earner of the free labour market is a free, independent agent, as the employer is. In classical and neo-classical economic theory, the haggling of the labour market containing many workers seeking work, and many employers seeking labour, not only coordinated dispersed wage bargains into an economic whole, but (by analogies and assumptions) provided perpetual full employment. In practice, this system can scarcely emerge or operate unless the demand for employment is usually greater

than the demand for workmen. A labour reserve is necessary for its sanctions to operate effectively. These sanctions are the simplified minimum of the attraction of wages and the repulsion of dismissal — i.e., the attraction of life and the repulsion of starvation. On the other hand, the system can only continue if the bulk of the labouring class, with their families, are able to retain life and some measure of working efficiency. A proportion of failure in this respect can be met very economically, and had been met, by the provision of relief agencies. However, the system would lose its logic if employers had to provide in relief, in order to retain their work force, as much as they gained by evading the overhead costs of the working class.

[2] The employer's position resembles that of the person who, to have use of a machine, finds he must buy rather than rent, and that he must buy a larger size of machine than he requires ordinarily. To the extent that the capitalist labour market involved a pooling of labour that could be shared by employers whose needs did not coincide, it presented a genuine economy.

[3] J.H. Clapham and E. Power, eds., *The Cambridge Economic History of Europe*, vol. I, *The Agrarian Life of the Middle Ages* (Cambridge, 1941), pp. 112-15, 234-43. (C.E. Stevens and Marc Bloch). See also, M. Dobb, *Studies in the Development of Capitalism* (London, 1946), pp. 33-70.

[4] H.C. Pentland, "Feudal Europe: An Economy of Labour Scarcity", *Culture*, September 1960, pp. 280-307.

[5] For opinions that New France benefited from government intervention, see Lunn, *op. cit.*, p. 437, and Reid, *op. cit.*, pp. 119-27.

[6] The system is described and analyzed in W.B. Munro, *The Seigniorial System in Canada* (New York, 1907). Munro argued that the Canadian system was much closer in operation to early classical feudalism than was the contemporary French system (p. 13).

[7] H.A. Innis, *Select Documents in Canadian Economic History, 1497-1783* (Toronto, 1929), pp. 301-2.

[8] Reid, *op. cit.*, p. 132. Miss Reid appears to mean that the small-scale and craft nature of colonial industry was perpetuated by craft tradition and apprenticeship — "still the royal road to mastership in Quebec as late as 1710" — but the argument is dubious. Apprenticeship training was supplemented by immigration and by a trades school given rather extravagant praise in Mason Wade, *The French Canadians 1760-1945* (Toronto, 1955), p. 26, though artisan training did not last long. There is no inherent relation between apprenticeship training and the scale of industry. Miss Reid herself notes the absence of legal impediments to the choice of industrial employments in New France.

[9] Gérard Tremblay, "L'Évolution des relations industrielles au Canada", *Rapport, Premier Congrès des relations industrielles de Laval, 1946* (Quebec, 1946), pp. 12-13.

[10] Lunn, *op. cit.*, pp. 255, 257, 348. Discussions of industry in New France here and elsewhere rest primarily on Lunn, *Economic Development in New France*, which is considered the best treatment of the subject, and on Fauteux, *Essai sur l'industrie*.

[11] Cf. Hocquart's defensive explanation that the wages of shipyard workers are very high, but they do begin work before sunrise and work until long past sunset. Fauteux, *op. cit.*, vol. I, p. 251.

[12] Costs of living seem to have been substantially higher than in France. Cf. Innis, *Select Documents...1497-1783*, pp. 412-13.

[13] Fauteux, *op. cit.*, vol. I, p. 232.

[14] Lunn, *op. cit.*, pp. 305-8.

[15] This section is not intended to add to the literature of the Canadian fur trade, but to point out features of its labour organization. The argument is substantially that of H.A. Innis, *The Fur Trade in Canada* (New Haven, 1930), in which work labour organization was given a position of importance.

[16] Innis, *op. cit.*, pp. 39, 58-61.

[17] The role of Canadian labour in the American fur trade was discussed by H.A. Innis in "Interrelations between the Fur Trade of Canada and the United States", *Mississippi Valley Historical Review*, vol. XX, no. 3 (Dec. 1933), pp. 328-29.

[18] The trade seems to have taken about 500 men in 1681 (population not quite 10,000); about 1,000 men in 1720 (population about 25,000); and 2,000 men or more in 1750 (population about 50,000). Cf. Innis, *op. cit.*, pp. 62-67, 103-4, 115; Lunn, *op. cit.* p. 109.

[19] Lunn, *op. cit.*, pp. 126-27. The data are not sufficient to be very conclusive, but the figure given fits well with wages shortly after the Conquest, and custom had a powerful influence on such wage rates.

[20] Skilled shipyard workers were paid 2 livres in winter and 2½ livres in summer per day. The number of days worked and the costs of subsistence can only be guessed, but it seems doubtful whether carpenters could have been much better off than canoemen.

[21] Innis, *op. cit.*, pp. 115-16.

[22] *Ibid.*, pp. 170-72, 194 ff.

[23] *Ibid.*, pp. 215, 220, 242-43. Innis gives wages of canoemen in 1767, following Ermatinger, as follows: guide (to Grand Portage), 350 l.; foremen and steersmen, 300-320 l.; middlemen, 250 l.; winterers, 300-400 l. Lawrence Ermatinger's "Engagement Book" for 1773-75 (P.A.C., M.G. 29/128) shows that some wages had risen substantially in the intervening seven years. Middlemen still get about 250 l.; foremen and steersmen are at about 350 l. — 10 per cent higher; but the pay of guides had doubled to 700 l., and winterers' pay was nearly doubled at about 600 l. Innis shows further wage increases by 1800, especially for skilled men (p. 242). The indefatiguable reporter, Lord Selkirk, writing in January 1804, indicates still higher wages and some other instructive points besides:

> The wintering men are bargained with for about $200 [1,000 livres if the livre remained the equivalent of a shilling, which it did so far as is known to the writer] — Bouts de Canot for $300 with equipments from $60 to 100 [the estimate is high or the kits had increased] — these wages are higher than they were before the competition, are doubly chargeable because formerly the engagés used to take a great part of their wages in articles of luxury for themselves or of finery for their Indian wives — this was encouraged by the Company so much that a great proportion of their men were in debt to them — this reduced their wages to little because these goods were charged at a very high rate — Of late however the men have become more saving partly because the wages are so immense that they are sufficient at once to enrich a man — partly because the competition of the Companies keeps them continually on the alerte following the Indians instead of waiting for them & leaves less time for idleness & dissipation. Tho' the Company have no regular civil authority yet they

keep discipline among these people (who are all French-Canadians) with a high hand — they have occasionally flogged and put them in irons — in general they find little difficulty in managing them — which certainly would not be the case with Americans who would know exactly how far their master's authority extended.... The Canadians however require a good deal of temper & attention in their leaders to be completely managed."

Earl of Selkirk, *Selkirk Papers* (P.A.C.), vol. 75 (Diary no. 3), pp. 120-21.

24 Innis, *op. cit.*, pp. 220-21, 245.

25 The leaders of the 1794 strike were sent back to Montreal. Cf. Selkirk's comments, note 23.

26 Innis, *op. cit.*, p. 245.

27 These strictures on the qualities of the Montreal traders receive support from the findings of Richard Glover. See E.E. Rich and A.M. Johnson, eds., *Cumberland House Journals and Inland Journal, 1775-82: First Series, 1775-79*, (London: Hudson's Bay Record Society, 1951), Introduction by Richard Glover, pp. xliv-li.

28 Innis, *op. cit.*, pp. 230-46.

29 *Ibid.*, p. 243.

30 *Ibid.*, p. 244.

31 *Ibid.*, pp. 140-41, 158-61.

32 *Ibid.*, pp. 164-68. Cf. Rich and Johnson, *op. cit.*, p. xxxvii.

33 Rich and Johnson, eds., *Cumberland House Journals and Inland Journal, 1775-82: First Series, 1775-79, loc. cit.*; *Second Series, 1779-82* (London, 1952), pp. xv-lii.

34 Innis, *op. cit.*, pp. 289-91, 312-18.

35 *Ibid.*, pp. 319-21, 360-62.

36 D.D. Calvin, *A Saga of the St. Lawrence* (Toronto, 1945).

37 *Ibid.*, pp. 33-34.

38 *Ibid.*, pp. 32, 84, 154.

39 *Ibid.*, pp. 145-50.

40 *Ibid.*, p. 156.

41 *Ibid.*, pp. 29, 44.

42 *Ibid.*, pp. 77-82.

43 *Ibid.*, pp. 68-69, 155, 160-61, 163-64.

44 The best account of the St. Maurice Forges (though with much extraneous material included) is Benjamin Sulte, "Les Forges Saint-Maurice", vol. 6 of *Mélanges historiques* (Montreal, 1920) (compiled, annotated and published by Gerard Malchelosse). Excellent discussions of the French period are found in Fauteux, *Essai sur l'industrie au Canada*, vol. I, ch. II, and in Lunn, *Economic Development in New France, 1713-1760*, chs. 11 and 12. Valuable material for the period 1767-1843 is found in Canada, *Sessional Papers, 1844-5*, appendix (O). The best short account is F.C. Wurtele, "Historical Record of the St. Maurice Forges, the Oldest Active Blast-Furnace on the Continent of America", *Transactions of the Royal Society of Canada*, IV (1886), sec. 2, pp. 77-90. A. Tessier, *Les Forges Saint-Maurice, 1729-1883* (Three Rivers, 1952) is a good general history.

45 Lunn, *op. cit.*, p. 290.

46 Lunn, *op. cit.*, p. 303, puts the Company's debt to the king at 193,000 livres

in 1738; but Sulte (*op. cit.*, p. 86) shows this debt at 139,000 livres in 1742. Lunn reports (p. 312) a total expenditure of 505,000 livres, 1737-41, of which 352,000 were not covered by receipts; but it is not clear what part capital expenditure played in these figures. Fauteux (*op. cit.*, p. 103) puts the Company's debts at 340,000 livres in 1742 in addition to 193,000 livres owed to the king. The Forges were valued in 1741 at 140,000 to 160,000 livres (Fauteux, *op. cit.*, p. 100). The Company bought the fief of St. Maurice in 1736 and other lands had been added (Sulte, *op. cit.*, pp. 51, 53, 56).

47 Four workmen were brought in 1736, about fifty-five in 1738, thirteen or fourteen in 1740, and two about 1746. The men talked at times of returning to France, but very few seem actually to have done this.

48 Sulte, *op. cit.*, p. 79. Peter Kalm thought the procedures identical with those in Sweden. Peter Kalm, *Travels in North America*, 3 vols. (Warrington and London, 1770-1), vol. III, p. 87. But in 1796 and 1817 techniques were held somewhat inferior to English ones (Sulte, *op. cit.*, pp. 172, 183).

49 Some seem to have thought St. Maurice iron superior to Swedish. Cf. Sulte, *op. cit.*, p. 179; Wurtele, *op. cit.*, p. 83.

50 Lunn, *op. cit.*, p. 312.

51 *Ibid.*, pp. 304, 312, 329.

52 *Ibid.*, pp. 328-29, 333-34; Fauteux, *op. cit.*, pp. 112-13.

53 It was claimed in 1762 that the Forges had produced 400 tons of cast iron to that date (R.C. Rowe, "The St. Maurice Forges", *Canadian Geographical Journal*, IX [1934], p. 21). Total output should have been about twenty times that amount. It is clear that cast products took a relatively small part of the output during the French regime, whatever the precise proportion.

54 Cf. Lunn, *op. cit.*, p. 329.

55 *Ibid.*, p. 335.

56 This is not readily explained, since it seems to have been generally considered that the Forges could be made very profitable. Hocquart evidently thought the lack of capital in the colony ruled out the possibility of local operation (Lunn, *op. cit.*, pp. 320-21).

57 *Ibid.*, pp. 329, 332, 335, 339-40.

58 Sulte, *op. cit.*, pp. 106, 118; Fauteux, *op. cit.*, pp. 121-22, Peter Kalm, *op. cit.*, p. 89.

59 Fauteux, *loc. cit.*

60 Lunn, *op. cit.*, p. 341.

61 The sawmill certainly operated in the 1770s, but the gristmill probably was not added until after 1796, by the Bell management. The mills are noted by Selkirk, 1804 (*Selkirk Papers*, vol. 75, p. 159).

62 The chief informant for this period is somewhat unreliable: Pierre de Sales Laterrière, a medical man who became Quebec sales agent for the iron works, assistant director of the Forges, and then director, 1776-78.

63 Laterrière's figures (Sulte, *op. cit.*, p. 159). Sales certainly were not so high a few years later, but the war period probably offered exceptional demand, and the claims of a very large labour force at this time support this view.

64 Mason Wade, *The French-Canadians, 1760-1945* (Toronto, 1955), p. 78.

65 *Ibid.*, p. 77.

66 Sulte, *op. cit.*, p. 147. Pelissier was evidently a man of strong liberal opinions.

67 Peter Force, ed., *American Archives . . . ., Fourth Series: A Documentary History of the*

English Colonies in North America from March 7, 1774 to the Declaration of Independence (Washington, 1839), vol. III, pp. 18, 135, 211, 743, 925, 962, 1342, 1396; vol. IV, pp. 170, 596 (this is a letter from Pelissier, director of the Forges, that reveals his attitudes well), 796, 854, and *passim*. These documents also demonstrate how the general sympathy in Lower Canada for the Americans was changed by the spring of 1776 to hostility, by expectations that the British would win, by the lack of hard money and by the continuous agitation of the clergy. Cf. vol. V, p. 751.

68 D.G. Creighton, *The Commercial Empire of the St. Lawrence, 1760-1850* (Toronto, 1937), pp. 9, 21, 30-31; H.A. Innis, *The Fur Trade in Canada*, pp. 122-23.

69 The precise course of events is hard to follow. Wurtele ascribes the 1887 transfer to Gugy's bankruptcy (*op. cit.*, p. 86), Sulte to his death (*op. cit.*, p. 171).

70 Sulte, *op. cit.*, pp. 171-72, 175.

71 There is conflicting information concerning the various leases and extensions in Sulte, *op. cit.*, p. 175; Wurtele, *op. cit.*, p. 86; and *Sessional Papers, 1844-5*, app. (O). The text is at least essentially correct.

72 Sulte, *op. cit.*, pp. 175, 176.

73 *Ibid.*, p. 180 cites John Lambert as stating (1808) that Munro & Bell would have bid £1,200 per year rather than let the Forges go.

74 *Selkirk Papers*, vol. 75, p. 159. Selkirk's information came from a man named Lees, formerly at the Forges.

75 Sulte, *op. cit.*, p. 174.

76 *Ibid.*, p. 179.

77 *Ibid.*, pp. 180, 186.

78 Part of the machinery for the *Accommodation*, 1809, was produced at the Forges. *Ibid.*, pp. 180-81.

79 *Selkirk Papers*, *loc. cit.*

80 Sulte, *op. cit.*, pp. 176-78, 182.

81 *Ibid.*, pp. 181, 184; Wurtele, *op. cit.*, p. 87.

82 *Sessional Papers, 1844-5*, app. (O); Sulte, *op. cit.*, pp. 185-86.

83 Aylmer's statement, *Sessional Papers, 1844-5*, app. (O).

84 Albert Tessier, "Une campagne antitrustarde il y a un siècle", *Les Cahiers des dix*, 1937, pp. 198-203.

85 Sulte, *op. cit.*, pp. 185-88; Canada, *Sessional Papers, 1852*, app. CCC.

86 Sulte, *op. cit.*, pp. 187-88. Tessier would have it that Bell frightened away buyers by spreading rumours of how bad the land was (Sulte, *op. cit.*, p. 202). There is an excellent discussion of the settlement question in *Sessional Papers, 1852*, app. CCC (by E. Parent, Asst. Prov. Sec.), which suggests that price, in the sense of some amount of cash payment, may have been the obstacle in 1843. According to Parent, the overflowing habitant population would take any kind of land, however bad, if offered under feudal tenure — i.e., with no cash down. But he asserted also that those who took bad lands in this way did not plan to settle permanently, but only to strip wood and exhaust the soil with a few crops. Parent also observed that settlement agitation from 1836 — the beginning of a genuine popular agitation — was aimed at Cap de la Madeleine rather than St. Maurice.

87 Sulte, *op. cit.*, p. 188.

88 *Sessional Papers, 1844-5*, app. (O); *Sessional Papers, 1852*, app. CCC.

89 Parent noted that the agitation from 1829 to 1836 was strictly a Three Rivers affair (*Sessional Papers, 1852*, app. CCC).

90 Very little cash circulated at the Forges, but rather purchases were debited against wages (*ibid*).

91 The habitants had perfected collusion in ore sales, wood sales and land purchases by 1851 (*ibid*).

92 *Ibid* and *Sessional Papers, 1844-5*, app. (O).

93 *Ibid.*

94 Wurtele, *op. cit.*, p. 88. Wurtele says Stuart rented the Forges to the Ferriers; but it seems clear that ownership passed to the Ferriers about 1848 (*Sessional Papers, 1852*, app. CCC). Sulte, *op. cit.*, p. 192, has a different sequence of events that appears to be in error.

95 Wurtele, *loc. cit.* Stuart & Porter seem to have faced expensive repairs for which they lacked capital (*Sessional Papers 1852*, app. CCC).

96 Wurtele, *op. cit.*, p. 89.

97 According to Laterrière (Sulte, *op. cit.*, p. 159).

98 *Ibid.*, p. 160. The hostile Committee of 1843 declared that the St. Maurice workmen were housed "in shanties or small log houses the construction of which is not intended for permanent residence" (*Sessional Papers, 1844-5*, app. [O]). The houses seem to have been in serious need of repairs in 1851 (*Sessional Papers, 1852*, app. CCC).

99 This number was also shown by a census of 1765 (Sulte, *op. cit.*, pp. 143, 184). Laterrière said that there were 130 houses in the 1770s, but he may have counted bunkhouses (*ibid.*, p. 160).

100 Lunn, *op. cit.*, p. 306.

101 *Ibid.*, pp. 307-11; Fauteux, *op. cit.*, pp. 95, 102, 111.

102 Lunn, *op. cit.*, p. 330. There is a graphic description of the inflation of 1744 and succeeding years in *Réflections sommaires sur le commerce qui s'est fait en Canada* ... (P.A.C., Pamphlets I, no. 281). Local produce doubled in price, imported goods rose much more than this.

103 Lunn, *op. cit.*, p. 330.

104 From about 1739. It is the parish records, transcribed by Sulte, that allow descent and intermarriage to be traced.

105 Sulte, *op. cit.*, pp. 63-79 and *passim*. Examples of intermarriages connecting skilled workmen are shown, pp. 64-65 and 67. On p. 65 appears the record of Michel Chaille who was born in Canada and presumably had no acquaintance with ironmaking, but who married the daughter of a master forgeworker, was employed as a fireman and eventually qualified as a master forgeman.

106 *Ibid.*, p. 6, Malchelosse's introduction; and the genealogical data.

107 The workmen were described as "European artists, distinguished for their ingenuity" by "A Citizen of the United States" (J.C. Ogden), *A Tour through Upper and Lower Canada* (Litchfield, 1799) (P.A.C., Pamphlets I, no. 852), p. 20. Ogden praised the Forges generally: "In every part, ability and enterprize are discovered, and a better regulated factory need not be sought for in North America."

108 Sulte, *op. cit.*, p. 190.

109 *Ibid.*, pp. 106n, 107.

110 *Ibid.*, p. 107.

111 *Ibid.*, p. 161.

112 Canada, *Sessional Papers, 1852*, app. CCC.

113 Three Welshmen were brought to the Forges about 1770. One, at least, married a Canadienne and was absorbed into the community. The given names in later lists of workmen indicate other cases of absorption. A census of 1820 numbered only five Protestants at the Forges. A list of 1842 shows twelve residents with British surnames (Sulte, *op. cit.*, pp. 169, 184, 189-90).

114 *Ibid.*, p. 179.

115 At the start, all workmen were evidently on annual salary. By 1752 (Franquet) some were on piece rates, but a large number were not. By the early nineteenth century, according to Selkirk and Lambert, nearly all workmen were paid by the piece.

116 Sulte, *op. cit.*, p. 121.

117 *Ibid.*, p. 173.

118 *Selkirk Papers*, vol. 75, p. 159.

119 Skilled shipyard workers during the French regime evidently received 2½ livres per day in summer and 2 livres in winter. Allowing 250 days of employment per year, their annual incomes would be a little less than 600 livres compared with the 400 livres (exclusive of rent and fuel) at the Forges. Kalm reported that journeymen got considerably more in 1749 — three to four livres per day — suggesting a much worse relative position for the forgeworkers (Kalm, *Travels in North America*, vol. III, p. 70). If the rate for skilled workmen around 1800 is taken at 6s. per day — and this is thought a fair and slightly conservative estimate — 250 days of employment would yield 1,500 livres, against 800 to 1,000 livres for the forgeworkers. The forgemen certainly failed to maintain parity with skilled canoemen in this period, but the competition of rival fur companies was a factor in that situation.

120 Lunn, *op. cit.*, pp. 338-39.

121 Sulte, *op. cit.*, p. 137. There was nothing peculiar to the Forges in this, for British military governors operated a kind of planned economy with rather general interference with prices and labour supply. *Add. MSS M377* (P.A.C.), p. 12, deals with the impressment of shipyard workers at Quebec, 1763. *S10* (*L.C. Sundries*, P.A.C.) covers the conscription of habitants to serve as bateaumen in 1765.

122 E.Z. Massicette, "Les Forges de Sainte-Géneviève de Bastican", *Recherches historiques*, vol. 41 (1935), pp. 564-67, 708-11. Batiscan evidently featured a company town, much like St. Maurice. Selkirk noted that the Batiscan works had proved very expensive to build and that Batiscan sales were only about half those of St. Maurice (in 1804). He also reported that the Batiscan management contracted with "American workmen" for their supply of charcoal, rather than hiring workmen to produce it as St. Maurice did, because they considered that method to be cheaper (*Selkirk Papers*, vol. 75, p. 159). The defalcations of John Craigie, one of the proprietors of Batiscan, and the subsequent actions of the Crown against him, may have played a part in the downfall of Batiscan.

123 Sulte, *op. cit.*, pp. 192-94. One of the Islet proprietors was "Robichon". This name was celebrated at the Forges from the beginning. There is some further information on the Radnor Forges in Benoit Brouillette, "Le Développement industriel de la Vallée du St. Maurice", in *Pages trifluviennes*, Series A, no. 2 (Trois Rivières, 1932), pp. 12-13. The Radnor Forges produced about 2,000 tons of iron per year and employed 200 to 400 workmen in 1863, according to Charles Robb in *Eighty Years Progress of British North America*, ed. H.Y. Hind (Toronto, 1863), p. 319.

124 British military authorities took numerous steps to strengthen Canada's military defences following the war of 1812-14, but the Kingston naval base and dockyard was an anchor of their system. It was considered that the base would be much more secure if it had access to a local supply of iron, and naval authorities sought to encourage the establishment of an iron-works to use the iron deposits known to be around Kingston by offering contracts for iron "ballast". Negotiations were carried on in 1816 with William Henderson, a Quebec merchant (*U.C. Sundries*, Jan. 31, 1816 [Owen to Gore]; Feb. 12 [Henderson to Gifford]; Feb. 12 [Gore to Owen]; Feb. 20 [Henderson to Gifford]; July 16 [Committee of Executive Council on Henderson's application for a grant of land]). The prohibition at this time on the grants of land in fee simple gave some difficulty, but the main trouble may have been that the land Henderson wanted (at Gananoque) had already been granted to one Sunderlin, who had complied with the conditions of his grant by erecting an ironworks in 1802. Other land was, however, offered. Henderson was quite aware of the Marmora deposits but probably decided they were too remote. Hayes was given a contract to supply pig iron to the dockyard to the extent of £13,000 sterling (Upper Canada, *Journals of the Assembly, 1839*, Appendix vol. II, part I, p. 239), probably at about 20s. currency per cwt. He was granted lands at Marmora forthwith though he may not have appreciated his location, for he referred to his grant as the "Gananoque ore" (*U.C.S.*, Nov. 3, 1820 [Hayes to Hillier]).

125 *U.C.S.*, Mar. 31, 1824 (Hayes to Hillier). The figure may be inflated by the inclusion of operating costs, since Hayes reported his investment at £10,000 in May 1822, by which time the main works at least were completed (*U.C.S.*, May 20 [Hayes to Hillier, enclosure]).

126 Upper Canada, *Journals of the Assembly, 1839*, Appendix vol. II, part II, pp. 236-37.

127 Hayes blamed his financial troubles on "the unexpected failure of my Agent in London". He tried to mortgage Marmora to the provincial government for £6,000 (*U.C.S.*, Mar. 31, 1824 [Hayes to Hilliar]). Next he tried to sell, and evidently approached the Carron Company (*U.C.S.*, Nov. 16, 1824 [Hayes to Maitland]). The creditors foreclosed sometime before August 1825. Hayes returned to England in 1826, still intent upon getting Marmora back (*U.C.S.*, Apr. 3, 1826 [Hayes to Hillier]). Peter McGill explained his succession to the Forges with delightful delicacy: "...on such suspension or failure in 1825, your Memorialist, as one of the principal creditors of Mr. Hayes, was appointed a Trustee for the said works, for the benefit of himself and all the other Creditors; but that finding such Trust incompatible with the interest and Security of your Memorialist, or the advantage of the establishment itself, the Deed of Trust was cancelled, and your Memorialist became possessed of the Property in fee simple..." (*U.C.S.*, Jan. 25, 1828 [McGill to Hillier]).

128 McGill filled a navy order for 750 tons of pig iron in 1826, and it may have operated at times, on some scale, in the 1830s (*U.C.S.*, Mar. 2, 1832. Return of Manufactures of Midland District, Aug. 13, 1832 [Smith to Rowan]). In 1828 he tried to obtain £10,000 from the government on a mortgage of the works. He must, as he claimed, have thought Marmora had prospects, to make this move; but he refused, on the other hand, to risk more of his own money in it. He offered the bland explanation that he did not feel he could withdraw more money from his own and his partners' business (*U.C.S.*, Jan. 25, 1828 [McGill to Hillier]). McGill tried to sell Marmora to the province in 1837, to become the provincial penitentiary, for £20,000 of debentures at 6 per cent or £25,000 at 5 per cent (Upper Canada, *Journals, 1839*, Appendix vol. II, part I, p. 241).

129 *Ibid.*, p. 244, shows the expenses in 1826. They amount to £19 per day of which £12 went for wages. The master founder was paid 15s. per day, skilled labour from 5s. to 7s. 6d., and labourers 3s. 3d. Hayes probably paid the same rates. He remarked that expenses could not be reduced below £100 per week (*U.C.S.*, Dec. 11, 1822 [Hayes to Maitland]).

130 Hayes made a great point from the beginning of insisting that no taverns should be licensed in the vicinity of Marmora (*U.C.S.*, Mar. 6, 1821). He argued similarly against taverns ("the very Pest of the Province") at the end of 1822, but said his workmen had so far been sober (*U.C.S.*, Dec. 19, 1822 [Hayes to Maitland]).

131 *U.C.S.*, May 20, 1822 (Hayes to Hillier). According to Hayes, there was a good deal of violence at this time, but its nature and cause are not clear. Hayes was at odds with local settlers, among others.

132 *U.C.S.*, Dec. 19, 1822 (Hayes to Maitland).

133 *U.C.S.*, Dec. 30, 1822 (Hayes to Maitland).

134 Hayes and his workmen encountered to begin with the hostility of local Indians who complained of his encroachment on their hunting grounds and of his parsimony with presents (*U.C.S.*, Mar. 6, 1821 [Hayes to Hillier]). Summer brought, of course, mosquitoes and black flies (*U.C.S.*, Aug. 2, 1821 [Hayes to Hillier]). Hayes had been allowed time to choose the precise lands he wanted, and he took a long time making up his mind. The chronic hostility shown by settlers was probably due to this delay and to the dispossession of squatters, and possibly also to Hayes's intense hostility to liquor sales (*U.C.S.*, Aug. 2, 1821 [Hayes to Hillier]; May 20, 1822 [Hayes to Hillier]; Mar. 31, 1824 [Hayes to Hillier]).

135 The King's corvée evidently was not usually onerous and it could be commuted for a money payment (W.B. Munro, *The Seigniorial System in Canada* [New York, 1907], pp. 132-33). The heavy calls upon the militia for transport service are noted in G.F.G. Stanley, *Canada's Soldiers, 1604-1954: The Military History of an Unmilitary People* (Toronto, 1954), p. 23. The construction of the fortifications of Quebec from 1701 onwards involved heavy corvées upon the people of Quebec and of its neighbourhood, which brought protests. Regular troops were also employed in this work and liked it no better. (Reid, *The Development and Importance of the Town of Quebec, 1608-1760*, p. 95.)

136 Stanley, *op. cit.*, p. 19. During 1665, part of the Carignan-Salières regiment was used to build several forts and a road in the Richelieu Valley (*ibid.*, p. 14).

137 Reinforcements were sent to the number of fifty to a hundred per year. This was a sufficient number to make up for settlers, but not for total

wastage. (Lunn, *Economic Development in New France, 1713-1760*, pp. 16-18.)

138 According to Lunn, the force varied between 600 and 800 men until it was sharply expanded in 1750 (*ibid.*, p. 15). While a great deal depends upon how far below strength the companies were kept, the description in Stanley, *op. cit.*, pp. 25-26, suggests a considerably larger force.

139 At Quebec, at Montreal (Lunn, *op. cit.*, p. 18) and presumably at the various outposts.

140 Fauteux, *Essai sur l'industrie au Canada*, vol. I, p. 232.

141 Lunn, *op. cit.*, pp. 235-36, 239-40.

142 Fauteux, *op. cit.*, vol. I, p. 24. Those employed were regular troops, but in this case presumably not Troupes de la Marine.

143 Lunn, *op. cit.*, p. 18.

144 *Ibid.*, p. 338.

145 *Ibid.*, pp. 236, 339.

146 There are numerous examples in W.O. 34, vols. 60, 61 and 64 (*Amherst Papers*, P.A.C.) pertaining to the recruitment of labour in Boston and New York, 1759-62. One of the groups involved was the seamen. The British military authorities were for solving their problems by impressment, but they were frustrated at Boston by the solid opposition of all the elements of the city (vol. 61, *passim*). The same seems to have been true at New York and Philadelphia. Amherst was reluctant, however, to obtain the required seamen by offering higher wages as "their Wages are already so Exorbitantly high, that I cannot possibly think of Encreasing it..." (vol. 60, p. 152 and *passim*). Much the same trouble arose in hiring carpenters at New York in 1759, Amherst and his officers being reluctant to meet the New York wage rates of 11s. 6d. per day, or better, plus provisions (vol. 64, p. 161). One consequence of this conflict was that the military authorities handled the artisans they did get with a brusqueness and lack of consideration that must have made it harder and harder for them to get men.

147 Stanley, *op. cit.*, p. 99.

148 At Montreal in 1763 and 1784 (*Add. MS, M377* [P.A.C.], p. 12 [Appeal of Knipe and Le Queene]).

149 Sulte, "Les Forges St. Maurice", p. 137.

150 Stanley, *op. cit.*, pp. 100-101, 122.

151 A determined resistance of the habitants about Montreal to bateau service in 1765 is shown in *S10* (P.A.C.), Oct. 12, 1765 (Christie to Gray); Oct. 13, 1765 (Christie to Burton); Oct. 13, 1765 (Le Goterie to Christie). A stubborn refusal of the habitants in 1810 to provide carts and horses to transport the baggage of troops employed in road making appears in *S85*, Nov. 30, 1810 (Bowen to Craig).

152 Lord Selkirk reported that Fort George was built by troops on a piece work basis, in the belief that the cost was only half of what it would have been with civilian labour (*Selkirk Papers*, vol. 75, p. 271).

153 William Claus's "Garden Book" (*William Claus Papers* [P.A.C.], vol. 20, no. 14) has entries of the labour which Claus hired at Niagara from 1809 to 1819. A substantial portion of the men hired were soldiers. The irregularity of their attendance suggests that they could only escape their duties at the neighbouring posts for a few days at a time.

154 Stanley, *op. cit.*, p. 145.

155 Scadding, *Toronto of Old . . .*, pp. 371-72.

156 Stanley, *op. cit.*, p. 146.

157 J.H. Aitchison, "The Development of Local Government in Upper Canada, 1783-1850" (Ph.D. thesis, University of Toronto, 1953), p. 330. Scadding's expression was that in 1794, the road was "half completed" to the Thames River (*op. cit.*, p. 306).

158 *Ibid.*, pp. 390, 415-17; Aitchison, *loc. cit.*

159 Scadding, *op. cit.*, pp. 306-7.

160 Aitchison, *op. cit.*, p. 331.

161 *U.C.S.*, April 2, 1816, (Owen to Gore). The temper of the period was also demonstrated in the treatment of some shipyard workers hired at Quebec for duty at Kingston in 1815. No work having been found for these men, they were nevertheless given an allowance of ten days' pay at wartime rates, the officer in charge stressing that he was anxious "not to offend any class of H.M. Subjects" (*U.C.S.*, Feb. 21, 1816 [Owen to Loring]).

162 *U.C.S.*, Aug. 19, 1830 (Gower to Mudge).

163 Land grants are noted in *U.C.S.*, June 12, 1816 (Owen to Gore); Aug. 19, 1822 (Yarwood to Hillier); June 17, 1834 (Petition of A. Manahan). However, it was asserted in 1835 that free grants were being denied to all claimants (*U.C.S.*, Oct. 30, 1835 [Rowan to Sughrue]).

164 The first of the Companies arrived on the Canal on June 10, 1827 (*U.C.S.*, July 7, 1827 [By to Mann]).

165 In 1826 and 1827, it was anticipated that skilled labour (in particular) would be hard to get and hold on the Rideau. One reason for the construction of a stone bridge across the Ottawa River during the winter of 1826-27 was to prevent masons from going elsewhere (John McTaggart, *Three Years in Canada: An Account of the Actual State of the Country in 1826-7-8 . . .*, 2 vols. [London, 1829], vol. I, pp. 337-47). The preparation of cut stone was under-taken for the same purpose of holding artisans (*U.C.S.*, Nov. 4, 1827 [By to Maitland]); and it seems likely that the expected shortage of labour provided much of the inspiration for other paternalistic measures, such as the assignment of town lots to artisans. By underestimated the labour that would prove to be available, and thereby got himself into trouble. He advertised more contracts than he desired actually to make, in the expecta-tion that few would be taken. But the contracts were taken up readily, to an extent much beyond the work that could be paid for out of the funds alloted to By at the time. The more By tried to restrain the work, so as to eke out his money, the more energetic each contractor became in the hope of reali-zing a larger share of whatever funds appeared (*U.C.S.*, April 24, 1828 [By to Hillier]; June 10, 1828 [By to Mann]).

166 By emphasized how expensive Sapper and Miner labour was, in refusing an offer of additional companies (*U.C.S.*, June 10, 1828 [By to Mann]).

167 Colonel Stanley gives the impression that grants of land were promised to the Sappers and Miners at the time of their recruitment (Stanley, *Canada's Soldiers . . .*, p. 187). In fact, By first proposed that land should be promised, as a means of checking desertion, in September 1828 (*U.C.S.*, Sept. 11, 1828 [By to Mann]; L. Breault, *Ottawa Old and New* [Ottawa, 1946], p. 49). The authorization of 100 acres for each of the Sappers and Miners, upon completion of the work, is noted in *U.C.S.*, Aug. 30, 1831 (Clegg to Felton, enclosure). Much earlier, land had been given to immigrat labourers em-

ployed on the Rideau Canal; but the grants failed either to get land settled or to keep labour available. The lands granted were too far from the works to allow labourers to live on their lots and do canal work too, and those who chose to give up their work to attempt settlement were driven out by starvation, according to By. (*U.C.S.*, April 29, 1829 [A.C. Buchanan to Colborne and enclosures]). While the Sappers and Miners were employed on the Rideau, a similar force called "Staff Corps Companies" was similarly occupied on the Grenville Canal. These men also were offered grants of land upon completion of satisfactory service (*Q199* [P.A.C.], pp. 60, 73, 77).

168 *U.C.S.*, July 7, 1827 (By to Mann); June 10, 1828 (By to Mann); April 4, 1829 (By to Mann).

169 *U.C.S.*, n.d. (Oct, 1) 1827. FitzGibbon's plan, like some others of this period was in the feudal military tradition. It is distinguishable in this way from other plans, notably those put forward in the 1840s, which were not in reality plans for military units but programs for the importation of indentured labour into Canada on a large scale, and for the enforcement of the indentures by military sanctions. These latter plans were discussed above in connection with slavery.

170 Stanley, *op. cit.*, p. 208.

171 Stanley notes the hostility of Maritime employers in 1793 to the raising of provincial military units, which depleted the local labour market (*ibid.*, pp. 134-35).

172 *U.C.S.*, Dec. 26, 1838 (Macaulay to Arthur, enclosure).

173 The distress that prevailed in 1837 is set out in *U.C.S.*, May 8, 1837 (Carey to Head); May 18, 1837 (Hagerman to Joseph). It is indicated that unemployed construction workers left for the United States in very large numbers in the summer of 1838 (*U.C.S.*, May 21, 1839 [Roy to Macaulay]; Aug. 31, 1839 [Hawke to Harrison]). There was complaint that harvest labour was short in 1838 (*U.C.S.*, Aug. 3, 1839 [Moyle to –]); and labour in general was scarce in 1839 (*U.C.S.*, Dec. 31, 1839 [Rubidge to Hawke]; Dec. 12, 1839 [Petition of Scott and others]).

174 *U.C.S.*, July 31, 1839 (Thompson to Harrison); Aug. 28, 1839 (Thomas to Bucknall); Civil Secretary's *Letter Book*, 1839 (*G 16C*): vol. 53, July 26, 1839 (to Birdsall), and vol. 54, Aug. 20, 1839 (to Battersby); *P.S.O., C.W.,* 1840-1, nos. 454, 533, 1373, 1715; 1841-3, nos. 197, 653, 783, 3334. Cf. Aitchison, *op. cit.*, p. 332.

175 The negro troops were ready to proceed with road work, as in previous years, in April 1842. No further evidence on road work has been found. It seems likely that it was the force which had been used hitherto on roads, which was dispatched to the Welland Canal when trouble arose there in August 1842. (*P.S.O., C.W.*, 1841-3, no. 4274.)

176 Troops were first called out in connection with the Welland Canal in July 1842. The use of troops at this work, and at others, is described in *C60* (P.A.C.), *passim*. The troops at the Welland Canal were supposed to be removed in July 1847 (*P.S.O., C.W.*, 1846-7, no. 16757; *C61* [P.A.C.], p. 170). However, if the troops left, they were soon back, for they were stationed on the canal in 1849 and 1850 (*C317* [P.A.C.], pp. 219-21, 234-37, 260; *C318*, p. 147). After 1850, it was the railway construction labourers (at Hamilton) who were deemed to require the oversight of troops (*C318*, pp. 188-94, 222, 278-80).

177 Calvin, *Saga of the St. Lawrence*, pp. 91-92.

178 *Selkirk Papers*, vol. 75, pp. 84-85, 103; John Howison, *Sketches of Upper Canada* (3rd ed., Edinburgh and London, 1825), pp. 41-52.

179 *Selkirk Papers*, vol. 75, pp. 5-6.

180 *Ibid.*, p. 103. A description of the raftsmen in the second half of the nineteenth century appears in Calvin, *op. cit.*, pp. 64-82.

181 (J.C. Ogden), *A Tour through Upper and Lower Canada* (Litchfield, 1799) (P.A.C., Pamphlets I, no. 852), p. 60.

182 *Selkirk Papers*, vol. 75, p. 121.

183 Selkirk said that "Seamen [on the lakes] must be paid all year tho' only 6 months work" (*Selkirk Papers*, vol. 75, p. 6); though at another place he reported that the men were paid £4 per month for six months (which he may have regarded as double wages) and were discharged in winter (*ibid.*, p. 86). According to another authority, "sailors had to be brought up from the ocean and retained on pay during the five or six months when the harbors are frozen up" (J.M. and Edw. Trout, *The Railroads of Canada for 1870-1* [Toronto, 1871], p. 21). These references apply to the beginning of the century. It was stated in reference to 1843 that sailors were hired only for the season, though wages were much as before, and that "about two-thirds of the seamen on these lakes are supplied from among the hardy west Highlanders, chiefly Argyle fishermen. They go to the lakes during the summer, and work usually upon a farm in the winter season..." (*Views of Canada and the Colonists, By a Four Year Resident* [Edinburgh, 1844], p. 141). The sailors of 1804 were said to be mostly deserters from English ships at Quebec (*Selkirk Papers*, vol. 75, p. 270). Some categories of skilled labour were evidently still more scarce on the lakes than sailors. There is an account of the extraordinary pains taken to retain a chief engineer, about 1827, in Scadding, *Toronto of Old*, p. 556.

184 At the beginning of the nineteenth century, carpenters were often brought to Upper Canada from New York (*Selkirk Papers*, vol. 75, p. 6; Trout, *op. cit.*, p. 21). Itinerant tradesmen were also available, but probably were not very skilled or reliable (W. Caniff, *History of the Settlement of Upper Canada* [Toronto, 1869], p. 216).

185 W.B. Munro, *The Seigniorial System in Canada* (New York, 1907), p. 13.

186 *Ibid.*, pp. 133, 228-33.

187 Andrew Picken, *The Canadas, as they at present commend themselves to the enterprize of Emigrants, Colonists, and Capitalists* (London, 1832), Appendix, paper C (pp. (xi-xxxii), "An account by P. Wright (1820) of his settlement and subsequent progress on the Ottawa", p. xxix.

188 *loc.cit.*

189 Norman MacDonald, *Canada, 1763-1841; Immigration and Settlement* (London, 1939), pp. 128-45, 186-201.

190 MacTaggart noted that "a settler of eminence is a kind of monopolist", setting out to own a grist mill, saw mill, fulling and carding mill, "smithery", and sometimes a distillery, tannery and a store, on all of which he expects to draw profits from his neighbours. (*Three Years in Canada...*, vol. I, pp. 197-98). Typical of a certain class of immigrant was one, Bastable, who wanted a township on the Ottawa to which he proposed to bring Irish settlers, and who considered that the government would welcome people like himself, "particularly in a Country where principles of democracy

inimical to our Establishments are to be guarded against..." (*U.C.S.*, Nov. 22, 1824 [Bastable to Maitland]; Feb. 16, 1825 [Bastable to Hillier]).

191 Sir Arthur Doughty, ed., *The Elgin-Grey Papers, 1846-1852*, 4 vols. (Ottawa, 1937), vol. III, pp. 1080, 1085-87, 1100-111; vol. IV, pp. 1431, 1447-57.

192 It failed in 1838. A large farmer, complaining in that year of the shortage of labour, said , "The success of the more respectable class of Farmers in this Country must very much depend on the annual arrival of labouring emigrants to supply the places of those Farm Servants who from the high wages obtained here, are soon enabled to be independent of their employers..." (*U.C.S.*, Aug. 3, 1838 [Moyle to P.S.]). Actually, it probably took a labourer at least four years to save enough to set up for himself. (R.L. Jones, *History of Agriculture in Ontario, 1613-1880* [Toronto, 1946], pp. 55n., 67).

193 *Selkirk Papers*, vol. 75, p. 111.

194 *U.C.S.*, July 9, 1839 (memorandum on the five-acre system by Hawke); Aug. 13, 1834 (petition of the inhabitants of Sunnivale); March 14, 1835 (Young to Hawke).

195 Canniff, *History of the Settlement of Upper Canada*, p. 591.

196 *Selkirk Papers*, vol. 75, p. 76.

197 *Ibid.*, p. 88.

198 B.P. and C.L. Davis, *The Davis Family and the Leather Industry, 1834-1934* (Toronto, 1934), pp. 87-88.

199 *Selkirk Papers*, vol. 75, p. 88; W.S. Fox, ed., *Letters of William Davies, Toronto, 1854-1861* (Toronto, 1945), pp. 55, 59.

200 H.A. Innis, *The Cod Fisheries; The History of an International Economy* (New Haven, 1940), pp. 278-86, 306, 356-57.

201 Howison, *Sketches of Upper Canada*, pp. 95-97, 127-29; C.W. Mixter, ed., *The Sociological Theory of Capital; Being a Complete Reprint of The New Principles of Political Economy, 1834* (original by John Rae, New York, 1905), pp. 119, 123-24; Elsie M. Murray, "An Upper Canada 'Bush Business' in the Fifties", *Papers and Proceedings of the Ontario Historical Society*, XXXVI (1944), pp. 41-47, which casts useful light on the downfall of this system.

202 R. Blanchard, *L'Est du Canada français*, 2 vols. (Paris and Montreal, 1935), vol. II, pp. 67-70.

Chapter III

1 *Census of Canada*, 1870-1, vol. IV, presents all the censuses taken before 1871 in considerable detail. *Census of Canada*, 1931, vol. I, presents summaries of population only for these.

2 What groups to include in a labour force is a matter of judgment. In current computations, great numbers are excluded because they do not offer themselves in the labour market, although under slightly different conditions than those that exist, they would do so. Any attempt to compute a labour force for the nineteenth century would have to face, in a far more acute form, the problem of how to handle persons not actively in the labour market, who nevertheless could be induced to take employment under some circumstances. The potential labour force, on the other hand, a fairly

stable conception, could be calculated pretty accurately for many dates, if there was any advantage in making the calculation. But general population figures seem to offer all the information required.

3 See Table I. The estimate that the French-Canadian population increased at 3 per cent per year is based on the fact that this rate gives close results when applied to years for which population figures are available. But the fit is far from perfect or consistent, which may indicate variations in the rate of population growth or errors in the censuses, or both. Cf., Jacques Henripen, "From Acceptance of Nature to Control", *Canadian Journal of Economics and Political Science*, February 1957, pp. 10-19, especially p. 14.

4 *Census of Canada*, 1870-1, vol. IV, p. 74.

5 See Table III, Immigration, pp. 82-84. The numbers coming to Canada from England became predominant in this decade, especially after 1854, but it may still be asked if these immigrants were English, or Irishmen sailing from Liverpool — as the immense "English" contribution of 1847 was (G.N. Tucker, *The Canadian Commercial Revolution, 1845-1851* [New Haven, 1936], p. 157n). There are good reasons for thinking that immigrants from England in the 1850s were mostly English. One is that the English immigration was particularly noticeable after 1854, when Irish migration is known to have dropped sharply, and especially so to Canada. Another is that great numbers of English are known to have come to Canada, to take the skilled wage-earning jobs that opened in the 1850s. Finally, the 1861 census of British-born in Canada supports this view.

6 Lunn, *Economic Development in New France, 1713-1760*, p. 109.

7 *Ibid.*, p. 440.

8 The extreme confinement of the French-Canadian population to agriculture was stressed by Georges Langlois, *Histoire de la population canadienne-française* (Montreal, 1934). Langlois cited data to show that about 22 per cent of the French population was urban in 1760, whereas only about 14 per cent was urban in 1861 — i.e., the rural proportion of a greatly enlarged population increased over this century from 78 per cent to 86 per cent (pp. 183-84). As a matter of fact, the dependence of the French on agriculture seems to have been greater still in 1851, when Montreal and Quebec held only 7½ per cent of the French population of Lower Canada, as against 8½ per cent in 1861, and 10½ in 1871. Simon Goldberg, in his "The French-Canadians and the Industrialization of Quebec" (M.A. thesis, McGill University, 1940), also stressed the degeneration of French society into one almost exclusively rural (p. 92), and provided an able analysis of the causes and significance of this phenomenon.

9 *Selkirk Papers*, vol. 75, p. 146.

10 *Baby Collection* (P.A.C.), vol. 3, no. 372, Viger to Berczy, Oct. 7, 1811.

11 (A Backwoodsman), *Statistical Sketches of Upper Canada for the Use of Emigrants* (3rd ed., London, 1833) (P.A.C., Pamphlet no. 1400), p. 63.

12 Q222 *pt. 1* (P.A.C.), Aylmer to Aberdeen, April 6, 1835: return of prices and wages in Lower Canada for 1834.

13 Earl of Durham, *Report on the Affairs of British North America* (4th ed., London, 1930), pp. 17-18.

14 John Howison, *Sketches of Upper Canada*, p. 18.

15 Joseph Bouchette, *The British Dominions in North America . . .*, 2 vols. (London, 1832), vol. I, p. 406.

16 W.S. Reid, "The Habitant's Standard of Living on the Seigneurie des Mille Isles, 1820-1850", *Canadian Historical Review*, September 1947, p. 275. In this same period, the French-Canadians on the Ottawa were reported to suffer from "Charbon", a deficiency disease like yaws which produced black spots (McTaggart, *Three Years in Canada...*, vol. II, p. 20).

17 Lower Canada, *Journal of the Assembly*, 1817, app. E; 1821-2, p. 44.

18 Durham, *Report*, p. 217.

19 *Report of the Select Committee of the Legislative Assembly appointed to Inquire into the Causes and Importance of the Emigration which takes place annually from Lower Canada to the United States* (Montreal, 1849) (P.A.C., Pamphlet I, no. 2142), p. 61.

20 Innis, *The Fur Trade in Canada*, p. 319.

21 Sir Arthur Helps, *Life and Labours of Mr. Brassey, 1805-1870* (3rd ed., London, 1872), p. 197. The enclosed quotation is presumably the statement of one, Rowan, who recruited French-Canadians under Brassey's instructions and attempted to work them. Helps added, "These men, however, though their powers were but feeble, proved to be of great use, inasmuch as their coming prevented the stalwart men from leaving" (*loc. cit.*). In Thomas Brassey, *Work and Wages* (New York, 1872), pp. 86-87, it is stated that French-Canadian labourers on the Grand Trunk were paid 3s. 6d. per day, while "Englishmen" received from 5s. to 6s. per day, "but it was found that the English did the greatest amount of work for the money".

22 Canada, *Sessional Papers, 1852*, app. CCC. Cf. Munro, *The Seigniorial System in Canada*, p. 218 and *passim*.

23 Munro, *op. cit.*, p. 218. Munro stated (p. 85) that the lands granted under seigneurial tenure averaged over 1,000 arpents per holding. This figure could only be obtained by including the seigneurial reserves, and refers to the year 1760. New English-speaking seigneurs practised various abuses (pp. 220, 233), including the diversion of what the habitants regarded as their patrimony to others. However, the abuses did not constitute a reason why habitants should desire a conversion of seigneurial lands to freehold, since even the slight existing restraints on the seigneurs would then be removed.

24 *Ibid.*, p. 227, quoting Lord Aylmer in 1830. Aylmer vigorously contested the anti-feudal views of James Stephen.

25 Canada, *Sessional Papers, 1852*, app. CCC. The statement is by Parent, Assistant Provincial Secretary, and a very intelligent observer.

26 *Q198 pt. 1*, p. 93. Stephen was commenting on an application for a new seigneury, March 3, 1832. Another of his objections was that the seigneuries could not be overpopulated, since they were so badly cultivated. It was in reply to views like this that men familiar with the province, like Aylmer, wrote to the contrary.

27 Lunn, *Economic Development in New France*, p. 111.

28 Mason Wade, *The French-Canadians, 1760-1945* (Toronto, 1955), p. 118.

29 V.C. Fowke, *Canadian Agricultural Policy: The Historical Pattern* (Toronto, 1946), pp. 79-82. For Canadian exports of wheat and flour, see Innis and Lower, *Select Documents in Canadian Economic History, 1783-1885*, p. 265.

30 Munro, *The Seigniorial System in Canada*, pp. 82-84.

31 *Ibid.*, p. 235.

32 Durham, *Report*, p. 217.

[33] See Table II, Lower Canada: Pressure on the Land. p. 71.

[34] For Church insistence on grain crops (which carried tithes) see Munro, *The Seigniorial System in Canada*, pp. 184, 234. Dr. Cooper believed that the seigneurs forced the habitants to concentrate on grain (J.I. Cooper, "French-Canadian Conservatism in Principle and in Practice, 1873-1891" [Ph.D. thesis, McGill University, 1938]). It seems to me that there is room to doubt whether the seigneurs were responsible to any great extent for the concentration on grain growing, and that peasant inertia was probably the chief factor in maintaining this emphasis.

[35] Fowke, *Canadian Agricultural Policy*, p. 88. Fowke, in his discussion of French agricultural methods (p. 82 ff.), noted complaints of deterioration of the soil in Lower Canada, but does not appear to have attached much importance to them.

[36] *Report of the Select Committee on Emigration* (1849), pp. 69-70.

[37] W.H. Parker, "The Geography of the Province of Lower Canada in 1837" (Ph.D. thesis, Oxford University, 1958), pp. 55-58; "A Revolution in the Agricultural Geography of Lower Canada, 1833-1838", *Revue canadienne de geographie*, 1957, pp. 189-94; "A New Look at Unrest in Lower Canada in the 1830s", *Canadian Historical Review*, September 1959, pp. 209-17.

[38] M.L. Hansen, *The Atlantic Migration, 1607-1860: A History of the United States* (Cambridge, Mass., 1940), pp. 9, 211-19 and *passim*; H.J. Habakkuk, "Family Structure and Economic Change in Nineteenth-Century Europe", *Journal of Economic History*, XV, no. 1 (1955), pp. 1-12.

[39] Durham, *Report*, p. 201. A considerable part of the early movement appears to have consisted of young men from Yamaska and Nicolet who took seasonal work in American brickyards. Cf. *Report of the Select Committee on Emigration* (1849), p. 6. For the permanent emigration of 1776 see E. Hamon, *Les Canadiens-Français de la Nouvelle-Angleterre* (Quebec, 1891), p. 164.

[40] M.L. Hansen and J.B. Brebner, *The Mingling of the Canadian and American Peoples* (New Haven, 1940), pp. 115-17 and ff.

[41] *Report of the Select Committee on Emigration* (1849), pp. 9, 61.

[42] Especially 1863-64 (Hansen and Brebner, *op. cit.*, pp. 151-52; Hamon, *op. cit.*, p.169).

[43] Hansen and Brebner, *op. cit.*, p. 168.

[44] Cited in Goldberg, "The French-Canadians and the Industrialization of Quebec", p. 131, citing Durand. The United States census of 1890, which was the first to distinguish French-Canadians, showed 302,000 of them. The census of 1900 showed 395,000.

[45] Cooper, "French-Canadian Conservatism", pp. 48-50; Wade, *The French-Canadians*, pp. 260, 336.

[46] Hansen and Brebner, *op. cit.*, pp. 124, 127, 141, 170-71.

[47] Cooper, "French-Canadian Conservatism", pp. 50-51.

[48] *Ibid.*, pp. 63-64.

[49] *Report of the Select Committee on Emigration* (1849), pp. 6, 7, 9.

[50] Hansen and Brebner, *op. cit.*, p. 169.

[51] C.E. Rouleau, *L'Emigration: ses principales causes* (Quebec, 1896), *passim*; Hamon, *Les Canadiens-Français de la Nouvelle-Angleterre*, p. 5.

[52] Bouchette, *The British Dominions in North America*, vol. I, p. 406.

[53] Habitant debts mounted in the 1840s in the form of cash loans, usually at

exorbitant rates of interest. Debt, and dispossession, increased also because of increasing demands made by the seigneurs. (Reid, "The Habitant's Standard of Living", pp. 276-77.)

54 *Report of the Select Committee on Emigration* (1849), p. 37.

55 Rouleau, *L'Emigration*, pp. 8-13, 24, 31, 37-38, 105-6.

56 Cooper, "French-Canadian Conservatism", p. 47.

57 Wade, *The French-Canadians*, pp. 335-36.

58 Rouleau, *L'Emigration*, p. 115 ff.

59 Wade, *The French-Canadians*, p. 189; S.B. Ryerson, *French Canada* (Toronto, 1943), pp. 115-16.

60 *Ibid.*, pp. 116-18. Munro stated that abolition harmed the seigneurs in the sense that the selling price of their land was lessened by it. From the French viewpoint — that seigneurs held their lands in trust for others — the seigneurs got too much. The biggest seigneur in the province by far, the Church, was neutralized by its exemption from the abolition.

61 *Report of the Select Committee on Emigration* (1849), p. 38.

62 Agitation was carried on in terms of the *censitaire's* right to use water power (Ryerson, *French Canada*, pp. 119-20), but in this respect cannot be taken seriously. The man who was uninterested in maximizing returns from his small farm would not have been capable of doing much with a mill site. What may have been behind the agitation was the desire of English-speaking capitalists to get power sites on easier terms than seigneurs offered. Some English-speaking seigneurs had promoted considerable industrial development on their estates (Wade, *The French-Canadians*, p. 116).

63 Cooper, "French-Canadian Conservatism", pp. 46, 48.

64 Goldberg, "The French-Canadians and the Industrialization of Quebec", p. 89.

65 Wade, *The French-Canadians*, pp. 336-37; Hansen and Brebner, *The Mingling of the Canadian and American Peoples*, p. 171.

66 Horace Miner, *St. Denis: A French-Canadian Parish* (Chicago, 1939); Everett Hughes, *French Canada in Transition* (Chicago, 1943).

67 A.L. Burt, *The Old Province of Quebec* (Minneapolis, 1933), p. 363.

68 Hansen and Brebner, *op. cit.*, pp. 69-70, 86-87; M.L. Hansen, *The Immigrant in American History* (Cambridge, Mass., 1940), pp. 179-82.

69 Hansen and Brebner, *op. cit.*, p. 90, cites an estimate that the population of Upper Canada was 80 per cent American. It is difficult to conceive how the English-speaking population of either Canada could have contained less than this proportion of Americans. There may have been as many as 150,000 Americans in Canada in 1815.

70 McTaggart, *Three Years in Canada*, vol. I, p. 329.

71 Many canal contractors in Canada, probably more than half the total, were Americans. The American willingness to dig canals, so long as the work was on a contract basis, was demonstrated in the construction of the Erie Canal in which a great number of small contractors were employed. See A.B. Hulbert, *The Erie Canal*, Historic Highways of America, vol. 14 (Cleveland, 1904), pp. 119-21.

72 Jones, *History of Agriculture in Ontario*, pp. 55, 60, 67.

73 *Selkirk Papers*, vol. 75, p. 76.

74 *Ibid.*, vol. 74, p. 256. The Genesee tract in New York is under discussion,

but presumably the statement would apply to Canada and to most Americans.

75 *Ibid.*, p. 197. The point was made by many other observers.

76 Hansen and Brebner, *The Mingling of the Canadian and American Peoples*, p. 77.

77 Canniff, *History of the Settlement of Upper Canada*, pp. 212-16.

78 McTaggart wrote about 1827, "Politics here are making a stir amongst the cobblers — I never mind them..." (*Three Years in Canada*, vol. I, p. 190). The position of the Toronto artisans is indicated in the solid support they gave Mackenzie politically and in Mackenzie's celebrated remark, in planning his rebellion, "that we should instantly send for Dutcher's foundrymen and Armstrong's axe-makers, all of whom could be depended on..." (Charles Lindsay, *The Life and Times of Wm. Lyon Mackenzie*, 2 vols. [Toronto, 1862], vol. II, pp. 19, 47, 55). There was a hubbub in 1816 when nine workmen at Gananoque provided a "riotous" celebration of July 4. They were required to promise not to repeat this gesture (*U.C.S.*, July 1816 [? to McMahon]). Also, "it is notorious that, a large proportion of disaffected persons forms the population of many large Towns in this, if not the other Province" (*U.C.S.*, Oct. 10, 1838 [Crooks to Macaulay]). Tanners are named particularly.

79 For the quality of Americans as pioneers and farmers, see Jones, *History of Agriculture in Ontario*, p. 54. Fowke, *Canadian Agricultural Policy*, pp. 82-88, argues the merit of extensive agriculture for Lower Canada. The argument would seem to apply to Upper Canada. Howison, *Sketches of Upper Canada*, p. 83, expressed his disapprobation of the "peasantry" of the Niagara Peninsula who were "indifferent to everything but field crops". See also p. 208.

80 *Selkirk Papers*, vol. 74. p. 250, refers to the houses of the settlers in northern New York, who had been there not more than a dozen years. On the accumulative powers of the American settlers, see Howison, *op. cit.*, p. 151.

81 Jones, *History of Agriculture in Ontario*, p. 40.

82 *Ibid.*, pp. 214-15; Scobie and Balfour, *Canadian Almanac*, 1857, p. 33.

83 Jones, *op. cit.*, p. 199.

84 Hansen and Brebner, *The Mingling of the Canadian and American Peoples*, pp. 79-80.

85 McTaggart, *Three Years in Canada*, vol. I, pp. 195, 204-7. The attitude of the pioneers, especially with respect to religion and clergymen, has been carefully explored by S.D. Clark. See his *Social Development of Canada*, pp. 284-86, 290-301.

86 McTaggart, *op. cit.*, vol. I, pp. 322-24; *Selkirk Papers*, vol. 74, p. 291. Selkirk also reported that all the Yankees, though sober people originally, took to drink when they went to the backwoods.

87 McTaggart, *op. cit.*, p. 195.

88 The role of Americans in the Canadian lumber industry has been noted in A.R.M. Lower, *The North American Assault on the Canadian Forest* (Toronto, 1933), pp. 93, 112-13, 140. Some of the Americans who came to Canada to establish manufacturing industries in Hamilton are noted in "Saturday Musings", *Hamilton Spectator* (Hamilton Public Library collection), especially n.d., 1917, "The First Industries in Canada". Americans were prevented from participating in Canada's first mining boom by legal restraints.

89 The literature of nineteenth-century Atlantic migration is very extensive.

The following works are relevant to the present study: Marcus Lee Hansen, *The Atlantic Migration, 1607-1860: A History of the Continuing Settlement of the United States,* ed. A.M. Schlesinger (Cambridge, Mass., 1940); W.F. Willcox, ed., *International Migrations,* 2 vols. (New York, 1931); S.C. Johnson, *A History of Emigration from the United Kingdon to North America, 1763-1912* (London, 1913); W.A. Carrothers, *Emigration from the British Isles, with Special Reference to the Development of the Overseas Dominions* (London, 1929); Helen I. Cowan, *British Emigration to British North America, 1783-1837* (Toronto, 1928); Brinley Thomas, *Migration and Economic Growth: A Study of Great Britain and the Atlantic Economy* (Cambridge, England, 1954); W.F. Adams, *Ireland and Irish Emigration to the New World from 1815 to the Famine* (New Haven, 1932); M.L. Hansen, *The Immigrant in American History* (Cambridge, Mass., 1940); L.G. Reynolds, *The British Immigrant: His Social and Economic Adjustment in Canada* (Toronto, 1935); R.T. Berthoff, *British Immigrants in Industrial America, 1790-1950* (Cambridge, Mass., 1953).

[90] Studies of emigration often devote a great amount of space to emigration movements sponsored by governments, possibly because these are well documented and because they illuminate government policy. Organized movements are useful to inaugurate emigration from areas previously unfamiliar with it. But machinery to cater to individual migration appears to have developed very quickly in European countries, and the individual method becomes the accepted one. Sponsored emigration accounts for only a tiny portion of the emigrants who went to America.

[91] See Adams, *Ireland and Irish Emigrants,* p. 415.

[92] Frances Moorhouse, "Canadian Migration in the Forties", *Canadian Historical Review,* Dec. 1928, p. 327.

[93] While the capitalistic labour market requires a surplus of labour large enough to meet upward fluctuations in demand, large numbers of chronically unemployed persons tend to be harmful to it. For example, the development of India is hampered by its chronic oversupply of labour, promoting low wage rates which hinder the introduction of new techniques in production and deny on the other side a mass market for all but a few articles. In Canada, the danger from the appearance of large numbers of unemployed persons, beyond the risk of political instability, has been that public welfare expenditures might become so burdensome as to neutralize the economies of the capitalistic market. See H.C. Pentland, "The Development of a Capitalistic Labour Market in Canada", *Canadian Journal of Economics and Political Science,* November 1959, especially p. 450-51.

[94] Hansen, *Atlantic Migration,* pp. 179-81; Adams, *Ireland and Irish Emigration,* pp. 162, 204, 306, 401-2. The ramifications of unused capacity were explored by H.A. Innis in "Unused Capacity as a Factor in Canadian Economic History", *Canadian Journal of Economics and Political Science,* Feb. 1936, pp. 1-15, and in "Transportation as a Factor in Canadian Economic History", *Problems of Staple Production in Canada* (Toronto, 1933), pp. 1-17.

[95] Hansen, *Atlantic Migration,* p. 9.

[96] H.J. Habakkuk, "Family Structure and Economic Change in Nineteenth-Century Europe", especially pp. 7-9.

[97] *Ibid.* pp. 11-12.

[98] "... in all the western European countries the industrial wage earners were the most prolific class" (*ibid.,* p. 11).

[99] Mediterranean and Slavic migration, 1890-1914 (or later), is not directly

relevant to this study; but it is noteworthy that these late peasant migrants resembled the early Celtic migrants in many respects. They were driven out by population and land pressures analogous to (but not the same as) those of Celtic lands, and their behaviour patterns in America were like those of Celts and French-Canadians. On the demand side, this migration may be supposed to reflect the expansion of unskilled and semi-skilled employment in mass production industries, superseding the stage of dependence on craftsmen which had supported Teutonic immigration.

[100] Hansen, *Atlantic Migration*, pp. 181-82.

[101] See Report of a Committee of the Executive Council of Canada, May 9, 1863, *State Book Y* (P.A.C.), pp. 470-74. The important section of the report is quoted in Fowke, *Canadian Agricultural Policy*, p. 136.

[102] Hansen, *Atlantic Migration*, p. 10; Habakkuk, "Family Structure and Economic Change in Nineteenth-Century Europe", pp. 2,10.

[103] Cowan, *British Emigration to British North America*, chs. 7 and 8, especially pp. 180-85. Norman Macdonald, *Canada, 1763-1841: Immigration and Settlement*, pp. 24-25, suggests a total number of about 35,000.

[104] Hansen, *Atlantic Migration*, pp. 89, 265; Carrothers, *Emigration from the British Isles*, pp. 50, 80-81; A.C. Buchanan's Report of Dec. 28, 1831 in *MG 22: 34* (P.A.C.).

[105] J.H. Clapham, "Irish immigration into Great Britain in the nineteenth century", *Bulletin of the International Committee of Historical Sciences*, vol. V, part III, no. 20 (July 1933), pp. 598-602.

[106] "In order to obtain information upon the subject of disturbances which occurred in a Cotton manufactory at Glasgow within the last eighteen months, it was found necessary to induce certain persons to give evidence upon the subject which rendered it unsafe for these persons, 44 in number, to remain in Scotland. They were therefore sent to Quebec...", where sums of money were paid them. (*C316* [P.A.C.], p. 170. The item is dated Oct. 22, 1838.)

On emigration societies among workmen see Hansen, *Atlantic Migration*, p. 143; Cowan, *British Emigration to British North America*, pp. 84-86.

[107] Reynolds, *The British Immigrant*, pp. 79-80, 90-96.

[108] Berthoff, *British Immigrants in Industrial America*.

[109] Note the easy adjustment of skilled British ironworkers into the upper part of the St. Maurice Forges hierarchy and the easy assimilation of non-Anglo-Saxon skilled workers to the predominantly British skilled group: some French from very early times, and later, Germans and Norwegians.

[110] There is an account of the first strike in the Hamilton and London shops of the Great Western Railway, in 1856, in the *Hamilton Herald*, July 5, 1902 (Hamilton Public Library). Most of the workers involved had recently come from Britain. Their strike was conducted in a very orderly manner and negotiated to a compromise agreement.

[111] Statistics of racial origin, and commentaries, indicate that very few Welsh came to Canada. A few came as miners or ironworkers; but the relatively small numbers of miners who came to Canada came more commonly from Cornwall. Presumably the opportunities for the Welsh in the United States, the precedents established of migration to that country, and the Welsh system of group emigration, worked to prevent the Welsh making any significant contribution to the people of Canada.

112 A.C. Buchanan estimated that Canada had received 5,000 overseas immigrants between 1790 and 1815 (*Q198 pt. 1* [P.A.C.], p. 59). Cowan, *British Emigration to British North America*, accounts for rather less than 5,000 to Canada before 1815, though for about double that number to the Maritime Provinces. J.T. Culliton, *Assisted Emigration and Land Settlement with Special Reference to Western Canada*, McGill University Studies no. 9 (Montreal, 1928), p. 13, asserted that 20,000 immigrants a year entered the St. Lawrence, 1769-74; but it appears that there is no basis for this claim and that it is quite wrong.

113 An excellent discussion of changes in the Scottish economy is furnished by Henry Hamilton, *The Industrial Revolution in Scotland* (Oxford, 1932). The early chapters deal with Scottish agricultural organization, and with the crucial linen trade. There is an analysis of Scottish emigration, pp. 68-75.

114 J.H. Clapham, *An Economic History of Modern Britain: The Early Railway Age, 1820-1850* (Cambridge, 1926), pp. 61-62.

115 Cowan, *British Emigration to British North America*, ch. 2, provides a clear account of Scottish immigration to Canada. See also Norman Macdonald, *Canada: Immigration and Settlement*, especially ch. I.

116 *Selkirk Papers*, vol. 75, pp. 103-4.

117 Howison, *Sketches of Upper Canada*, p. 188.

118 *Ibid*, p. 69. See Clark, *The Social Development of Canada*, p. 264, "These Highlanders are a stiff-necked race. They will not understand anything ... that requires thought."

119 *Selkirk Papers*, vol. 75, p. 108.

120 *Ibid.*, p. 109.

121 Selkirk offered a similar judgment of the Highlanders in Prince Edward Island (*ibid.*, vol. 74, p. 109).

122 Robert Gourlay, who had farmed both in Scotland and England, discusses the great superiority of the Scottish over the English farm labourer in *Statistical Sketches of Upper Canada*, vol. III, pp. civ, cvii and ff.

Chapter IV

1 George O'Brien, *The Economic History of Ireland in the Seventeenth Century* (Dublin and London, 1919), pp. 122-23, citing Petty's estimate. But the most careful student of Irish population proposes a larger total population for 1687 of 2,167,000, which may imply a smaller proportion of Protestants (K.H. Connell, *The Population of Ireland, 1750-1845* [Oxford, 1950], p. 25).

2 *Ibid.*, pp. 25, 238-40.

3 *Ibid.*, pp. 240-41.

4 Conrad Gill, *The Rise of the Irish Linen Industry* (Oxford, 1925), pp. 157-58, 233-35, 237-43.

5 An analysis of modern Irish society, which suggests a great deal also about early nineteenth-century society, is provided by Conrad M. Arensberg, *The Irish Countryman* (New York, 1937) and C.M. Arensberg and S.T. Kimball, *Family and Community in Ireland* (Cambridge, Mass., 1940).

6 Gill, *The Irish Linen Industry, passim.*

7 Gill, a judicious scholar, believed land tenure to be the only explanation

required (*op. cit.*, p. 23). But the evidence seems to support a contrary view. See Adams, *Ireland and Irish Emigration*, p. 37-38.

[8] Hansen, *Atlantic Migration*, p. 49; H.J. Ford, *The Scotch-Irish in America* (2nd ed., New York, 1941), pp. 186-87.

[9] *Ibid.*, p. 50.

[10] Gill, *The Irish Linen Industry*, pp. 178-80; Adams, *Ireland and Irish Emigration*, pp. 72-75.

[11] Ford, *The Scotch-Irish in America*, pp. 179-80; C.K. Bolton, *Scotch Irish Pioneers in Ulster and America* (Boston, 1910) p. 35.

[12] *Q vol 198*, pt. 1, p. 63.

[13] For example, Ford, *The Scotch-Irish in America*, pp. 198, 523, and *passim*.

[14] Adams, *Ireland and Irish Emigration*, pp. 69-70, particularly contesting the data offered by C.A. Hanna, a "Scotch-Irish" historian.

[15] Berthoff, *British Immigrants in Industrial America*, p. 5.

[16] Hansen, *Atlantic Migration*, p. 15, 24, 49; Hansen, *The Immigrant in American History*, pp. 60-67; F.J. Turner, *The Frontier in American History* (New York, 1920), pp. 24, 347.

[17] Ford, *The Scotch-Irish in America*, on pioneering in Ulster, pp. 115-16; in America, pp. 199-200, 211, 216, 220. Ford also argued that the stern hostility of Ulstermen to British government, and the cohesiveness and activity of this group in the revolutionary cause, was crucial to the success of the American Revolution, p. 458 ff.

[18] Except insofar as Loyalists were of this extraction, as some certainly were. But the McNutt settlement in the 1760s put about a thousand Ulster settlers in the Maritimes.

[19] Adams, *Ireland and Irish Emigration*, p. 352.

[20] Hamilton, *The Industrial Revolution in Scotland*, pp. 120-21.

[21] According to Buchanan in 1828, "A great proportion of those [immigrants] from Ireland are from the Province of Ulster, and their feelings strongly attach them to prefer a British Colony where they could still enjoy the blessings of our Constitution..." (*MG 22: 34*, Dec. 1828, Buchanan's evidence to a Committee on Roads and Internal Communications).

[22] (Dr. Dunlop), *Statistical Sketches of Upper Canada*, p. 8: "the linen weavers from the north of Ireland make the best choppers, native or imported, in the province..." Nicholas Flood Davin, *The Irishman in Canada* (Toronto, 1877), though a partisan writer, is reasonably judicious in exhibiting the evidences of Irish success, e.g. pp. 63-64, 245.

[23] Adams, *Ireland and Irish Emigration*, p. 222n.

[24] Davin, with some justice, called the early period of Canadian constitutional development "the Irish period" (*op. cit.*, ch. XIII). But the efforts of Irish Protestants to establish working arrangements with French Canada, within the Macdonald coalitions, was at least as important. See D. Creighton, *John A. Macdonald: The Young Politician* (Toronto, 1952), pp. 195-96.

[25] O'Brien, *The Economic History of Ireland in the Seventeenth Century*, p. 214.

[26] *Ibid.*, pp. 102, 214.

[27] Clapham, "Irish immigration into Great Britain in the nineteenth century", pp. 596-98.

[28] *Ibid.*, p. 601. Also J.H. Clapham, *An Economic History of Modern Britain: The Early*

Railway Age, 1820-1850 (Cambridge, 1926), pp. 57-66.

29 Adams, *Ireland and Irish Emigration*, p. 60; R.G. Lounsbury, *The British Fishery at Newfoundland 1634-1763* (New Haven, 1934), pp. 245-50.

30 Migration to New England, which was early and substantial, is noted in Lounsbury, *op. cit.*, pp. 172, 178-79, 253, 259, 325. Selkirk reported in 1803 that there were about 800 Irish fisherman in Nova Scotia, mostly from Newfoundland. Arrivals were occasional: "there is no regular migration" (*Selkirk Papers*, vol. 74, p. 174).

31 *Ibid.*, vol. 75, pp. 216-17. It would appear that a great part of the Irish in New York were indentured at this time. Selkirk also found a sizable group of Irish Catholics in Albany, a good part of whom, with their priest, had emigrated because they were implicated in the rebellion of 1798 (*ibid.*, vol. 74, p. 227).

32 Hansen, *Atlantic Migration*, p. 97; Adams, *Ireland and Irish Emigration*, p. 111 and map facing p. 1.

33 *Ibid.*, map facing p. 158; Hansen, *Atlantic Migration*, pp. 249-51.

34 Adams, *Ireland and Irish Emigration*, pp. 94, 160-62; Hansen, *Atlantic Migration*, p. 181.

35 *Q198, pt. 1*, p. 59.

36 *Q198, pt. 2*, pp. 358, 368; *MG 22: 34*, Buchanan statement, Dec. 1828. It is clear that a substantial body of Irish Catholics had settled in Quebec by 1830, and probably also in Montreal; but it is implied that this group was not numerous elsewhere.

37 P.A.C., *Public Works I, Vol 16, Lachine Canal Accounts: Labourers 1819-1821*. The volume is misdated. The accounts run to Oct. 1824.

38 *MG 22: 34*, Petition of the President of the Montreal General Hospital, Jan. 25, 1826.

39 "There is scarcely a Hut, or Log-house here [Kingston] but is filled with Sick and Needy, who are suffering, not only from Disease, but also from Hunger, & from almost every other Misery concomitant upon the want of the *common Necessaries of Life*... principally occasion'd, I understand, by the numerous Labourers returning from the Rideau Canal, where they have fallen victims of Disease..." (*U.C.S.*, Oct. 24, 1827 [Tunney to Hillier]).

"One thing I cannot omit to mention, which is, the slovenly dilapidated appearance of many of the houses [in Kingston], so familiar to those who have travelled in Ireland...houses in a half unfinished state... half tumbling to the ground, for want of timely repairs" (A Canadian Settler, *The Emigrant's Informant: or, A Guide to Upper Canada* [London, 1834] [P.A.C., Pamphlets I, no 1477], p. 129).

40 In 1826, Bishop Macdonell asked for more money to support more Irish priests — he had three in a total of eight — because Irish labourers and artisans were spread over "the whole of this tract of country" (Toronto eastward?) (*U.C.S.*, Mar. 9, 1826 [Macdonell to Maitland]). Maitland, however, declared that Roman Catholics then formed no more than one-fifteenth, at the most, of the population of Upper Canada (*Q340*, p. 539). This would only allow 11,000 Catholics to Upper Canada, and allow no room for Irish whatever. There is a statistic that reports 36,435 Roman Catholics in Upper Canada, in 1829. Allowing for Scottish and French Catholics, this might leave room for 10,000 Irish. There were reported to be 413,000 Catholics in Lower Canada in 1831, when French-Canadians must have numbered about 400,000; but the defect of this census is that

68,000 people did not state a religious denomination. However, there were more Irish Catholics in Lower than in Upper Canada, and a total of about 25,000 in 1830 seems possible. The number certainly increased rapidly in the 1830s. By 1842, there were 65,000 Catholics in Upper Canada, the additions certainly being mostly Irish. In 1844, Catholics exceeded French-Canadians in Lower Canada by 48,000. The "effectively Irish" in Canada are therefore estimated at 85,000 for 1842-44. The phrase "effectively Irish" indicates those Irish Catholic in culture and behaviour, including those born outside Ireland. The concept was used by Clapham, in consideration of Irish resistance to assimilation.

[41] The number of persons in Ontario and Quebec who confessed to Irish origin in 1871 was 683,000. This total is estimated to consist of 260,000 Irish Catholics and 425,000 Irish Protestants. The estimates, which can only be approximate, are reached by balancing origins, religious denominations and places of birth.

[42] Cowan, *British Emigration to British North America*, pp. 103-7, 111-14.

[43] But it appears that the bulk of unemployed Irish labourers simply moved into Chicago or to other works (many hundreds from here appeared in Canada), and that some of the Irish who settled on the land did so only until new employment opportunities appeared. See W.V. Pooley, *The Settlement of Illinois from 1830 to 1850*, Bulletin of the University of Wisconsin, no. 220, History Series, vol. I, no. 4 (Madison, 1908), pp. 492-501; P.W. Gates, *The Illinois Central Railroad and Its Colonization Work* (Cambridge, Mass., 1934), p. 89; B.L. Pierce, *A History of Chicago*, vol. I (New York, 1937), pp. 70-71, 151, 166.

[44] In Nova Scotia, "few take up farming as they are not well received in the interior part of the country by the Methodists & New light people, who are there prevalent & dislike them as Catholics..." (*Selkirk Papers*, vol. 74, p. 174). Protestant hostility was a factor in Selkirk's abortive plan to settle these Irish, and others from New York, in Upper Canada (*ibid.*, vol. 75, pp. 300-301).

[45] An excellent description and analysis of the position of the Irish in Boston is provided in Oscar Handlin, *Boston's Immigrants, 1790-1865* (Cambridge, Mass., 1941). Descriptions to the same effect may be found for other cities in North America and Great Britain.

[46] Adams, *Ireland and Irish Emigration*, pp. 358-86, offers a useful analysis of the impact of Irish immigration on American social life and politics. For Great Britain, see Clapham, "Irish immigration into Great Britain in the nineteenth century".

[47] Adams, *op. cit.*, p. 206, ascribes Irish crowding into the large cities of America to poverty, to a "manufacturing background" — an absence of this background is more evident — and to the "natural gregariousness of the Irish people". Hansen, besides noting that a good many Irish did become farmers, and perhaps exaggerating their numbers, credits urban crowding to the attractiveness of a steady job, of the church and of good Irish company (*Atlantic Migration*, p. 302; *The Immigrant in American History*, pp. 159, 165). Hostile writers made still less effort to understand the Irish position. Wakefield said they made themselves "virtually slaves by means of their servile, lazy, reckless habit of mind, and their degradation in the midst of the energetic, accumulating, prideful, domineering, Anglo-Saxon race..." (*A View of the Art of Colonization*, p. 175). McTaggart, who is quoted below, had a somewhat similar, though less savage, view.

⁴⁸ The fact that the Irish would accept very inferior conditions to win employment, and sometimes act as scabs in order to replace their predecessors, has obscured the fact that the Irish were even more readier than others in the British Isles to combine and to form unions, which were often very successful. See Adams, *op. cit.*, pp. 56-57, 169; Gill, *The Irish Linen Industry*, pp. 106-15, 140-48, 226, 241-42. The remarkable and fairly successful trade union movement of north Britain of the late 1820s and early 1830s, led by John Doherty (a Roman Catholic Ulsterman) must have depended to a considerable extent on the ubiquity of Irishmen for its cohesion (S. and B. Webb, *History of Trade Unionism* [London, 1920], pp. 117-124). Wakefield blamed Chartism mostly on the Irish (*op. cit.*, p. 68).

⁴⁹ *Statistical Sketches of Upper Canada*, p. 99. Dunlop wrote of the Robinson settlers that, "having got quit of the feeling of hopelessness and despair of ever bettering their condition, that weighs down and paralyzes the Irish peasant in his own country, they have acquired the self-respect essential to respectability" (*ibid.*, p. 73).

⁵⁰ Thomas McKay, who built the masonry of the Lachine and Rideau Canals, is quoted as writing in 1837 "that during the first five years he employed from 100 to 300 men, 2/3 of whom were from Ireland, & that he never experienced any difficulty in keeping them quiet. During the second five year period he employed from 40 to 50 men & never saw or heard a quarrel among them" (L. Breault, *Ottawa Old and New* [Ottawa, 1946], p. 63). However, these may have been predominantly Ulstermen. Employers and officials who gave evidence in connection with the strikes of Irish Catholic labourers in Canada, in 1843, were inclined to say that the Irish were unmanageable by nature. However, the Paymaster of the Board of Works testified that during the previous year when the labourers had been employed directly by the Board of Works rather than by contractors, had been paid 3s. a day, and that in cash rather than store pay, and when some provision was made for shanties to house the men, the workmen were extremely well behaved. Two priests gave evidence to the same effect (*Sessional Papers, 1843*, app. T). Clapham noted that in 1826 the Irish in England were held to be "most exemplary" labourers, even when starving (*Economic History of Modern Britain: The Early Railway Age*, p. 58).

⁵¹ McTaggart, *Three Years in Canada*, vol. II, pp. 243, 245, 248. McTaggart was Chief Clerk on the Rideau Canal until June 1828. He did not like the Irish, but his description seems accurate enough. The "houses" pictured on the Ottawa tally closely with Irish huts of the nineteenth, and also of the seventeenth, century. (O'Brien, *Economic History of Ireland in the Seventeenth Century*, pp. 137-42). The man in charge of the Rideau Canal, Colonel John By, was a paternalistic (and dictatorial) employer who tried to provide services for his workmen. On the other hand, Rideau conditions were especially conducive to disease and accidents. Several hundred labourers died there of malaria.

⁵² Adams, *Ireland and Irish Emigration*, p. 33.

⁵³ *Sessional Papers, 1844-5*, app. U and app. Y, are collections of evidence of alleged outrages near the Canadian public works. There is a great deal more of such evidence in various other documents.

⁵⁴ On attempts to win religious democracy in the United States, see Adams, *op. cit.*, p. 345. The Rev. W.J. O'Grady came to the Roman Catholic community of Toronto in 1828. Some parts of the unsuccessful battle of O'Grady and his flock for congregational autonomy and liberal political

views appear in *U.C.S.*, Jan. 8, 1833 (Macdonell to P.S.); Jan. 29, 1833 (Memorial of R.C. Churchwardens); June 30, 1833 (Petition of McDougall and others). See J.J. Lepine, "The Irish Press in Upper Canada and the Reform Movement, 1828-1848" (M.A. thesis, University of Toronto, 1946).

55 *U.C.S.*, May 7, 1834 (Mackenzie to Rowan, enclosure).

56 *Loc. cit.* The quotation in Mackenzie's own report after investigating complaints.

57 *U.C.S.*, May 21, 1834 (Mackenzie to Rowan, enclosure).

58 *U.C.S.*, May 14, 1831 (Baby to Mudge, enclosure). There is a useful study of the provisions made in Canada to deal with destitution, and the relation of this to immigration, in Aitchison, "The Development of Local Government in Upper Canada, 1783-1850", ch. 36 (Poor Relief), pp. 644-60; also ch. 37 (Boards of Health), pp. 661-89.

59 *U.C.S.*, Nov. 28, 1831 (Annual Report of the York Hospital).

60 *U.C.S.*, Aug. 9, 1834 (Leslie to Rowan).

61 *U.C.S.*, n.d. (Jan.) 1838 (Petition of the Committee of the House of Industry).

62 *Loc. cit.*, enclosure. In 1839 there were reported to be 1,967 Roman Catholics in Toronto in a total population of 12,153 (*U.C.S.*, Aug. 21, 1839 [Daly's return]). The proportion of Irish Catholics assisted by the House of Industry was much smaller in 1838 than in 1837, probably because immigration was very small (*U.C.S.*, Mar. 13, 1839 [Second Annual Report of the House of Industry]). In subsequent years, the House of Industry often reported that it could not care for all indigents and appealed for more funds.

63 *U.C.S.*, Oct. 1, 1838 (Smith to Macaulay, enclosing Return of Convicts Discharged from the Penitentiary). Nearly all the Irish Catholic convicts were from the Home and Midland Districts — i.e., Toronto and Kingston. See Clark, *The Social Development of Canada*, pp. 247-50. However, Adams asserted (*Ireland and Irish Emigration*, p. 364) that the Irish share in crime in the United States was less than might have been expected and that only one-fifth of the convicts in Auburn Penitentiary were Irish.

64 Upper Canada, *State Papers*, vol. 24 (P.A.C.). Statements by R.B. Sullivan and A. Baldwin (forming one of several statements by members of the Executive Council on the best way to handle the immigrants expected in 1840), June 1840.

65 *U.C.S.* July 9, 1939 (Memorandum of A.B. Hawke). This experiment was not very successful. Labourers were unable to get enough work to provide for their support.

66 Upper Canada, *State Papers*, vol. 24. Statement by W. Allen and R.A. Tucker.

67 *Loc. cit.*

68 *Loc. cit.* Cf. Marx in the *Communist Manifesto:* "The bourgeois is a bourgeois — for the benefit of the working class."

69 Lillian F. Gates, "The Decided Policy of William Lyon Mackenzie", *Canadian Historical Review*, September 1959, p. 203.

70 *Ibid.*, pp. 186-87. The quotation is not from Mackenzie but represents the author's opinion of what he stood for.

71 Upper Canada, House of Assembly, *First Report of the Select Committee Appointed*

to inquire into the state of Trade and Commerce of the Province of Upper Canada (1835), pp. 9-11. These remarks occur in the course of an argument against tariffs on articles bought by farmers.

72 Upper Canada, *State Papers*, vol. 24. Statement of Sullivan and Baldwin.

73 *Loc. cit.*

74 "As long as emigration was restricted, wages for unskilled labor in Canada were high — four shillings a day or more in Quebec — but with the coming of large numbers they sank to an average of between 2s. 6d. and 3s. 6d., and only two shillings on government contracts, which was somewhat lower than the average in the United States" (Adams, *Ireland and Irish Emigration*, p. 179). The principle is right. The wage rates cited cannot be supported very far, for one thing, because there was considerable variation from time to time and from place to place. However, it would not be far out to say that the typical daily wage for labourers in Canada was 3s. 6d. before immigration became heavy, and 2s. 6d. afterwards.

75 Norman Ware, *The Industrial Worker, 1840-1860* (Boston and New York, 1924).

76 E. Wiman, "Annual Report of the Board of Trade with a review of the Commerce of Toronto for 1861" (uncatalogued pamphlet, P.A.C.).

77 For the importance of the Irish to the rise of manufacturing in Montreal in the 1840s and 1850s see Reynolds, *The British Immigrant*, p. 92.

78 J.F. Maguire, *The Irish in America* (London, 1868), pp. 5-6, 92-93, 98. Similarly in cities in the United States, it was declared that "Irish labourers, who save a few pounds, enter into some street trading, take a store, and their sons become respectable merchants, a process we never observe in Manchester" (Sidney Smith, *The Settler's New Home or the Emigrant's Location* [London, 1849], p. 77).

79 The wave-like impact of new sets of immigrants was more continuous and perhaps more obvious in the United States than in Canada. See Hansen, *The Immigrant in American History*, pp. 157-73, 201. Ethnic distinctions permit interesting permutations in labour practice. Occasionally an ethnically unified working class can exploit ethnic or other differences among employers. A situation in which employers and employees were of the same minority group facilitated mature collective bargaining in the garment trades. French-Canadian workmen have occasionally been able to force employers of their own ethnic group to be wage leaders; but more commonly it is employers who exploit ethnic ties to remain wage laggards.

80 Hansen struck at the superficiality of ethnic characteristics by comparing behaviour of identical immigrants in different overseas environments (*ibid.*, pp. 23-25). The remarkably different behaviour of seemingly identical immigrant groups which hit upon different occupations deserves equal attention.

81 Montreal artificers at Kingston in 1816 (about half French) and Kingston shipwrights in 1819 (about 20 per cent French) appear to have got on together harmoniously (*U.C.S.*, June 12, 1816 [Owen to Gore, petition of artificers] and June 1, 1819 [petition of shipwrights]). Kingston shoemakers who petitioned in 1822 included, apparently, one Italian, at least one Frenchman and half a dozen southern Irish (*U.C.S.*, Oct. 1, 1822). Even in 1842, when Kingston's mechanics (primarily carpenters and joiners) petitioned against the displacement of men by machinery, the names were half "Anglo-Saxon", half French and Irish (*P.S.O.*, *C.W.*, 1841-3, no. 3226).

Petitions do not guarantee against ethnic cleavage; but they imply the power of occupational interest to overcome it.

82 Some writers have remarked the reinforcement of Canadian Reformers by radical British immigrants, which no doubt occurred. But the Rebellion of 1837 made plain how profoundly conservative most immigrants were. Immigrants, especially primitive ones, seem always to have been conservatives. See Hansen, *The Immigrant in American History*, p. 83 ff.

83 As early as 1820 an immigrant accused the Americans of Upper Canada of showing hatred for English and Irish immigrants. It was for this reason, and to defend themselves, that the immigrants "begin to form societies amongst themselves..." (An English Settler, *A Few Plain Directions for Persons Intending to Proceed as Settlers to...Upper Canada* [London, 1820] [P.A.C., Pamphlets I, no. 1084], p. 98).

84 *Ibid.*, p. 100; *MG 22: 34*, A.C. Buchanan to Horton, May 27, 1827; *Q198, pt. 1*, p. 235; *pt. 2*, pp. 356-57.

85 An English Settler, *A Few Plain Directions* ... pp. 45-46.

86 Howison, *Sketches of Upper Canada*, p. 77.

87 *Q198, pt. 2*, p. 429. Buchanan, 1831. These labourers were predominantly Irish Catholics. Buchanan praised their church-going habits.

88 A Canadian Settler, *The Emigrant's Informant: or a Guide to Upper Canada* (London, 1834) (P.A.C., Pamphlets I, no. 1477), p. 203.

89 *Q198, pt. 2*, p. 358.

90 *Ibid.*, p. 376. According to Buchanan, the poor Irish worked on annual engagements that terminated May 12, and the rush of these could not arrive until July, while Quebec's labour market was flooded by June 4 (1831?). However, it seems doubtful whether many Irish were fortunate enough to have an annual engagement.

91 *MG 22: 34*, Feb. 13, 1830, evidence of William Lyons, M.D., of the Quebec Emigrant Hospital, who wanted some power to inspect and regulate boarding houses in order to control epidemics.

92 *Ibid.*, Dec. 31, 1831, evidence of Shaw, McMahon. However, Buchanan thought this society too generous and that many received its charity under false pretences (*ibid.*, Dec. 28, 1831). The Quebec and Montreal Emigrant Societies, though praised as splendid institutions, "were instituted for the main purpose of relieving the inhabitants of the two cities from the miserable spectacle of crowds of unemployed and starving emigrants..." (E. Abbott, *Immigration: Select Documents and Case Records* [Chicago, 1924], p. 24).

93 A Canadian Settler, *The Emigrant's Informant* ..., p. 8.

94 A Four Year Resident, *Views of Canada and the Colonists* (Edinburgh, 1844) (P.A.C., Pamphlets I, no. 1964), pp. 42-44.

95 A Seven Years' Resident of North America, *A Practical Guide for Emigrants to North America including the United States, Lower and Upper Canada, and Newfoundland* (London, 1850)(P.A.C., Pamphlets I, no. 2193), p. 16.

96 *Report of the Select Committee on Emigration* (1849), pp. 7, 8, 10.

97 *Ibid.*, p. 10; (A Ten Years' Resident), *Emigrants' Guide: being Information Published by His Majesty's Commissioners for Emigration, respecting the British Colonies of Upper and Lower Canada, and New Brunswick* (Devonport, n.d. [1832]) (P.A.C., Pamphlets I, no. 1356), pp. 14, 17; (Dunlop), *Statistical Sketches of Upper Canada*, p. 63.

98 All the guides and descriptions cited above make this point.

99 It was claimed that 30,000 immigrants were settled in Upper Canada in 1831 (*Q198, pt. 2*, p. 391). This was an exceptional year for immigration and employment. But for the population of Upper Canada to expand at the rate it evidently did, about twenty thousand immigrants a year must have settled in the 1830s and 1840s, on average.

100 (Dunlop), *Statistical Sketches of Upper Canada*, p. 63; *U.C.S.*, Oct. 24, 1837 (Tunney to Hillier).

101 *U.C.S.*, July 4, 1832 (Macaulay to Rowan). Probably the same difficulty existed in 1834, but there was much better provision that year for public relief and more construction work available.

102 *U.C.S.*, May 8, 1837 (Carey to Head); May 14, 1839 (Roy to Macaulay); Nov. 30, 1839 (Roy to Poulett Thompson); Aug. 31, 1839 (Hawks to Harrison).

103 *U.C.S.*, Aug. 3, 1838 (Moyle to P.S.); Dec. 31, 1839 (Rubidge to Hawks).

104 References to the number of labourers arriving on various public works suggest that at least ten thousand labourers entered from the United States, going mostly to the Welland and Lachine-Beauharnois Canals. Moorhouse, "Canadian Migration in the Forties", p. 327, offers estimates of the gross and net flows across the Canadian-United States border in the 1840s. She shows only 6,000 entering Canada in 1842, few in 1843, but 5,000 in 1844. Since she shows substantial movements in the other direction, she calculated a net movement to Canada of only 2,500, even in these years (1842-44). The significance of this movement does not lie so much in numbers involved, as in the reversal of a flow on which Canada normally depended to match its immigration with the demand for labour.

105 For example, *P.S.O., C.W., 1841-3*, no. 712 (Report of the House of Industry, Toronto). A common Irish expedient to fend off starvation, much used in these years, was to sell liquor without a licence to do so (e.g. *P.S.O., C.W.,1840-1*, nos. 3162, 3195, 3163).

106 *P.S.O., C.W., 1847-8*, nos. 18122, 18433, 19549.

107 Vere Foster, *Work and Wages; or, the Penny Emigrant's Guide to the United States and Canada* ... (5th ed., London, n.d.) (P.A.C., Pamphlets, no. 2452), p. 16.

108 Hansen and Brebner, *The Mingling of the Canadian and American Peoples*, pp. 113-18. Statistics on immigration are consistent with the "congestion" thesis. Canada took over 250,000 immigrants during the 1830s. The number for the 1840s is only 100,000 larger, at 350,000, which should have been an easy number for a larger economy to handle if it was in good health. Yet evidences of difficulty in absorption are more common in the 1840s, quite aside from the famine migration.

109 *Ibid.*, p. 135. It is extremely difficult to account for the divergent growth patterns of the United States and Canada without assigning high importance to the continuous agricultural frontier in the United States as against the interrupted one in Canada. Cf. John H. Young, "Comparative Economic Development: Canada and the United States", *American Economic Review, Papers and Proceedings*, vol. XLV, no. 2 (May 1955), especially pp. 83-84.

110 The Irish rated below the negroes in Boston (Handlin, *Boston's Immigrants, 1790-1865*, pp. 67-68).

111 *Sessional Papers, 1843*, app. Q (Report of the Board of Works — i.e., Hamilton Killaly).

[112] *P.S.O., C.E.,* 1843, no. 1200 (Killay to P.S.).

[113] For a more detailed discussion on Orangeism see pp.

[114] *Q198, pt. 2,* p. 358. Buchanan certainly knew better, but he was engaged in defending immigration against the charge that it drove the native population out of employment.

[115] McTaggart, *op. cit.,* vol. I, p. 290.

[116] A.D. Ferrier, *Reminiscences of Canada, and the Early Days of Fergus* (Guelph, 1866) (P.A.C., Pamphlets I, no. 3394), p. 5.

[117] Durham, *Report,* p. 22.

[118] See Mason Wade, *The French Canadians,* pp. 333-57. J.I. Cooper, "The Quebec Ship Labourers' Benevolent Society", *Canadian Historical Review,* December 1949, pp. 342-43, provides a spirited description of the transition from Irish to French employment in Quebec longshoring. Though it featured some pitched battles, this change was managed with great diplomacy. It may be wondered whether so much consideration would have been shown if it had been the Irish who were on the way in.

[119] *Sessional Papers, 1843,* app. (T), evidence of Laroque. See H.C. Pentland, "The Lachine Strike of 1843", *Canadian Historical Review,* September 1948, p. 266n.

[120] Pentland, *loc. cit.*

[121] *P.S.O., C.E.,* 1844, no. 1799 (May 23, 1844, Killaly to P.S. and enclosures). These men were employed directly by the Board of Works. Killaly wrote, somewhat inaccurately, "the novel feature in this case [is] that it originated with the Canadian labourers, and not with the Strangers, and that the ring-leaders are known." Killaly thought it improper to grant the demands of the strikers, but impolitic to prosecute the strike leaders, since Irish strike leaders had never been prosecuted because the authorities had never discovered their identity. What the Board of Works did to deal with this strike was to suspend the work indefinitely.

[122] Montreal *Transcript,* Aug. 4, Aug. 11, 1842.

[123] *Ibid.,* Aug. 20, Aug. 23, 1842.

[124] Quebec *Gazette,* March 31, 1843; also, Montreal *Transcript,* April 1, April 6, 1843. The Irish had just won higher wages by means of a hard strike, but besides employing strike-breakers, the contractor involved had in mind to replace all the Irish with French labourers. The Irish had every reason to think, therefore, that they were being robbed of their victory and defrauded under their agreement with the contractor. The opinion of a police officer on these tactics was:

> This, as will be found, will also fail, from the circumstance of the Canadians, who are a quiet, timid, people, being alarmed at coming into collision with the discharged Irish, of whom they have a great dread; thus the work will be stopped. Even supposing the Irish should leave the work quietly, it will only be to proceed to Beauharnois to get work on the Canal there; and where the same scenes will be re-enacted *de novo* ...(*C60,* p. 63).

There was, indeed, another ethnic clash at Beauharnois in August 1843. The police were supposed to prevent further ones (Quebec *Gazette,* Aug. 23, 1843; Montreal *Transcript,* Aug 1843).

[125] *U.C.S.,* June 1, 1835 (Hamilton to Colborne).

[126] *Ibid.;* see also *U.C.S.,* June 15, 1835 (Baker to Rowan).

127 *C317*, pp. 108-13 (attempt of the Sheriff to seize a raft at Bytown, July 22, 1847).

128 J.L. Gourlay, *History of the Ottawa Valley* (n.p., 1896), pp. 52, 142-43; L. Breault, *Ottawa Old and New*, pp. 65-68.

129 Reynolds, *The British Immigrant*, pp. 13-15, 93-110, 180-82.

130 W.J.S. Mood, "The Orange Order in Canadian Politics, 1841-1867" (M.A. thesis, Toronto, 1950), contains a useful summary of the origins and structure of the Orange Order.

131 Hansen, *Atlantic Migration*, pp. 132-34; Adams, *Ireland and Irish Emigration*, p. 129.

132 *Selkirk Papers*, vol. 75, p. 301.

133 Leslie R. Saunders, *The Story of Orangeism* (Toronto, 1941), pp. 18, 21; E.C. Guillet, *Early Life in Upper Canada* (Toronto, 1933), p. 328.

134 Most of the information about Orangeism in this area appears in letters from William Morris, who constantly urged more drastic action against it. In August 1825, when he first wrote, he stated that the Perth lodge was established "about two years ago" (*U.C.S.*, Aug. 13, 1825 [Morris to Hillier]). A letter of May 1830 said it was formed "about 6 years" before (*U.C.S.*, May 27, 1830 [Morris to Colborne]). Long afterwards, Morris suggested that Orangeism arose partly in consequence of a fight among militia companies at Ramsay in 1824 (*P.S.O., C.W.*, 1840-1, no. 1792 [Morris to P.S., Aug. 10, 1840; with no. 2899?]). There was indeed a fight at Ramsay in April 1824 between the "Scotch" and "Irish" (mostly Robinson settlers?) companies but the investigation of it implies that the Irish party included both Protestants and Catholics (*U.C.S.*, May 20, 1824 [statement of Fitzmaurice and others; enclosure]).

135 *U.C.S.*, May 27, 1830 (Morris to Colborne). Matheson had been removed from his office on Morris's complaints.

136 This circular urged unity between Protestant and Catholic Irish. It is dated June 18, 1826, but appears in *U.C.S.*, April 18, 1826; and again July (n.d.) 1836 (Fitzgibbon to Joseph, enclosure). Fitzgibbon had a part in the inquiry into the Ramsay riot.

137 *U.C.S.*, April 4, 1827 (Tully to Maitland).

138 *Ibid.*

139 *U.C.S.*, July 27, 1827 (Morris to Hillier).

140 *U.C.S.*, Sept. 17, 1827 (Macdonell to Hillier); Jan. 21, 1828 (Fraser to Hillier). The second letter reported that a new Orange lodge had been formed by a person who had recently come from Ireland expressly to organize lodges. It is not clear whether this refers to a first or second lodge in Kingston.

141 The circular seems to have been issued at the instigation of Morris (*U.C.S.*, May 27, 1830 [Morris to Colborne]) and one Maguire, a Roman Catholic magistrate in the Robinson settlement in Cavan (*U.C.S.*, July 5, 1830 [Maguire to Bergin]). Subsequently, Maguire was accused of having fomented most of the Orange-Green trouble (*U.C.S.*, June 1 [should be July 1] 1831 [Huston to McMahon]; June 29, 1831 [Elliott to McMahon]; July 7, 1831 [Brown to McMahon]). On the consolidation of Orange lodges in the east, see Mood, "The Orange Order in Canadian Politics", p. 9.

142 Saunders, *The Story of Orangeism*, p. 22.

[143] *U.C.S.*, Sept. 3, 1833 (Covert to Rowan: Address of Orangemen to Lieutenant Governor, Aug. 18, 1833).

[144] Most of these figures are given by Mood, "The Orange Order in Canadian Politics", pp. 10n., 98, 165. The figure for 1846 was given in George Benjamin's Belleville *Intelligencer* & Victoria *Advertiser* (*P.S.O., C.W.*, 1846, no. 14635, enclosure).

[145] Mood, *op. cit.*, p. 9.

[146] As a government project, the Robinson settlements were well documented, and accounts appear in all works on emigration and settlement. The Catholics were settled in the rear concessions of the township of Emily, with wild land between them and Protestant Irish who had settled before 1820 in Cavan and the first four concessions of Emily. (J.H. Aitchison, "The Development of Local Government in Upper Canada, 1783-1850", p. 197.)

[147] *U.C.S.*, June 15, 1831 (Maguire to O'Grady); July 7, 1831 (Brown to McMahon). There are supposed to have been a thousand Orangemen in York in 1833 .(Saunders, *The Story of Orangeism*, p. 22); but Toronto Orangeism does not seem really to have flourished until after an election battle between Orange and Green in 1841. Early apathy is indicated in H. Lovelock, "Reminiscences of Toronto Orangeism", *Official Orange Souvenir*, Toronto, July 12, 1902 (pamphlet).

[148] Mood, "The Orange Order in Canadian Politics", p. 9.

[149] *Ibid.*, pp. 103, 106, 165, 168. It is asserted that even in 1860 less than a quarter of the Orangemen in Canada West were west of Oshawa.

[150] For example, *P.S.O., C.W.*, 1850, no. 620 (also C317, pp. 219-21, 234), July 12, 1849 clash between Orangemen and canal labourers in St. Catharines; nos. 1344 and 1378, July 12, 1850 clash between Orangemen and labourers at Brantford, and subsequent prosecution of Orangemen under the law of 1843. In *P.S.O., C.W.*, 1851, no. 788, Benjamin proposed repeal of the law against Orange processions on the ground that it had provoked the holding of more processions than before.

[151] Opponents, of course, used the same device and maximized the threat of Orangeism. Orangemen were frequently accused of keeping their opponents from the polls by force at elections (*P.S.O., C.W.*, 1840-1, nos. 3323, 3339, 3345) but others did this also when they could. With respect to local offices, "the truth is that the Orange Society are endeavouring to control all our Township affairs. No appointment can take place, even as Teachers of Common Schools, or any Township Office where trust and responsibility is required, but their Secret influence is immediately called into Action and the Nominees of their Leaders, are supported and carried in, whether qualified by education fitness or otherwise — frequently the exclusion of many good and loyal subjects..." (*P.S.O., C.W.* 1843-4, no. 7062. Memorial of S. Walford, village of Bolton, Jan. 21, 1844). Even while local government was carried on entirely by appointed magistrates, there was a great deal of trouble, especially because governors did not appoint enough magistrates and showed a marked partiality for recent British immigrants (Aitchison, "The Development of Local Government in Upper Canada, 1783-1850", pp. 60-68). It is notable that Gowan came out in 1839 in favour of the Reform objective of elected local officers; for one thing, no doubt, because he was confident that the Orange Order could elect a great many of them (*ibid.*, p. 82). Tory opposition to the secret ballot presumably

reflected, in part, the fact that the Orange machine functioned better with open elections (*ibid.*, p. 577).

152 What the Orangemen seem to have meant was that they were prepared to have their organized Protestant body coexist peacefully alongside an organized Catholic body; but they wanted Catholics kept in their place. They did not propose to have Catholics overrun and dominate them. Where the Catholics' place was well defined, as in Lower Canada, this arrangement worked well enough. It was the undetermined position of Irish Catholics in Upper Canada that gave trouble.

153 *U.C.S.*, July 1, 1834 (Gowan to Rowan). According to Gowan, the population at the front of Leeds consisted of disloyal Americans; but the four-fifths in back locations were loyal British immigrants, mostly disfranchised because polling places for elections were at the front. There was a vendetta between Gowan and local "American" officials: Gowan charged systematic persecution by prosecutions of himself and his friends on various trumped-up charges (*U.C.S.*, July 4, 1833 [Gowan's memorial]; Feb. 10, 1834 [Gowan to Colborne]).

154 Mood, "The Orange Order in Canadian Politics", p. 11.

155 *U.C.S.*, May 9, 1836 (Stewart to Joseph, and enclosures).

156 Mood, *op. cit.*, p. 17.

157 The Irish Catholics offered as a reason for their refusal to come forward in 1837 the partiality of the government towards Orangemen. The indefatiguable Morris used this as a further reason why Orangeism should be discouraged. The government proceeded to discourage Orangeism quietly — i.e., by refusing offices to known Orangemen (*P.S.O.*, *C.W.*, no. 1792 [Morris to P.S., Aug. 10, 1840 and enclosure]). The Irish Catholics also maintained neutrality in 1849 (Doughty, *Elgin-Grey Papers*, vol. I. p. 412).

158 Mood, "The Orange Order in Canadian Politics", pp. 19-21.

159 *Ibid.*, p. 172, and *passim* for the political development of Orange coalitions. Also Creighton, *John A. Macdonald; The Young Politician*, pp. 92, 142-45, 195-96, 301-4.

160 The predominantly Orange "British American League" of 1849 had proposed union of British North America, and perhaps made this an Orange objective (Mood, *op. cit.*, pp. 67-72).

161 *Ibid.*, pp. 104-5, and the statement of Lord Elgin, who disliked the Order, that "the Orangemen are powerful only in the towns, where their organization enables them to act at once" (May 1849) (Doughty, *Elgin-Grey Papers*, vol. I, p. 352).

162 *C316*, p. 213 (March 22, 1841); also *P.S.O.*, *C.W.*, 1840-1, nos. 3345, 3360, 3387.

163 *P.S.O.*, *C.W.*, 1840-1, no. 758 (Bridges to P.S., April 10, 1840).

164 Doughty, *Elgin-Grey Papers*, vol. I, pp. 409-12.

165 Mood, "The Orange Order in Canadian Politics", p. 87.

166 *P.S.O.*, *C.W.*, 1846, no. 14758; *C317*, pp. 44-45.

167 *C317*, p. 48.

168 *P.S.O.*, *C.W.*, 1843-4, no. 6184.

169 *Ibid.*, no. 7363.

170 Doughty, *Elgin-Grey Papers*, vol. I, p. 523. Montreal Orangemen were sup-

posed to have been annexationists in 1849. However, there were not a great many lodges in the lower province — a spokesman in 1839 said there were "several" (*P.S.O.*, *C.W.*, 1840-1, no. 758).

Chapter V

[1] For a description of how "almost completely self-supporting" rural economy was in Lower Canada, and of the striking effect of railways, see J.I. Cooper, "French-Canadian Conservatism in Principle and in Practice, 1873-1891" (Ph.D. thesis, McGill University, 1938), pp. 47-48.

[2] The total population of the Canadas multiplied two and a half times between 1830 and 1850 and more than three times between 1830 and 1860. However, it only grew by 12 per cent between 1861 and 1871. City populations from 1851 appeared in the 1871 Census, Table E. Cf. Innis and Lower, *Select Documents, 1783-1885*, p. 618. Data cited for years before 1851 come from a wide variety of sources.

[3] Ottawa, London, Brantford, St. Catharines, Guelph, Chatham, Port Hope, Brockville, Belleville.

[4] Bert F. Hoselitz, "The City, the Factory, and Economic Growth", *American Economic Review, Papers and Proceedings*, May 1955, pp. 166-84, especially 173, 177-79.

[5] Cf. Innis and Lower, *Select Documents, 1783-1885*, pp. 590-617.

[6] Hamilton *Spectator*, "Saturday Musings", various issues (Hamilton Public Library collection).

[7] J.M. Clark, "Common and Disparate Elements in National Growth and Decline" (mimeo., 1948).

[8] These places tended to leap to an economy of service industries without much of a manufacturing stage either. Manufactures rose in New England with the support of commerce; in the Netherlands as a means of livelihood when service industries declined; while in Nova Scotia they have not had a strong development at any time.

[9] Limitations of this kind appear to form the basis of Professor W.L. Morton's concept of "northern" economies.

[10] N.S.D. Gras produced and generalized a theory of economic development not far removed from the present one in his studies of metropolitan evolution. The starting point for his conception of metropolitan growth (of London) was his *Evolution of the English Corn Market* (1915), in which he stressed the strength of local agriculture as the first condition for a metropolis. This same consideration promoted his *History of Agriculture in Europe and America* (1925) and several other works relating to agriculture. Cf. also Gras, *An Introduction to Economic History* (1922).

[11] The trade seems to have taken about 500 men in 1681 (population about 10,000); about 1,000 men in 1720 (population about 25,000); and 2,000 men in 1750 (population about 50,000); though one estimate of 1758 put the number drawn by the fur trade at 4,000 (H.A. Innis, *The Fur Trade in Canada*, pp. 62-67, 103-4, 115; Lunn, "Economic Development in New France, 1713-1760", p. 109. Cf. A.L. Burt, *The Old Province of Quebec* (Toronto, 1933), p. 447: "...the fifteen hundred men whom the business drew from the agricultural life of the colony were five hundred too many."

12 It is true that agriculture served as an auxiliary and reserve supplying food and men for the fur trade. However, state policy was directed to make agriculture the primary and independent economic foundation of New France and, on the whole, succeeded.

13 Upper Canada House of Assembly, *First Report of the Select Committee appointed to inquire into the state of Trade and Commerce of the Province of Upper Canada, 1835:* "On the shores of Lake Ontario it [the cost of producing wheat] ranges from half a dollar to three shillings: Eastward, in Glengarry and the Ottawa, it rises from 3s. to 3/9; and westward, in the Gore and London districts, falls to between half a dollar and 2 shillings in our currency."

14 *C108* (P.A.C.), p. 127 ff., R. Hamilton to Major James Green, June 11, 1803.

15 *Loc. cit.*

16 *C105*, p. 11, indicates that in 1791 it was already the established custom, in buying local flour for the up-country garrisons, to pay a premium over the Lower Canadian price reflecting at least a part of the cost of transport. *C105*, pp. 133, 143, 150, 154, 173, are correspondence concerning the feeding of the garrisons in 1797. In that year, the army tried to exploit the position of a monopsonist, with some success against the farmers around Kingston, but was soundly defeated on the Niagara and Detroit frontiers by the competitive buying of the Americans, who also had garrisons on the frontiers to feed and paid up to 50 per cent more than the highest price that the British purchasing agent was authorized to pay. *C105*, p. 146, indicates that there were 2,200 troops in Upper Canada to be fed and 1,000 more at "outposts" out of a total force in Canada of 5,800, and also sets out the daily ration of flour, pork, peas, beef, etc. that had to be found for them. *C105*, pp. 186-87, sets out a purchasing policy that concedes that prices have to be somewhat higher as one proceeds westward. *C109*, pp. 47-48, is a statement of the prices paid for supplies to the army of flour, peas and pork, 1799-1803, in Lower and Upper Canada (Kingston, York, Fort George, Chippewa, Fort Erie, Amherstburg). They indicate maintenance of the principle that prices were progressively higher in Upper Canada as one went westward, and show substantially higher prices for all points in Upper Canada than in Lower Canada in 1799. However, John Craigie, Deputy Commissary General, appears to have achieved his goal of driving down Upper Canadian prices from 1800 to 1803. The prices reported to be paid in Upper Canada were somewhat lower than Lower Canadian prices in 1800 and 1801, and substantially lower in 1803 — perhaps because (according to the blanks in the table) the army refrained from buying anything in Upper Canada in 1802, with the exception of one purchase.

17 In 1842 it was stressed that to support the timber trade it was necessary for supplies to be drawn from (at least) the Kingston area (*P.S.O., C.W., 1841-3, no. 4860*). *Sessional Papers, 1847*, app. QQ, indicates that in 1846-47 (a bigger year than hitherto) the Ottawa Valley trade employed 11,000 men, 8,600 horses, 2,200 oxen, and required 30,000 bbls. of flour, 27,000 bbls. of pork, 450,000 bus. of oats and 11,000 tons of hay.

18 Cf. D.G. Creighton, "The Economic Background of the Rebellion of Eighteen Thirty-Seven", *Canadian Journal of Economics and Political Science*, Aug. 1937, pp. 322-34.

19 *Ibid.*, especially pp. 324-26.

20 Discussed below.

21 "A new state of affairs has arisen in Western Canada during the past two

years. High prices have had a marvellous effect... As an illustration we
may take Toronto markets, and trace the effects of high prices among
many of the smaller farmers of that neighborhood. The illustration will
hold good for every other town and village in the country... Again it may
be observed that during the past two years the value of every kind of farm
produce has risen in the same proportion as products adapted for exporta-
tion..." (Scobie and Balfour, *Canadian Almanac*, 1856). The 1857 *Almanac*
contained a glowing account of the effect of railways. It stated that lumber
could not be carried economically by road for more than 40 miles, but that
railways made it pay. Also, "Now... at every station, fruit, butter, eggs and
vegetables command a ready cash sale. The price of firewood has risen
considerably..." Adam Shortt stated that production could not satisfy
home demand in this period, and that food was imported heavily from the
United States ("Railway Construction and National Prosperity: An Historic Parallel", *Transactions of the Royal Society of Canada*, 1914, pp. 303-4).

[22] H.C. Pentland, "The Role of Capital in Canadian Economic Development
before 1875", *Canadian Journal of Economics and Political Science*, November 1950,
pp. 457-74, deals with the changes described in this section.

[23] "In the Mercantile System surplus value is only relative; what one wins the
other loses. 'Profit upon alienation', 'oscillation', or 'vibration of the
balance of wealth between different parties'..." (Karl Marx, *Theories of
Surplus Value* [New York, 1952], p. 55).

[24] Cf. the trouble in industrial ventures of the Hancocks. W.T. Baxter, *The
House of Hancock: Business in Boston, 1724-1775* (Cambridge, Mass., 1945), ch.
XI.

[25] With the qualification that the merchant made every effort to channel
trade through his city (until he abondoned it) rather than any competing
city.

[26] The outlook of the merchants who came to Canada after 1759 is revealed in
"A Friend of Trade", *The True Interest of Great-Britain in Regard to the Trade and
Government of Canada, Newfoundland, and the Coast of Labrador* (London, 1767)
(P.A.C., Pamphlet I, no. 415) and "An Old and Experienced Trader"
(Alexander Cluny), *The American Traveller: or, Observations on the Present State,
Culture, and Commerce of the British Colonies in America* (London, 1769) (P.A.C.,
Pamphlet no. 427).

[27] Merchants of the American colonies generally opposed local development,
even though they were its beneficiaries in many ways. Even in so diversified an economy as that of Massachusetts, merchants were against home
manufactures and measures to encourage them. On the other hand, there
were always some Massachusetts merchants who joined with farmers and
artisans to promote manufactures — sometimes, no doubt, because they
intended to invest in them — and doctrines of the "economy of high
wages", the reflection of an integrated economy primarily dependent on
the home market, were advanced periodically. Such views were conspicuously absent from the St. Lawrence region. Cf. O. and M. Handlin, *Commonwealth: A Study of the Role of Government in the American Economy: Massachusetts,
1774-1861* (New York, 1947), especially pp. 38-39, 111, 131, 197.

[28] V.C. Fowke, *Canadian Agricultural Policy; The Historical Pattern* (Toronto, 1946),
p. 106.

[29] French policy in New France probably was fostered also by the traditional
self-sufficiency of France, the desire in France to maintain it and, not

unrelated to this, the persistence of feudal traditions.

30 Government conscription of shipyard workers at Quebec in 1763 is noted in *British Museum, Additional MSS, M377* (P.A.C.), p. 12 (Christie vs Knipe and Le Queusne). Conscription and other means of regulation were used to re-establish the St. Maurice Forges, 1760-63 (Sulte, "Les Forges Saint-Maurice", p. 130-39). Conscription of men to serve on bateaux was common and the cause of bitter clashes between habitants and the government (e.g., *S Series* [Lower Canada Sundries] Oct. 12, 1765 [Christie to Burton]; Oct. 13, 1765 [Christie to Burton]; Oct. 13, 1765 [La Goterie to Christie]). Regulation of the price of bread was long continued in Quebec and Montreal and the Legislature wanted regulation extended in area and scope (Lower Canada, *Journals of the Assembly*, 1814, pp. 12, 24, 26, 110-12; 1815, pp. 152, 322-26, 440). The persistence of support for regulation is suggested by the occurrence in Quebec in 1855 of "bread riots" against a combine that had raised bread prices by 20 per cent (Montreal *Gazette*, Jan. 11, 1855).

31 Military authorities opposed canals along the exposed portions of the St. Lawrence and the Welland Canal, as well as industrial development near the frontier, all of which they thought too exposed to American attack, in spite of mercantile desires to strengthen the position of the St. Lawrence in the interior market. Cf. J.L. McDougall, "The Welland Canal to 1841" (M.A. thesis, University of Toronto, 1923), pp. 7-9.

32 Peter Force, ed., *American Archives: consisting of a Collection of Authentick Records, State Papers, Debates, and Letters ... Fourth Series: A Documentary History of the English Colonies in North America from March 7, 1774 to the Declaration of Independence* (Washington, 1839). Vols. 2-6 contain a wealth of information on Canada and Canadian attitudes.

33 The Force documents, aside from revealing the activity of some merchants who were wholehearted supporters of the Revolution, suggest that the rest of the mercantile community would have been ready enough to support it too: "Their [the Canadians'] minds are all poisoned by emissaries from New-England and the damned rascals of merchants here and at Montreal..." (Gamble to General Gage, Sept. 6, 1775, Force, *American Archives*, vol. III, p. 963). But cf. D.G. Creighton, *The Commercial Empire of the St. Lawrence, 1760-1850* (Toronto, 1937), pp. 60-67.

34 E.C. Kirkland, *Men, Cities, and Transportation: A Study in New England History, 1820-1900*, 2 vols. (Cambridge, Mass., 1948), vol. I. A great deal of information on the various competitive transportation routes of the United States was collected in a series of studies, "Historic Highways of America" by A.B. Hulbert (Cleveland, 1904).

35 Mulhall's *Dictionary of Statistics* (London, 1892) gives the following populations:

	New York	Philadelphia	Boston	Baltimore
1730	8,600	12,000	11,500	—
1750	10,000	18,000	14,000	—
1790	33,000	44,000	18,000	13,800
1810	96,000	95,000	33,000	36,000
1820	124,000	113,000	43,000	63,000
1830	203,000	161,000	61,000	81,000
1840	313,000	220,000	93,000	102,000

36 "New York City" in *Encyclopaedia Britannica* (1960 ed.)

37 L. Hartz, *Economic Policy and Democratic Thought: Pennsylvania, 1776-1860* (Cambridge, Mass., 1948); A.B. Hulbert, *The Chesapeake and Ohio Canal and the Pennsylvania Canal*, Historic Highways of America, vol. 13 (Cleveland, 1904).

38 A.B. Hulbert, *The Cumberland Road*, Historic Highways of America, vol. 10 (Cleveland, 1904).

39 Kirkland, *Men, Cities, and Transportation*, vol. I, pp. 66 ff., 187-90, 208.

40 The federal government of the United States did give particular assistance to New York by breaking the Iroquois Confederacy, and gave assitance to its cities in general by its land policies and subsidization of canals. Cf. J.B. Rae, "Federal Land Grants in Aid of Canals", *Journal of Economic History*, Nov. 1944, pp. 167-77.

41 A.B. Hulbert, *The Erie Canal*, Historic Highways of America, vol. 14 (Cleveland, 1904).

42 *Ibid.*, p. 54.

43 The improvement of Montreal's harbour is examined in L.C. Tombs, *The Port of Montreal* (Toronto, 1926). In H.Y. Hind, ed., *Eighty Years' Progress of British North America* ... (Toronto, 1863), Thomas Keefer discussed the problem of dredging Lake St. Peter (pp. 165-66) and the attempt to strengthen Montreal by subsidization of the Allan line (pp. 142-45), a step that reflected contemporary Canadian annoyance with the British government's subsidization of New York. The same author noted in another work that in spite of all the difficulties, St. Lawrence traffic increased more than New York's between 1820 and 1840, as measured by tonnage of vessels (T.C. Keefer, *The Canals of Canada: their Prospects and Influence* [Toronto, 1850], pp. 30-31, 44). New York, of course, was also improving its canals, and had the power to lower its canal tolls when competition threatened to become serious, so that rates on both lines fell considerably (G.P. de T. Glazebrook, *A History of Transportation in Canada* [Toronto, 1938], pp. 89-97). The net effect was that rates on the St. Lawrence were less than half of Erie rates in 1849. (G.N. Tucker, *The Canadian Commercial Revolution, 1845-1851* [New Haven, 1936], pp. 55, 58).

44 L.H. Jenks, "Railroads as an Economic Force in American Development", *Journal of Economic History*, May 1944, pp. 12-13.

45 W.V. Pooley, *The Settlement of Illinois from 1830 to 1850* (Madison, 1908), pp. 482-85; P.W. Gates, *The Illinois Central and Its Colonization Work* (Cambridge, Mass., 1934), p. 87. Similarly, lumber, and a good proportion of immigrants, continued to move to Chicago by water.

44 L.H. Jenks, "Railroads as an Economic Force in American Development", *Journal of Economic History*, May 1944, pp. 12-13.

45 W.V. Pooley, *The Settlement of Illinois from 1830 to 1850* (Madison, 1908), pp. 482-85; P.W. Gates, *The Illinois Central and Its Colonization Work* (Cambridge, Mass., 1934), p. 87. Similarly, lumber, and a good proportion of immigrants, continued to move to Chicago by water.

46 It was New York's power to engross the wheat of the Midwest through its command of the water route to Chicago that led to its own local agricultural base being turned to dairying and stock-raising (Pooley, *op. cit.*, pp. 339-43). The St. Lawrence valley did not complete this transition until about a half-century later.

47 Ocean freight rates on flour to Liverpool were reported to be about twice as much from Montreal as from New York — 4s. per bbl. from Montreal,

2s. from New York (Doughty, *Elgin-Grey Papers*, vol. III, p. 1282).

48 Cf. Tucker, *The Canadian Commercial Revolution*, pp. 58-62.

49 It is interesting to compare the rivalry between New York and Montreal with that between Montreal and Hudson's Bay — Montreal in the second case taking the role of New York as the port with a well-developed hinterland, by which it was able to command better shipping services and rates, notwithstanding its greater distance from Europe. In respect to the rivalry of the fur trade, Hudson's Bay was able to win. This may mean that ocean shipping costs were of little consequence in the fur trade as against interior shipping costs; though it may be that Montreal could have continued to compete strongly after 1821 if monopoly had not intervened. In the twentieth century, the Hudson's Bay route has been unable to offer serious competition to Montreal precisely because of those economic disadvantages of a port catering to staple exports, and the political weakness of a port with a weak hinterland, that defeated Montreal in its contest with New York.

50 There has been a decided movement to reinterpret British policy in the nineteenth century in respect to free trade and to laissez faire in general. On free trade, see J. Gallagher and R. Robinson, "The Imperialism of Free Trade", *Economic History Review*, Aug. 1953, pp. 1-15. In respect to general policy, it has been argued that public policy in nineteenth-century Britain was interventionist and hostile to an earlier laissez faire attitude. Cf. E. Lipson, *The Growth of English Society* (New York, 1953), ch. X. Cf. also the reinterpretation of Bentham in J.B. Brebner, "Laissez-Faire and State Intervention in Nineteenth Century Britain," *Tasks of Economic History* (Supplement to the *Journal of Economic History*), no. VIII (1948), p. 59-73.

51 L. Hartz, *Economic Policy and Democratic Thought: Pennsylvania, 1776-1860*; O. and M. Handlin, *Commonwealth: A Study of the Role of Government in the American Economy: Massachusetts, 1774-1861*.

52 The ancestry of the National Policy is traced briefly in V.C. Fowke, "The National Policy — Old and New", *Canadian Journal of Economics and Political Science*, Aug. 1952, especially pp. 271-72. The events of Canadian tariff history are set out in O.J. McDiarmid, *Commercial Policy in the Canadian Economy* (Cambridge, Mass., 1946).

53 D.G. Creighton, "British North America at Confederation", Royal Commission on Dominion-Provincial Relations, Appendix 2 (Ottawa, 1939), pp. 5-6 (Editorial Foreword).

54 Tucker, *The Canadian Commercial Revolution*, p. 46.

55 J.I. Cooper, "Some Early French-Canadian Advocacy of Protection: 1871-1873", *Canadian Journal of Economics and Political Science*, Nov. 1937, p. 534.

56 The rise of competition on the basis of transport, which allowed some producers to conquer wide markets, but which was generally distasteful and led to combinations to restrict it, is discussed for the United States by T.C. Cochran and W. Miller, *The Age of Enterprise, a Social History of Industrial America* (New York, 1942), pp. 57-62. John R. Commons considered the new competition to be a vital factor in the growth of unions, especially of national unions (*History of Labour in the United States*, 2 vols. [New York, 1918], vol. I, pp. 438-53).

57 Fowke, *Canadian Agricultural Policy*, p. 68. General Murray is quoted on the merits of hemp raising, among others, that it "will divert them from manu-

facturing coarse things for their own use, as it will enable them to purchase those of a better sort manufactured and imported from Great Britain" (Murray's Report, 1762).

[58] In 1843, the Colonial Secretary had opposed duties that provided protective differentials and forbidden governors to assent to them. In 1846, Secretary Gladstone urged that Canadian duties be lowered, Liverpool interests protested the Canadian duties of 1845 which favoured raw as against refined sugar imports, and the British government demanded that no goods which Canada had usually imported from Britain should be taxed more than formerly (*E.C.O.*, 1846, nos. 530, 1584, 1615). *Sessional Papers, 1849*, app. N, covers among other things numerous protests from Britain against Canada's "prohibitive" new tariff of 12½ per cent on many manufactured products. In respect to money, following the attempt to keep Canadian banks to gold payments in 1837, there may be noticed the Colonial Secretary's opposition to a Provincial Act to raise £100,000 in small notes in 1839 (Upper Canada, *Journals of the Assembly*, 1839, Appendix vol. II, part II, p. 553), the extremely restrictive advice of Gladstone upon the proper principles of banking (*E.C.O.*, 1846, no. 1733, and *Sessional Papers, 1847*, app. W), and British objection to any extension of the power granted in 1842 to Canadian banks to issue 5 shilling notes (*E.C.O.*, 1847, no. 94). Cf. Innis and Lower, *Select Documents, 1783-1885*, pp. 638-42; R.M. Breckenridge, *The Canadian Banking System, 1817-1890* (New York, 1895), pp. 92, 106, 113-14, 122-26.

[59] Baxter, *The House of Hancock: Business in Boston, 1724-1775*, ch. II, "Trade without Money": an excellent account of all the ingenious devices used by Boston merchants to carry on business with a minimum of cash.

[60] R.A. Lester, "Currency Issues to Overcome Depressions in Pennsylvania, 1723 and 1729", *Journal of Political Economy*, June 1938, pp. 324-75, and "Currency Issues to Overcome Depressions in Delaware, New Jersey, New York and Maryland, 1715-37", *Journal of Political Economy*, April 1939, pp. 182-217.

[61] *Reflections Sommaires sur Le Commerce qui s'est fait en Canada; d'après un manuscrit à la bibliothèque du roi à Paris"* (Pamphlet, n.d., about 1759) (P.A.C., Pamphlet no. 281). This is a vivid account of the difficulties in New France when it was isolated by war, 1744-48 and 1755-59. It indicates that the supply of industrial products was not elastic enough for New France to achieve self-sufficiency. With the disruption of imports, industrial products became scarcer and scarcer until the farmers, the only persons who had real goods, refused any longer to take paper money which could not buy the articles they wanted.

[62] S. Reznick, "The Rise and Early Development of Industrial Consciousness in the United States, 1760-1830", *Journal of Economic and Business History*, Supplement to vol. 4, no. 4, Aug. 1932, pp. 784-811.

[63] Cooper, "Some Early French-Canadian Advocacy of Protection, 1871-1873", p. 530. It is noted in the same place that the Rouges were protectionists in the 1840s. Cooper has treated protectionism, among other things, more extensively in his "French-Canadian Conservatism in Principle and Practice, 1873-1891" (Ph.D. thesis, McGill University, 1938).

[64] Fowke, *Canadian Agricultural Policy*, p. 92.

[65] *Ibid.*, p. 95.

[66] Lillian F. Gates, "The Decided Policy of William Lyon Mackenzie", *Canadian*

Historical Review, Sept. 1959, pp. 203-4, 208.

67 *Ibid.*, pp. 204-6; R.A. MacKay, "The Political Ideas of William Lyon Mackenzie", *Canadian Journal of Economics and Political Science*, Feb. 1937, pp. 14-18.

68 Fowke, *op. cit.*, pp. 112, 251-52, 259. Cf. R.L. Jones, *History of Agriculture in Ontario, 1613-1880* (Toronto, 1946), especially p. 307.

69 *Address delivered before the Provincial Agricultural Association at the Thirteenth Annual Exhibition at Toronto, 1858, by the President, William Ferguson, Esq., of Kingston, C.W.* (Toronto, n.d.) (uncatalogued pamphlet, P.A.C.).

70 Lake shippers were occasionally protected against American competitors, as in 1823 (*U.C.S.*, July 15, 1823 [Crooks to Hillier]; July 26, 1823 [John B. Robinson to Hillier]), but their interest was usually ignored. Shipowners and sailors asked protection in 1845 (*P.S.O., C.W.*, 1844-5, no. 9420). Professional men and beneficiaries of patent grants had special protectionist interests. The Canadian government sought and received power over copyright in 1846 (*E.C.O.*, 1846, no. 2799).

71 W.H. Merritt sought protection for his new salt works in 1818 (*U.C.S.*, Oct. 23, 1818). In 1819, Kingston shipwrights asked protection from persons who bought American vessels and used them in the Canadian coasting trade (*U.C.S.*, June 1, 1819). However, the "inhabitants of Kingston" objected to the high prices of foods occasioned by the agricultural protection of 1822 (*U.C.S.*, n.d. [Nov.], 1823). A bounty appears to have been used to assist a new paper mill in 1826 (*U.C.S.*, June 6, 1826).

72 The history of Canadian tariffs in this period is set out in McDiarmid, *Commercial Policy in the Canadian Economy*, p. 28 ff. The philosophy of tariff legislation, in particular reference to a 30 per cent duty on paper, was expressed in 1840 as "intended to check foreign importations prejudicial to our domestic manufacture, and the consumption of British paper" (*P.S.O., C.W.*, 1840, no. 1173 [the opinion of J.W. Macaulay, Inspector General]). Canadian and British interests were thus harmonized.

73 Sir Arthur Helps, *Life and Labours of Mr. Brassey, 1805-1870* (3rd ed., London, 1872), pp. 186-95. McDiarmid expressed the contrary opinion that British manufactures were too competitive to be much affected by Canadian tariffs of the 1840s (*op. cit.*, p. 67).

74 *P.S.O., C.W.*, 1841-3, nos. 1438 and 4304. *Ibid.*, no. 4597, from Hamilto ironfounders, asked for retention of free importation of coal from the United States, instead of the 15 per cent duty proposed to 1842, alleging strong competition from American iron producers.

75 McDiarmid, *op. cit.*, pp. 51-56. Agriculture seems to have won a clear victory in this instance, contrary to Fowke, *Canadian Agricultural Policy*, p. 92.

76 *P.S.O., C.E.*, 1842, no. 2555; 1843, no. 1920. Also *Journals of the Assembly*, 1842, p. 56.

77 *P.S.O., C.W.*, 1844-5, no. 9127 (Kingston shoemakers, Dec. 1844); no. 9382 (Belleville shoemakers, Jan. 1845). The Kingston petition also urged that most shoes imported from the United States were made by prison labour.

78 *Ibid.*, no. 9129 (Montreal sawmill operators for an equalized duty of 25 per cent, instead of the present 5 per cent, on American lumber, Dec. 1844); no. 9390 (inhabitants of Brockville for higher, and specific rather than *ad valorem*, duties on American iron castings).

79 *Ibid.*, no. 9673 (Kingston merchants especially on this point).

80 McDiarmid, *Commercial Policy in the Canadian Economy*, p. 55. Commercial

interests were divided at this time. Montreal favoured, while Toronto opposed, differential duties to promote importation by the St. Lawrence rather than the American border, and won its point at the price of general protection. Montreal was provoked by American drawback legislation, which Toronto did not find objectionable (*P.S.O.*, *C.W.*, 1844-5, nos. 10116 and 10121). The British government had vetoed protective differentials in 1843, and demanded in connection with the 1845 tariffs that British goods customarily imported into Canada should not be taxed more than formerly (*E.C.O.*, 1846, no. 1615). It is worth observation that the British government, Montreal and Toronto were all concerned with special advantage, and not at all with freer trade or any other principle.

81 McDiarmid, *op. cit.*, p. 56. *E.C.O.* 1846, no. 530 contains Gladstone's objections to protection. *Ibid.*, no. 1584 conveys the protest of Liverpool refiners to a Canadian duty of 2d. per lb. on refined sugar, but only 1d. on raw sugar. *Ibid.*, no. 2698 contains Grey's objections to discrimination.

82 There is a valuable discussion of the British American League in Donald Creighton, *John A. Macdonald: The Young Politician* (Toronto, 1952), pp. 142-45, and a great deal of information about it is scattered through Doughty, *Elgin-Grey Papers*. For R.B. Sullivan's speech at Hamilton in November 1847 "on the connection between agriculture and manufactures of Canada", sometimes credited with launching the protectionist movement as a serious force in Canada, see E. Porritt, *Sixty Years of Protection in Canada, 1846-1907* (London, 1908), pp. 199-201. Elgin reported the great growth of protectionist sentiment in Canada under "British party" auspices at least as early as Jan. 4, 1849 — a development used by Hincks to argue for final repeal of the Navigation Acts (*Elgin-Grey Papers*, vol. I, pp. 280, 285). The Annexationist Manifesto appears in *ibid.*, vol. IV, pp. 1487-94 and as an appendix to Tucker, *The Canadian Commercial Revolution*, p. 227. Its author dismissed "the protection of home manufactures" as a remedy: "Although this might encourage the growth of a manufacturing interest in Canada, yet, without access to the United States market, there would not be a sufficient expansion of that interest, for the want of consumers, to work any result that could be admitted as a 'remedy' for the numerous evils of which we complain." Cf. also Samuel Thompson, *Reminiscences of a Canadian Pioneer* (Toronto, 1884), pp. 245-51.

83 *Sessional Papers, 1849*, app. N. The complaints, from interested parties in Britain, were against the increase of Canadian duties in 1849 to 12½ per cent on leather and leather goods, paper, hardware (scythes, axes, castings), cottons, linens, woollens and silks. But there were interested producers of all these in Canada, except for linens and silks.

84 *P.S.O.*, *C.W.*, 1849, no. 426 (Johnstown District Council, Feb. 1849). *Ibid.*, nos. 489 (Gananoque public meeting) and 522 (Colborne District Council) are similar and appeared at the same time. Probably all were inspired by Ogle R. Gowan and connected with the agitation that produced the British American League. However the Colborne resolution, unlike the others, seemed to favour Canadian protection primarily as a bargaining weapon to win reciprocity from the United States.

85 French-Canadian thought and policy received a careful and illuminating treatment in J.I. Cooper, "French-Canadian Conservatism in Principle and in Practice, 1873-1891". On vacillations of policy, see especially pp. 530-33.

86 Cf. C.E. Rouleau, *L'Emigration: ses principales causes.*

87 Colonization plans of 1848 appear in *Elgin-Grey Papers*, vol. IV, pp. 1361-72. Colonization more generally was examined by Cooper, *op. cit., passim*. Cooper makes the point that industrialization was also seen not as an alternative but as a necessary base (market) for colonization (p. 539). This would imply a goal of a self-contained economy, with rural and urban parts interdependent. However, there was also desire to preserve the structure portrayed in Horace Miner, *St. Denis: A French-Canadian Parish* (Chicago, 1939).

88 The emphasis on sacrifice may be explained by the fact that those who had recently moved in Lower Canada, to the Eastern Townships and to Gaspe, were especially likely to emigrate. Reluctance to accept hardship thus struck at the base of the whole colonization program.

89 *Report of the Select Committee of the Legislative Assembly appointed to Inquire into the Causes and Importance of The Emigration which takes place annually, from Lower Canada to the United States.* (Montreal, 1849)(P.A.C., Pamphlet I, no. 2142), pp. 38, 45. This attack on the seigneurs appears to foreshadow the end of Seigneurial Tenure.

90 *Ibid.*, pp. 30-31.

91 The Canadian mining boom of the 1840s was promoted by a large-scale search for copper in the Lake Superior region of the United States which began in 1842, and by information gathered by the Canadian Geological Survey, which also began in 1842. The first licence to explore was granted in 1845, and a rush soon followed. By 1847, the government had worked out a policy for handling mining claims. By the end of 1848, twenty-two mining locations had been taken up, and the most important one, Bruce Mines (Montreal Mining Company), was employing 163 men. See *Sessional Papers, 1847*, app. AAA and *E.C.O.*, 1849, no. 80; also Innis and Lower, *Select Documents, 1783-1885*, pp. 580-81, 585-86.

92 This transformation was discussed in Pentland, "The Role of Capital in Canadian Economic Development before 1875," especially pp. 466-67. A specific example of a manufacture (of cement) directly promoted and financed by canal construction appears in the questionable record of one, Norton, a Commissioner for the Cornwall Canal (*U.C.S.*, Sept. 11, 1838 [Hume to McDonald]).

93 In November 1844, the Executive Council concurred with the opinion of the President of the Board of Works that "water privileges" created by canal construction ought to be leased for periods up to 50 years (*P.S.O., C.W.*, 1844-5, no. 8872). Some leases are noted in *P.S.O., C.W.*, 1846-7, no. 16785 (Cornwall Canal, May 1847); no. 16846 (Williamsburg Canal, June 1847); no. 17122 (Welland Canal, June 1847). It is stated that mill sites are being eagerly sought in *P.S.O., C.W.*, 1847-8, no. 18393 (November 1847).

94 *E.C.O.*, 1847, no. 237, A Report by Samuel Keefer on the Water Power of the Welland Canal, January 1847. Keefer set out the amount of water power available, but also urged that protection was needed to promote industry, that industry would develop better on a small rather than large scale, but that existing mills were inefficient and ought to be forced to modernize.

95 Presumably copper ore from the new mines of the Upper Lakes was concentrated here.

96 From *Sessional Papers, 1851*, app. T, a complete listing of mills on all the canals.

[97] *E.C.O.*, 1847, no. 237. Report by Samuel Keefer.

[98] The application to import cotton manufacturing machinery from the United States for the Thorold Joint Stock Cotton Factory was from Jacob Keefer, Chairman, October 1846 (*P.S.O., C.W.*, 1846, no. 15354).

[99] Another cotton mill was established at Sherbrooke in 1845 or 1846. See Innis and Lower, *Select Documents, 1783-1885*, pp. 299-305.

[100] *Sessional Papers, 1851*, app. T.

[101] Innis and Lower, *Select Documents, 1783-1885*, pp. 603-4. By 1853 the canal supported also a woollen mill, a chair factory, an axe factory, a sash and blind factory, a locomotive factory, a rubber factory, a stave and barrel factory and a cotton mill (St. Catharines *Journal*, June 2, 1853).

[102] *Sessional Papers, 1849*, app. (BB); *1850*, app. H.

[103] S.D. Clark, *The Canadian Manufacturers' Association* (Toronto, 1939), p. 6.

[104] In 1811, it was reported that a tenth of the persons in Boucherville were available for hire as servants at 7s., 6d. per month (*Baby Collection* [P.A.C.], vol. 3, no. 372 [Viger to Berczy, Oct. 7, 1811]). Generally in the nineteenth century, the towns of the lower province seem to have more and more become repositories for useless persons, rather than active economic entities.

[105] Scobie and Balfour, *Canadian Almanac*, 1859.

[106] The machinery question in the 1850s centred around the sewing machine, which Toronto tailors drove out of their city in 1851. The *Globe* nevertheless did not hesitate to point out that it was "impossible to prevent the encroachment of machinery", that employers must use the machines to meet competition, that those displaced by machines faced a world full of opportunities, and so forth (Toronto *Globe*, Dec. 7, 1852). The Toronto *Leader*, not always on the same side, had much the same opinion, that machines were inevitable and desirable, and that the "crusade against sewing machines" of the Hamilton tailors in 1854 was mad, evil and doomed to failure (Toronto *Leader*, July 12, 1853; Aug. 30, 1853; Feb. 15, 1854). The less sophisticated were deeply concerned about machines as a cause of unemployment (not employment). The displacement of farm labourers by threshing machines since about 1840 was given as an important cause of emigration in the 1849 *Report*. One of the pleas for protection of the 1840s was directed against the importation of American machinery which was alleged to be putting building tradesmen out of employment in Kingston (*P.S.O., C.W.*, 1841-3, no. 3226).

[107] H.C. Pentland, "The Development of a Capitalistic Labour Market in Canada", *Canadian Journal of Economics and Political Science*, November 1959, pp. 450-61.

[108] *Loc. cit.*; and ch. V above.

[109] Porritt, *Sixty Years of Protection in Canada*, pp. 205-6. Clark has pointed out that the commercial groups present wanted differential duties to favour the St. Lawrence route, not protection for manufacturers (*The Canadian Manufacturers' Association*, p. 1n). The effect was nevertheless to promote higher tariffs.

[110] Porritt, *loc. cit.*; R.S. Longley, *Sir Francis Hincks* (Toronto, 1943), pp. 125-26.

[111] *Report of the Committee on Trade*, Montreal *Gazette*, June 18, 1855.

[112] Porritt, *op. cit.*, p. 218; McDiarmid, *Commercial Policy in the Canadian Economy* reviews this period in ch. IV and considers that "the stability of the trend in

tariff rates is remarkable in view of the chaotic state of politics during this period" (p. 74).

113 *"Labour's Political Economy, or the Tariff Question Considered", by Horace Greeley, to which is added the Report of the Public Meeting of Delegates, Held in Toronto on the 14th April, 1858. Published by the Association for the Promotion of Canadian Industry* (Toronto, 1858) (pamphlet). The Association was supported by the Tariff Report Association of Montreal, formed in 1856. Representatives were present from Montreal, London, Hamilton, Port Hope, Louth, Gananoque, Georgetown, Dundas, Ottawa, Kingston, Brockville, Lyn, Galt, Belleville, Colborne, Merrickville, Niagara, Oshawa, Newcastle and Caledonia. In spite of a later rhetorical remark by Buchanan that the meeting was to bring manufactures into existence, those who met in 1858 were mostly manufacturers and covered a considerable range of products.

114 Porritt, *op. cit.*, pp. 229, 238.

115 *Address of the Hon. John A. Macdonald to the Electors of the City of Kingston, with Extracts from Mr. Macdonald's Speeches, Delivered on different occasions in the Years 1860 and 1861...* (Pamphlet, P.A.C.), pp. viii-ix, 70.

116 The effect on some was a heightened Canadian nationalism involving plans for further industrial diversification and expansion into Rupert's Land. Cf. *Lectures on Canada, illustrating Its Present Position, and shewing forth Its Onward Progress, and predictive of Its Future Destiny. By the Late Mr. Charles Bass* (Hamilton, 1863) (Pamphlet, P.A.C.); Fowke, *Canadian Agricultural Policy*, p. 140.

117 *Association for the Promotion of Canadian Industry, Its Formation, By-Laws, &c.* (Toronto, September 1866)(pamphlet).

118 Historians have inclined to take as evidence of disinterest in protection what appears to be a deliberate tactic. Porritt claimed that there was little support for protection between 1859 and 1876 (*op. cit.*, pp. 261, 266). Clark noted that tariffs were lowered to bring the Maritimes into Confederation, and takes this to mean that "Confederation, by emphasizing the demands of the trading interests and railway promoters, temporarily halted the movement towards protection. On questions of trade policy, the importing and transportation enterpreneurs of Montreal combined with the frontier agricultural vote to resist the imposition of a protective tariff" (*op. cit.*, pp. 3n, 4). Creighton has stated the view that few wanted a protective policy in 1867, though he also cited examples of the premonition that it would come ("British North America at Confederation", pp. 42-43). W.A. Mackintosh took much the same view of prevailing opinion, but treated protection as implicit in the Confederation decision ("The Economic Background of Dominion-Provincial Relations", Royal Commission on Dominion-Provincial Relations, appendix 3 [Ottawa, 1939], p. 17). It is difficult to imagine that intelligent observers were unaware that an east-west economic structure implied protection, or that protection would be resumed.

119 Porritt, *op.cit.*, pp. 263, 275-76. Protectionist sentiment was not only rising in Canada West at this time, but became very strong in Canada East because of the rapid increase of emigration. Cf. Cooper, "Some Early French-Canadian Advocacy of Protection, 1871-1873".

120 There is a great mass of Buchanan's papers in the Public Archives of Canada. Published sources of length include Buchanan's *Address to the Free and Independent Electors of Hamilton* (Hamilton, 1861) and H.J. Morgan, ed., *The*

Relations of the Industry of Canada, with the Mother Country and the United States (Montreal, 1864).

121 Hamilton *Spectator*, Extra, June 22, 1861.

122 *Loc. cit.*

123 "A Jotting", 14 June, 1876 (*Buchanan Papers*). Some of Buchanan's views on trade were published in *The Patriotic Party versus the Cosmopolite Party; or, in other words, Reciprocal Free Trade versus Irreciprocal Free Trade* (Toronto, 1848) and *British American Federation a Necessity, its Industrial Policy a Necessity* (Hamilton, 1865).

124 The phrase, "full employment" was used both by Buchanan and by Horace Greeley, whom he much admired. In a letter to the president of the Hamilton Cooperative Association (Feb. 8, 1866), Buchanan proposed to relieve unemployment by issuing money "enough to enable Employers to have the means of employing every man able and willing to work." A particularly far-reaching monetary program appeared in Buchanan's one-page address "To the Members of the Currency Convention, Hamilton, 28th October, 1879" (*Buchanan Papers*, printed).

125 Soap and sugar were new interests of the 1850s, oil in 1864, cotton in 1879. For the variety of new interests of 1900 and the problem of maintaining unity, cf. Clark, *Canadian Manufacturers' Association*, pp. 13-14.

126 *Ibid.* pp. 3n, 4.

127 *Ibid.*, p. 6; H.A. Innis, *Problems of Staple Production in Canada* (Toronto, 1933), pp. 10-13; Fowke, *Canadian Agricultural Policy*, pp. 261-62.

128 Cf. Porritt, *Sixty Years of Protection in Canada*, pp. 336-39.

129 Clark, *Canadian Manufacturers' Association*, pp. 27-46.

130 D.G. Creighton, "George Brown, Sir John Macdonald, and the 'Working-man'", *Canadian Historical Review*, December 1943, pp. 362-76.

131 Cf. J.M.S. Careless, "The Toronto Globe and Agrarian Radicalism, 1850-67", *Canadian Historical Review*, March 1948, pp. 14-39. It was actually the *Globe* whose care for the common man "it seems" was excluded from city dwellers (p. 34), and Careless makes clear that Brown (or the *Globe*) was not much bothered about common farmers either. Some examples of the *Globe's* view are:

April 9, 1850 — urged abolition of usury laws.

April 6, 1850 — against any interference by law with the truck system, a matter of individual contract. Besides, the mechanic "is the most independent man in the country and gets the 'cream of all the good things'."

July 10, 1852 — crowed over loss of the Amalgamated Engineers' strike in England. The moral is that workers were seeking an "artificial price for labor" and trying to run the trade.

May 11, 1854 — on a Grand Trunk strike: "Supply and demand must regulate the price of labor as of everything else."

Sept. 12, 1867 — applauded a court decision in Britain that outlawed peaceful picketing.

And still in the 1880s:

March 11, 1882 — on demands that women should receive equal pay with men for the same work: the law of supply and demand lies behind lower rates for women, and there is nothing else that can be done about it.

But there was variation in:

July 5, 1853 — welcoming a rubber factory in Montreal: "The raw material is not bulky, labour is cheap, and we have the protection of a small duty."

132 Ottawa *Citizen*, Aug. 6, 1874. This report was "highly approved" by the appropriate committee. Free Traders were not usually found in labour gatherings, but one or more appeared at the 1877 meeting of the Canadian Labour Union, and stoutly opposed protection. However his (their) eloquence did not prevail, and protection was approved by (the *Globe* said) "a majority of one or two" (Toronto *Globe*, Aug. 9, 1877).

Chapter VI

1 Karl F. Helleiner, "Moral Conditions of Economic Growth", *Journal of Economic History*, Spring 1951, pp. 97-116, is concerned with the transformation of European societies in general.

2 V.H. Jensen, *Lumber and Labor* (New York and Toronto, 1945), p. 3. This work deals most competently with the social conditions of lumber workers and the economics of their industry.

3 An important factor causing returns to staple producers to be low is the extreme seasonality of their occupations. Lumbering permitted some to overcome this disability, by working in the woods in winter and elsewhere in summer, but most staple producers, even in lumbering, remained unemployed in their off season. The effect of this has been a relatively low *annual* productivity. This disadvantage can be overcome by the introduction of industries that can operate during all seasons — manufactures — and it is significant that industrialization has developed chiefly in temperate regions.

4 The replacement on the Ottawa in the second half of the nineteenth century of human, equalitarian employers by wealthy, impersonal, grasping ones who provoked strikes by their meanness and won them by use of their wealth, is pictured in J.L. Gourlay, *History of the Ottawa Valley* (n.p., 1896), p. 113.

5 *U.C.S.*, 1838. Report on the State of the Province by R.B. Sullivan (n.d., 1838).

6 On the great difficulty of inducing French workers to adopt an outlook suited to industrial capitalism, cf. A.L. Dunham, "Industrial Life and Labor in France, 1815-1848", *Journal of Economic History*, Nov. 1943, pp. 128-29.

7 (Halifax) *Free Press*, May 6, May 13, May 20, 1817; (Halifax) *Acadian Recorder*, Feb. 1, Feb. 14, March 13, 1818. These references were kindly provided by Miss Norah Storey.

8 Lower Canada, *Journals of the Assembly, 1812*, pp. 556, 558. Other references are at *ibid., 1812*, pp. 236, 388, 554; *1817*, pp. 38, 146, 252, 312, 442, 944; *1820-1*, pp. 36, 155, 205, 220, 256.

9 *Ibid., 1820-1*, p. 256.

10 Hartz, *Pennsylvania, 1776-1860*, pp. 208-12, 295, 305-6.

11 Montreal Mechanics' Institute, *The Mechanics' Institute of Montreal, 1840-1940* (n.p., n.d. [Montreal, 1940]) (Hundredth Anniversary brochure).

12 References to the various Institutes appear in *U.C.S.* and *P.S.O.* correspondence: e.g., *P.S.O., C.W.*, 1841-3, no. 4049 (Dundas); no. 6208 (Peterboro); *P.S.O., C.W.*, 1849, no. 564 (Brantford); no. 830 (Niagara); no. 1063 (Carle-

ton Place). Other Institutes whose applications for grants appear at the beginning of *P.S.O., C.W.*, 1856, were at Ayr, Belleville, Chatham, Dunn-ville, Galt, L'Original, Ottawa, "Port Sarnia", Prince Edward, Richmond, Sydenham, St. Mary's, Whitby and Watertown. *Rules of the Quebec Mechanics' Institute, founded February 1, 1831* (Quebec, 1832), appears as Pamphlet I, no. 1392 (P.A.C.).

[13] *P.S.O., C.W.*, 1849, no. 564.

[14] *P.S.O., C.W.*, 1841-3, no. 1557. The Kingston Institute was established in 1834. The apparent deficiencies, on the other hand, of most Institutes may be indicated by a petition of Toronto mechanics in January 1849 for a "School of Art and Design", on the ground that, "but little has been done in behalf of the numerous class with which they stand connected and, hence, to a certain extent, they are led to attribute the comparative backward state of our manufactures and mechanical sciences..." (*P.S.O., C.W.*, 1849, no. 193). The Toronto Mechanics' Institute had at this time been operating for eighteen years and had accumulated large supplies of books and equip-ment, but probably conducted no regular courses. Montreal's Institute seems to have been unusual in doing so.

[15] Montreal Mechanics' Institute, *op. cit.*, pp. 5-8.

[16] *An Opening Address Delivered at the first meeting of the Halifax Mechanics' Institute on Wed., Jan. 11, 1832, by Joseph Howe* (Halifax, 1832)(P.A.C., Pamphlet I, no. 1391).

[17] Montreal Mechanics' Institute, *op. cit.*, p. 8.

[18] The patents referred to in this section appear from time to time in the correspondence of the Provincial Secretaries, and information is drawn largely from this source — i.e., *U.C.S.*; *S* Series; *P.S.O., C.W.*; *P.S.O., C.E.* Notice of patents of invention also appeared in contemporary newspapers, and some information has been drawn from that source.

[19] This conclusion is drawn on the basis of correspondence concerning dis-puted patents, and suits for infringement. This indicates that the common farm machines were made by numerous village smiths, with more or less variation in design according to individual taste and capacity.

[20] Industrial society also achieved notable economies of scale and "external" economies other than the external economy of the labour market; but these economies were not unknown to staple production, and therefore are not distinctive.

[21] These were the chief issues raised by the first and possibly the only number of "The People's Magazine and Workingman's Friend", published in Quebec, March 16, 1842 (with *P.S.O., C.E.*, 1842, no. 794).

[22] The high valuation placed on skilled labour, and the need to keep the flow of it from Britain open, is given as the basic reason for the Trade Unions Act of 1872 being proposed and passed in B. Ostry, "Conservatives, Liber-als, and Labour in the 1870s", *Canadian Historical Review*, June 1960, pp. 106-10.

[23] The proceedings of the Commissioners are in *Public Works I*, vol. 13 (P.A.C.). A strike of carpenters (the only group that did not strike in 1823) against the Commission in 1825 turned out in much the same way — that is, the Commissioners took a strong line against yielding to their workmen, but the carpenters got what they wanted in a round-about way.

[24] *Ibid.*, vol. 16.

25 The Toronto (semi-weekly) *Leader* was especially willing to consider the labour viewpoint. The issue of May 9, 1854, noted that rising food prices were the cause of the current increases in wages and of the wave of strikes, and regretted only that so often the wage increases called for were won by strikes without previous notice to the employer: "The price of labour is a subject of bargain, requiring mutual agreement..." The issue of May 29, 1854, in reporting a strike in Toronto for the 10-hour day, said that 10 hours was certainly long enough, that it produced more work than a 12-hour day, and that the employers should yield. (The Toronto *Globe*, May 11, 1854, also approved of the 10-hour day "in principle" — if in accord with supply and demand, by which the *Globe* meant, if the employer is agreeable to it.) Cf. also a laconic announcement in the St. Catharines *Post* (Toronto *Leader*, May 3, 1854): "The Shoemakers' strike in this town has been amicably arranged, the employers agreeing to pay the wages demanded by the Journeymen."

26 A strike of Grand Trunk employees (and others) in Montreal in 1855 for higher wages produced a compromise settlement (Montreal *Gazette*, May 5, 1855). In a strike of carters against the Grand Trunk in 1864 in Montreal, the company offered a compromise but apparently was defeated by public opinion and legal rulings (Toronto *Globe*, Oct. 3, Oct. 5, 1864). When Grand Trunk engineers struck on Dec. 29, 1876 for the principle of seniority in lay-offs, they won their point. The Toronto *Globe*, which had been violently opposed to any acknowledgement that seniority established a claim to anything, then denounced the General Manager of the Grand Trunk as weak, undetermined, changeable, etc. (Toronto, *Globe*, Jan. 7, 1877).

27 Workingmen may have achieved some immunity to charges of conspiracy also because lawyers, as a class, were engaged themselves in a conspiracy at mid-century. Lawyers opposed popular proposals that their fees should be set by the courts, and the lawyers of Quebec City went on strike in 1850 against the setting of their fees (Toronto *North American*, Jan. 3, 1851).

28 *Sessional Papers, 1843*, app. Q. Another argument was that even the lowest Canadian (daily) wage rates were higher than rates in Europe, on which people somehow survived. Those who argued this way seem never to have included comparisons of the cost of living, and of annual incomes, in their demonstrations.

29 *P.S.O., C.E.*, 1843, no. 1200. It was true that many of the Irish labourers in Canada in 1843 had come over recently from the United States, but usually the movement was the other way, and Killaly here twisted the facts to win the argument. But it was also true that Canada had the easy way of getting rid of surplus labourers by letting them go south — the province had been cleared of its surplus in this way in 1838.

30 *U.C.S.*, May 15, 1827 (By to Maitland); Aug. 8, 1829 (Cooper to Colborne, enclosure); May 28, 1831 (Hagerman to Colborne); McTaggart, *Three Years in Canada*, passim.

31 *U.C.S.*, April 4, 1827 (Wilson to Hillier); May 18, 1827 (Wilson to Hillier).

32 *U.C.S.*, June 10, 1828 (By to Mann).

33 *U.C.S.*, Dec. 22, 1834 (Hume to Rowan, enclosure).

34 *U.C.S.*, Dec. 26, 1834 (Robinson to Colborne).

35 For example, *C*60, pp. 86-87 (General I.A. Hope, April 3, 1843).

36 *Sessional Papers, 1843*, app. T, Evidence of John Falvey, R.J. Begley.

[37] H.C. Pentland, "The Lachine Strike of 1843", *Canadian Historical Review*, Sept. 1948, pp. 255-77.

[38] *P.S.O., C.E.*, 1843, no. 524 (March 8, 1843); no. 626 (March 23, 1843, enclosure); no. 733 (April 3, 1843).

[39] *P.S.O., C.E.*, 1843, no. 626 (March 23, 1843, enclosure: Executive Council Report).

[40] Provincial Secretary, C.E., *Letter Book*, March 25, 1843 to Rev. Archambault; March 23, 1843 to Ermatinger.

[41] *P.S.O., C.E.*, 1843, no. 2297 (Oct. 28, 1843).

[42] *P.S.O., C.E.*, 1843, no. 2618 (Dec. 12, 1843); 1844, no. 2955 (Sept. 9, 1844).

[43] *P.S.O., C.E.*, 1844, no. 729 (Feb. 20, 1844).

[44] *P.S.O., C.E.*, 1844, no. 1045 (March 14, 1844, enclosure); no. 1505 (April 21, 1844); no. 1606 (April 29, 1844).

[45] *P.S.O., C.E.*, 1844, no. 3212 (Oct. 4, 1844); no. 3310 (Oct. 26, 1844).

[46] *P.S.O., C.E.*, 1844, no. 3336 (Oct. 31, 1844).

[47] *P.S.O., C.E.*, 1844, no. 3453 (Nov. 17, 1844).

[48] 8 Vic., c.6.

[49] The Report of the Executive Council that foreshadowed the Act, Jan. 3, 1845. With *P.S.O., C.E.*, 1844, no. 3534 (Nov. 28, 1844).

[50] For example, troops were maintained in the vicinity of the Welland Canal, along with police, at least until May 1847 (*P.S.O., C.W.*, 1846-7, no. 16757 [May 15, 1847]). Priests were kept on the Welland and Cornwall Canals, e.g., *P.S.O., C.W.*, 1843-4, no. 8168 (March 29, 1844); 1846-7, no. 17399 (Executive Council Report, July 23, 1847).

[51] The revolt of the Rev. J.C. Clark, who immediately went, or was shipped, back to Ireland, appears in *P.S.O., C.W.*, 1844-5, nos. 10145 (April 14, 1845); 11362; 11462; 11611; 11635 (Aug., Sept., 1845).

The Report of a Committee of Executive Council, July 31, 1844, on the request of a Rev. M. Brethour for payment for spiritual services rendered to canal labourers (*E.C.O.*) contains some direct statements: "the employment of a Roman Catholic clergyman (at Beauharnois) took place on Mr. Killaly's recommendation as a measure of Police and for the preservation thro' his influence of peace and order on the line of this Canal, not for the purpose of furnishing religious instruction to the labourers. It having been deemed inexpedient to Station Troops along the line of the Canal and the large expense which would have attended the establishment of an effective Police force, combined to render the course adopted desirable as economic and effectual for the object designed. If the employment of any person other than a Roman Catholic Clergyman would have been more likely to gain the object, such a course would have been taken."

[52] C317, p. 278 ff.; C318, p. 79 ff.

[53] *P.S.O., C.W.*, 1851, no. 284 (Feb. 12, 1851).

[54] *Loc. cit.*, enclosures: Executive Council, Feb. 17, 1851; June 17, 1851; July 7, 1851.

[56] *P.S.O., C.W.*, 1853, no. 1050 (July 7, 1853).

[57] However, there was apprehension that July 12 might bring a clash between Orangemen and labourers, and there was a rush to get the detachment of troops to Hamilton before that date.

58 14 & 15 Vic., c. 76.

59 14 & 15 Vic., c. 77.

60 *P.S.O., C.W.,* 1853, no. 1050 (July 7, 1853) and enclosure.

61 *P.S.O., C.W.,* 1853, no. 1107 (Aug. 8, 1853).

62 *P.S.O., C.W.,* 1854, no. 99 (Jan. 12, 1854).

63 *P.S.O., C.W.,* 1854, no. 111 (Jan. 21, 1854), enclosure; C318, pp. 392-97.

64 *P.S.O., C.W.,* 1854, no. 111, enclosure.

65 *Ibid.,* enclosure.

66 The contemporary achievement of dominance by the Pennsylvania Railroad over the State of Pennsylvania is well analyzed by Hartz, *Pennsylvania, 1776-1860,* especially pp. 310-20. How the officers of state learned to jump in Canada appears from an episode of April 1858, when the provincial authorities were called on to send troops to deal with "a serious riot" on the Grand Trunk line at St. Mary's. The troops were dispatched from Toronto immediately and arrived in St. Mary's by 4 o'clock on the following morning (C319, p. 151).

67 For Cork-Connaught strife at Lachine, cf. Pentland, "The Lachine Strike of 1843", pp. 263-64; on the Welland Canal, C60, pp. 130, 213, 217, 220. Cf. also C60, p. 80.

68 *P.S.O., C.W.,* 1844-5, no. 8689 (Sept. 12, 1844), enclosure.

69 These were prime issues in the Beauharnois strikes of 1843 (C60, p. 83; *Sessional Papers, 1843,* app. T, Evidence of P.J. Begley), and the investigation that followed them implied that remedies were provided, by writing stricter provisions into contracts. At any rate, the issue seems not to have been raised after this.

70 Pentland, "The Lachine Strike of 1843", p. 266; *P.S.O., C.W.,* 1844-5, no. 8939 (Jan. 18, 1845).

71 *U.C.S.,* May 18, 1837 (Hagerman to Joseph); Aug. 18, 1837 (Jones to Joseph).

72 *P.S.O., C.E.,* 1844, no. 1045 (March 14, 1844).

73 *P.S.O., C.E.,* 1844, no. 1505 (April 21, 1844), enclosure.

74 *P.S.O., C.E.,* 1844, no. 1799 (May 23, 1844) and enclosures.

Bibliography

A. Manuscript Sources
(Public Archives of Canada)

Correspondence of the Provincial Secretaries (originals)
Incoming correspondence:
    Lower Canada Sundries (S series), 1765-1840.
    Upper Canada Sundries (U.C.S.), 1816-40.
    Provincial Secretary, Canada East (P.S.O., C.E.), 1842-45.
    Provincial Secretary, Canada West (P.S.O., C.W.), 1840-56.
Outgoing correspondence:
    Provincial Secretary's Letter Books.
Executive Council Office (E.C.O., originals)
    Executive Council Papers, 1846-49.
    Executive Council Papers "Put By", 1844-66.
    Executive Council, Additional Papers, 1841-67.
    Also:
    State Papers, Upper Canada. vol. 24 (1840: Opinions of various members of Council on the state of the province and the best method of handling immigration).
Military Affairs (C series, originals): C60-1 (Canals); C105, 108-9 (Supplies); C316-19 (Military Aid at Riots).
State Books, Province of Canada, A to Y (1841-63, originals).
Public Works (originals): I, vols. 5-22, 80-81; V, vols. 1-6.
Colonial Office (transcripts from the Public Record Office, Great Britain)
    Q series nos. 198, 222, 238, 337-40, 344.
    Also: C.O. 43.
Miscellaneous and Private Series
    M377, Add. MSS (British Museum, Additional Manuscripts, transcript).
    W.O. 34, vols. 59-64 (War Office, transcripts, Amherst Papers, 1758-63).
    B21 (Emigration Returns, 1841-44, Memoranda).
    M.G. 22:34 (Immigration, Miscellaneous MS., 1792-1863).
    M.G. 29:128 (Lawrence Ermatinger's Engagement Book, 1773-75).
    Upper Canada Land Petitions, A-A1 (1792-1840).
    Lord Selkirk, *Papers* (vols. 74-75 are Selkirk's diaries, Aug., 1803- Sept., 1805).
    Baby Collection.
    William Claus Papers.
    Mitchell Collection.
    William Dummer Powell MS., "First Days in Upper Canada".
    Isaac Buchanan Papers.

B. Government Reports, Printed

Lower Canada, *Journals of the Assembly*, and Appendices, 1809-23.

Upper Canada, *Journals of the Assembly*, and Appendices, 1833-34, 1835, 1839.

Canada, *Journals of the Assembly*, and Appendices (Sessional Papers), 1841-59.

Canada, *Census*, 1871 (vol. IV contains all earlier censuses), 1881, 1891.

Report of the Royal Commission on Dominion-Provincial Relations. 3 vols. Ottawa, 1939. And Appendices.

Great Britain, *Imperial Blue Books* relating to Canada (Parliamentary Papers)
Emigration, 1829-39.
Emigration to Canada, 1841-43.
Emigration to Canada, 1844-48.
Emigration, 1848-50.
Emigration, 1851-72.

C. Newspapers

Halifax *Acadian Recorder*, 1818.

Ottawa *Citizen*, 1874.

Halifax *Free Press*, 1817.

Montreal *Gazette*, 1855.

Quebec *Gazette*, 1840, 1841, 1843.

Toronto *Globe*. Daily: 1850, 1852-54, 1864, 1867, 1875-77, 1882, 1885. Semi-weekly: 1854.

St. Catharines *Journal*, 1851, 1853.

Toronto *Leader*. Semi-weekly: 1853. Daily: 1854-55, 1858.

Toronto *Mail*, 1873.

Toronto *North American*, Semi-weekly: 1850. Weekly: 1850-51.

Hamilton *Spectator*, "Saturday Musings", 1908-20 (Hamilton Public Library collection); July 15, 1946.

Ottawa *Times*, 1874.

Montreal *Transcript*, 1842-43, 1851, 1853, 1855-56.

Hamilton *Herald*, July 5, 1902; Nov. 7, 1903 (Hamilton Public Library).

D. Reference Works and Collections of Documents

Abbott, E. *Immigration: Select Documents and Case Records.* Chicago, 1924.

Canada and Its Provinces. Edited by Adam Shortt and Arthur G. Doughty. 23 vols. Toronto, 1913-17.

Canadian Archives, Reports, 1899, 1904.

Dictionnaire général du Canada. Edited by R.P.L. LeJeune, 2 vols. Ottawa, 1931.

Doughty, Sir. A.G., ed. *The Elgin-Grey Papers, 1846-52.* 4 vols. Ottawa, 1937.

Encyclopaedia Britannica. Edited by W. Yust. 1960.

Force, Peter, ed. *American Archives: consisting of a Collection of Authentick Records, State Papers, Debates, and Letters... Fourth Series: A Documentary History of the English Colonies in North America from March 7, 1774 to the Declaration of Independence.* Washington, 1839.

Innis, H.A. *Select Documents in Canadian Economic History, 1497-1783.* Toronto, 1929.

Innis, H.A., and Lower, A.R.M. *Select Documents in Canadian Economic History, 1783-1885.* Toronto, 1933.

Mulhall, M.G. *The Dictionary of Statistics.* London, 1892.

Almanac

Canadian Almanac and repository of useful knowledge. Toronto: Scobie and Balfour, 1850; H. Scobie, 1851-54; Maclear & Co., 1855-59.

Minutes

Minutes of the Toronto Trades Assembly, 1871-78 (typescript).

E. Theses

Aitchison, J.H. "The Development of Local Government in Upper Canada, 1783-1850". Ph.D. thesis, University of Toronto, 1953.

Camm, R.W. "History of the Great Western Railway of Canada". M.A. thesis, University of Western Ontario, 1947.

Cooper, J.I. "French-Canadian Conservatism in Principle and in Practice, 1873-1891". Ph.D. thesis, McGill University, 1938.

Goldberg, Simon A. "The French-Canadians and the Industrialization of Quebec". M.A. thesis, McGill University, 1940.

Lepine, J.J. "The Irish Press in Upper Canada and the Reform Movement, 1828-48". M.A. thesis, University of Toronto, 1946.

Lunn, Alice J.E. "Economic Development in New France, 1713-1760". Ph.D. thesis, McGill University, 1942.

Mood, W.J.S. "The Orange Order in Canadian Politics, 1841-1867". M.A. thesis, University of Toronto, 1950.

Parker, W.H. "The Geography of the Province of Lower Canada in 1837". Ph.D. thesis, Oxford University, 1958.

Reid, Allana G. "The Development and Importance of the Town of Quebec, 1606-1760". Ph.D. thesis, McGill University, 1950.

F. Books and Monographs

Adams, W.F. *Ireland and Irish Emigration to the New World from 1815 to the Famine.* New Haven, 1932.

Arensburg, Conrad M. *The Irish Countryman.* New York, 1937.

Arensberg, C.M., and Kimball, S.T. *Family and Community in Ireland.* Cambridge, Mass., 1940.

Baxter, W.T. *The House of Hancock: Business in Boston, 1724-1775.* Cambridge, Mass., 1945.

Berthoff, R.T. *British Immigrants in Industrial America, 1790-1950.* Cambridge, Mass., 1953.

Blanchard, R. *L'Est du Canada Français.* 2 vols. Paris and Montreal, 1935.

Bolton, C.K. *Scotch-Irish Pioneers in Ulster and America.* Boston, 1910.

Bouchette, Joseph. *The British Dominions in North America...* 2 vols. London, 1832.

Brassey, Thomas. *Work and Wages.* New York, 1872.

Breault, L. *Ottawa Old and New.* Ottawa, 1946.

Breckenridge, R.M. *The Canadian Banking System, 1817-1890.* New York, 1895.

Bridenbaugh, Carl. *The Colonial Craftsman.* New York and London, 1950.

Brouillette, Benoit. "Le Développement industriel de la Vallée du St. Maurice". Pages trifluviennes, Series A, no. 2. Trois Rivières, 1932.

Burt, A.L. *The Old Province of Quebec.* Toronto, 1933.

Calvin, D.D. *A Saga of the St. Lawrence.* Toronto, 1945.

Canniff, William. *History of the Settlement of Upper Canada.* Toronto, 1869.

Carrothers, W.A. *Emigration from the British Isles, with Special Reference to Development of the Overseas Dominions.* London, 1929.

Clapham, J.H. *An Economic History of Modern Britain: The Early Railway Age, 1820-1850.* Cambridge, 1926.

Clapham, J.H., and Power, Eileen, eds. *The Cambridge Economic History of Europe.* Vol. I, *The Agrarian Life of the Middle Ages.* Cambridge, 1941.

Clark, S.D. *The Canadian Manufacturers' Association.* Toronto, 1939.

—. *The Social Development of Canada.* Toronto, 1942.

Cochran, T.C., and Miller, W. *The Age of Enterprise: A Social History of Industrial America.* New York, 1942.

Coleman, McAlister, *Men and Coal.* New York, 1943.

Commons, John R., and associates. *History of Labour in the United States.* 2 vols. New York, 1918.

Conant, Thomas. *Upper Canada Sketches.* Toronto, 1898.

Connell, K.H. *The Population of Ireland, 1750-1845.* Oxford, 1950.

Cowan, Helen I. *British Emigration to British North America, 1783-1837.* Toronto, 1928.

Creighton, D.G. *The Commercial Empire of the St. Lawrence, 1760-1850.* Toronto, 1937.

—. *John A. Macdonald: The Young Politician.* Toronto, 1952.

Culliton, J.T. *Assisted Emigration and Land Settlement with Special Reference to Western Canada.* Montreal, 1928.

Davin, Nicholas Flood. *The Irishman in Canada.* Toronto, 1877.

Davis, B.P. and C.L. *The Davis Family and the Leather Industry, 1834-1934.* Toronto, 1934.

Dobb, M. *Studies in the Development of Capitalism.* London, 1946.

"A Backwoodsman" (Dr. William Dunlop). *Statistical Sketches of Upper Canada for the Use of Emigrants.* 3rd ed. London, 1833.

Durham, Earl of (J.G. Lambton). *Report on the Affairs of British North America.* 4th ed. London, 1930.

Fauteux, J.N. *Essai sur l'industrie au Canada.* 2 vols. Quebec, 1927.

Finch, I. *Travels in the United States and Canada.* London, 1833.

Ford, H.J. *The Scotch-Irish in America.* 2nd ed. New York, 1941.

"Four Year Resident". *Views of Canada and the Colonists, By a Four Year Resident.* Edinburgh, 1844.

Fowke, V.C. *Canadian Agricultural Policy: The Historical Pattern.* Toronto, 1946.

Fox, W.S., ed. *Letters of William Davies, Toronto, 1854-1861.* Toronto, 1945.

Gates, P.W. *The Illinois Central and Its Colonization Work.* Cambridge, Mass., 1934.

Gill, Conrad. *The Rise of the Irish Linen Industry.* Oxford, 1925.

Glazebrook, G.P. de T. *A History of Transportation in Canada.* Toronto, 1938.

Gourlay, J.L. *History of the Ottawa Valley.* n.p., 1896.

Gourlay, Robert. *Statistical Account of Upper Canada.* 3 vols. London, 1822.

Gras, N.S.D. *Evolution of the English Corn Market.* Cambridge, Mass., 1915.

—. *History of Agriculture in Europe and America.* New York, 1925.

—. *An Introduction to Economic History.* New York and London, 1922.

Greaves, Ida. *The Negro in Canada.* Orillia, n.d. (1930).

Greene, L.J. *The Negro in Colonial New England.* New York, 1942.

Guillet, E.C. *Early Life in Upper Canada.* Toronto, 1933.

Hamilton, Henry. *The Industrial Revolution in Scotland.* Oxford, 1932.

Hamon, E. *Les Canadiens-Français de la Nouvelle-Angleterre.* Quebec, 1891.

Handlin, Oscar. *Boston Immigrants, 1790-1865.* Cambridge, Mass., 1941.

Handlin, O. and M. *Commonwealth: A Study of the Role of Government in the American Economy: Massachusetts, 1774-1861.* New York, 1947.

Hansen, M.L. *The Atlantic Migration, 1607-1860: A History of the Continuing Settlement of the United States.* Edited by A.M. Schlesinger. Cambridge, Mass., 1940.

—. *The Immigrant in American History.* Cambridge, Mass., 1940.

Hansen, M.L. and Brebner, J.B. *The Mingling of the Canadian and American Peoples.* New Haven, 1940.

Hartz, L. *Economic Policy and Democratic Thôught: Pennsylvania, 1776-1860.* Cambridge, Mass., 1948.

Helps, Sir Arthur. *Life and Labours of Mr. Brassey, 1805-1870.* 3rd ed. London, 1872.

Hind, H.Y., ed. *Eighty Years' Progress of British North America . . .* Toronto, 1863.

Howison, John. *Sketches of Upper Canada.* 3rd ed. Edinburgh and London, 1825.

Hulbert, A.B. *Historic Highways of America.* Cleveland, 1904.

 — Vol. 10, *The Cumberland Road.*

 — Vol. 13, *The Chesapeake and Ohio Canal and the Pennsylvania Canal.*

 — Vol. 14, *The Erie Canal.*

Innis, H.A. *The Cod Fisheries: The History of an International Economy.* New Haven, 1940.

—. *The Fur Trade in Canada.* New Haven, 1930.

—. *Problems of Staple Production in Canada.* Toronto, 1933.

Jensen, V.H. *Lumber and Labor.* New York and Toronto, 1945.

Johnson, S.C. *A History of Emigration from the United Kingdom to North America, 1763-1912.* London, 1913.

Jones, R.L. *History of Agriculture in Ontario, 1613-1880.* Toronto, 1946.

Kalm, Peter. *Travels in North America.* Translated by J.R. Forster. 3 vols. Warrington and London, 1770-71.

Keefer, T.C. *The Canals of Canada: Their Prospects and Influence.* Toronto, 1850.

Kirkland, E.C. *Men, Cities, and Transportation: A Study in New England History, 1820-1900.* 2 vols. Cambridge, Mass., 1948.

Langlois, Georges. *Histoire de la population canadienne-française.* Montreal, 1934.

Laval University. *Rapport, Premier Congrès des relations industrielles de Laval, 1946.* Quebec, 1946.

LeMoine, J.M. *Picturesque Quebec.* Montreal, 1882.

Lindsay, Charles. *The Life and Times of Wm. Lyon Mackenzie.* 2 vols. Toronto, 1862.

Lipson, E. *The Growth of English Society.* New York, 1953.

— Logan, Harold A. *The History of Trade-Union Organization in Canada.* Chicago, 1928.

Longley, R.S. *Sir Francis Hincks.* Toronto, 1943.

Lounsbury, R.G. *The British Fishery at Newfoundland, 1634-1763.* New Haven, 1934.

Lower, A.R.M. *The North American Assault on the Canadian Forest.* Toronto, 1933.

McDiarmid, O.J. *Commercial Policy in the Canadian Economy.* Cambridge, Mass., 1946.

MacDonald, Norman. *Canada, 1763-1841: Immigration and Settlement.* London, 1939.

McTaggart, John. *Three Years in Canada: An Account of the Actual State of the Country in 1826-7-8...* 2 vols. London, 1829.

Maguire, J.F. *The Irish in America.* London, 1868.

Marx, Karl. *Theories of Surplus Value.* New York, 1952.

Mathews, Hazel C. *Oakville and the Sixteen: The History of an Ontario Port.* Toronto, 1933.

Miner, Horace. *St. Denis: A French-Canadian Parish.* Chicago, 1939.

Mixter, C.W., ed. *The Sociological Theory of Capital: Being a Complete Reprint of the New Principles of Political Economy, 1834.* New York, 1905.

Morgan, H.J., ed. *The Relations of the Industry of Canada with the Mother Country and the United States.* Montreal, 1864.

Munro, W.B. *The Seigniorial System in Canada.* New York, 1907.

Nussbaum, F.L. *History of the Economic Institutions of Modern Europe.* New York, 1935.

O'Brien, George. *The Economic History of Ireland in the Seventeenth Century.* Dublin and London, 1919.

— Ogden, J.C. ["A Citizen of the United States"]. *A Tour through Upper and Lower Canada.* Litchfield, 1799.

Picken, Andrew. *The Canadas, as they at present commend themselves to the enterprize of Emigrants, Colonists, and Capitalists.* London, 1832.

Pierce, B.L. *A History of Chicago.* Vol. I, New York, 1937.

Pooley, W.V. *The Settlement of Illinois from 1830 to 1850.* Madison, 1908.

Porritt, E. *Sixty Years of Protection in Canada, 1846-1907.* London, 1908.

Ragatz, L.J. *The Fall of the Planter Class in the British Caribbean, 1763-1833.* New York and London, 1928.

Reynolds, L.G. *The British Immigrant: His Social and Economic Adjustment in Canada.* Toronto, 1935.

Rich, E.E. and Johnson, A.M., eds. *Cumberland House Journals and Inland Journal, 1775-82:*
— First Series, 1775-79. London: Hudson's Bay Record Society, 1951.
— Second Series, 1779-82. London: Hudson's Bay Record Society, 1952.

Rolph, Thomas. *Emigration and Colonization.* London, 1844.

Rouleau, C.E. *L'Emigration: ses principales causes.* Quebec, 1896.

Ryerson, S.B. *French Canada.* Toronto, 1943.

Scadding, H. *Toronto of Old: Collections and Recollections.* Toronto, 1873.

Stanley, G.F.G. *Canada's Soldiers, 1604-1954: The Military History of an Unmilitary People.* Toronto, 1954.

Sulte, Benjamin. "Les Forges Saint-Maurice". Vol. 6. *Mélanges historiques.* Edited by Gerard Malchelosse. Montreal, 1920.

Tanser, H.A. *The Settlement of Negroes in Kent County, Ontario.* Chatham, 1939.

Tessier, A. *Les Forges Saint-Maurice, 1729-1883.* Three Rivers, 1952.

Thomas, Brinley. *Migration and Economic Growth: A Study of Great Britain and the Atlantic Economy.* Cambridge, 1954.

Thompson, Samuel. *Reminiscences of a Canadian Pioneer.* Toronto, 1884.

Tombs, L.C. *The Port of Montreal.* Toronto, 1926.

Trout, J.M. and Edw. *The Railroads of Canada for 1870-1.* Toronto, 1871.

Tucker, G.N. *The Canadian Commercial Revolution, 1845-1851.* New Haven, 1936.

Turner, F.J. *The Frontier in American History.* New York, 1920.

Wade, Mason. *The French Canadians, 1760-1945.* Toronto, 1955.

Wakefield, Edward Gibbon, *A View of the Art of Colonization.* Collier ed. Oxford, 1914.

Ware, Norman, *The Industrial Worker, 1840-1860.* Boston and New York, 1924.

Webb, S. and B. *History of Trade Unionism.* London, 1920.

Willcox, W.F., ed. *International Migrations.* 2 vols. New York, 1931.

Williams, Eric. *Capitalism and Slavery.* Chapel Hill, 1944.

G. Pamphlets

(Most of the pamphlets cited here are available in the Public Archives of Canada)

Anonymous. *Reflections Sommaires sur Le Commerce qui s'est fait en Canada: d'après un manuscript à la bibliothèque du roi à Paris.* n.p., n.d. (about 1759).

Association for the Promotion of Canadian Industry, Its Formation, By-Laws, &c. Toronto, Sept. 1866.

(Backwoodsman, A). *Statistical Sketches of Upper Canada for the Use of Emigrants.* 3rd ed. London, 1833.

Bass, Charles. *Lectures on Canada, illustrating Its Present Position, and shewing forth Its Onward Progress, and predictive of Its Future Destiny. By the Late Mr. Charles Bass.* Hamilton, 1863.

Buchanan, Isaac. *Address to the Free and Independent Electors of Hamilton.* Hamilton, 1861.

—. *British American Federation a Necessity, its Industrial Policy a Necessity.* Hamilton, 1865.

—. *The Patriotic Party versus the Cosmopolite Party; or, in other words, Reciprocal Free Trade versus Irreciprocal Free Trade.* Toronto, 1848.

(Canadian Settler, A). *The Emigrant's Informant; or, A Guide to Upper Canada.* London, 1834.

Clark, J.M. *Common and Disparate Elements in National Growth and Decline.* Mimeo., n.p., 1948.

"An Old and Experienced Trader" (Alexander Cluny). *The American Traveller: or, Observations on the Present State, Culture, and Commerce of the British Colonies in America.* London, 1769.

(English Settler, An). *A Few Plain Directions for Persons Intending to Proceed as Settlers... to Upper Canada.* London, 1820.

Ferguson, William. *Address delivered before the Provincial Agricultural Association at the Thirteenth Annual Exhibition at Toronto, 1858, by the President, William Ferguson, Esq., of Kingston, C.W.* Toronto, n.d. (1858).

Ferrier, A.D. *Reminiscences of Canada and the Early Days of Fergus.* Guelph, 1866.

Foster, Vere. *Work and Wages; or, the Penny Emigrant's Guide to the United States and Canada.* 5th ed. London, n.d.

(Four Year Resident, A). *Views of Canada and the Colonists.* Edinburgh, 1844.

(Friend of Trade, A). *The True Interest of Great-Britain in Regard to the Trade and Government of Canada, Newfoundland, and the Coast of Labrador.* London, 1767.

(Great Western Railway). *Report to the Directors of the Great Western Railway to the Shareholders, upon the Report made by the Commission appointed to enquire into certain accidents upon the Great Western Railway.* n.p., April 1855.

Greeley, Horace. *"Labour's Political Economy, or the Tariff Question Considered" by Horace Greeley, to which is added the Report of the Public Meeting of Delegates, Held in Toronto on the 14th April, 1858. Published by the "Association for the Promotion of Canadian Industry".* Toronto, 1858.

Howe, Joseph. *An Opening Address Delivered at the first meeting of the Halifax Mechanics' Institute on Wed., Jan. 11, 1832, by Joseph Howe.* Halifax, 1832.

LaFontaine, L.H., and Viger, M.J. *"De l'esclavage en Canada". Mémoires et documents relatifs à l'histoire du Canada publiés pas la Société historique de Montréal.* Montreal, 1859.

Lovelock, H. *Official Orange Souvenir,* Toronto, July 12, 1902.

Macdonald, John A. *Address of the Hon. John A. Macdonald to the Electors of the City of Kingston, with Extracts from Mr. Macdonald's Speeches, Delivered on different occasions in the Years 1860 and 1861.* n.p., June 10, 1861.

Montreal Mechanics' Institute. *The Mechanics' Institute of Montreal, 1840-1940.* Montreal, n.p., n.d. (1940).

Quebec Mechanics' Institute, *Rules of the Quebec Mechanics' Institute, founded February 1, 1831.* Quebec, 1832.

Saunders, Leslie R. *The Story of Orangeism.* Toronto, 1941.

(Seven Years' Resident of North America, A). *A Practical Guide for Emigrants to North America including the United States, Lower and Upper Canada, and Newfoundland.* London, 1850.

Smith, Sidney. *The Settler's New Home or the Emigrant's Location.* London, 1849.

(Ten Year Resident, A). *Emigrants' Guide: being Information Published by His Majesty's Commissioners for Emigration, respecting the British Colonies of Upper and Lower Canada, and New Brunswick.* Devonport, n.d. (1832).

Wiman, E. *Annual Report of the Board of Trade with a review of the Commerce of Toronto for 1861*. Toronto, n.d. (1861).

H. Periodical Articles

Brebner, J.B. "Laissez-Faire and State Intervention in Nineteenth Century Britain". *Tasks of Economic History*. Supplement VIII to the *Journal of Economic History* (1948), pp. 59-73.

Careless, J.M.S. "The Toronto Globe and Agrarian Radicalism, 1850-67". *Canadian Historical Review*, March 1948, pp. 14-39.

Clapham, J.H. "Irish immigration into Great Britain in the nineteenth century". *Bulletin of the International Committee of Historical Sciences*, V, part III, no. 20 (July 1933), pp. 598-602.

Cooper, J.I. "Some Early French-Canadian Advocacy of Protection: 1871-1873". *Canadian Journal of Economics and Political Science*, Nov. 1937, pp. 530-40.

—. "The Quebec Ship Labourers' Benevolent Society". *Canadian Historical Review*, Dec. 1949, pp. 336-43.

Creamer, Daniel, "Recruiting Laborers for Ameskeag Mills". *Journal of Economic History*, May 1941, pp. 42-48.

Creighton, D.G. "The Economic Background of the Rebellion of Eighteen Thirty-Seven". *Canadian Journal of Economics and Political Science*, Aug. 1937, pp. 322-34.

—. "George Brown, Sir John Macdonald, and the 'Workingman'". *Canadian Historical Review*, Dec. 1943, pp. 362-76.

Dunham, A.L. "Industrial Life and Labor in France, 1815-1848". *Journal of Economic History*, Nov. 1943, pp. 117-51.

Fowke, V.C. "The National Policy — Old and New". *Canadian Journal of Economics and Political Science*, Aug. 1952, pp. 271-86.

Gallagher, J., and Robinson, R. "The Imperialism of Free Trade". *Economic History Review*, Aug. 1953, pp. 1-15.

Gates, Lillian F. "The Decided Policy of William Lyon Mackenzie". *Canadian Historical Review*, Sept. 1959, pp. 185-208.

Habakkuk, H.J. "Family Structure and Economic Change in Nineteenth-Century Europe". *Journal of Economic History*, XV (1955), no. 1, pp. 1-12.

Helleiner, Karl F. "Moral Conditions of Economic Growth". *Journal of Economic History*, Spring 1951, pp. 97-116.

Henripen, Jacques. "From Acceptance of Nature to Control". *Canadian Journal of Economics and Political Science*, Feb. 1957, pp. 10-19.

Hoselitz, Bert F. "The City, the Factory, and Economic Growth". *American Economic Review, Papers and Proceedings*, May 1955. pp. 166-84.

Innis, H.A. "Interrelations between the Fur Trade of Canada and the United States". *Mississippi Valley Historical Review*, Dec. 1933, pp. 321-32.

—. "Unused Capacity as a Factor in Canadian Economic History". *Canadian Journal of Economics and Political Science*, Feb. 1936, pp. 1-15.

Jenks, L.H. "Railroads as an Economic Force in American Development". *Journal of Economic History*, May 1944, pp. 1-17.

Landon, Fred. "Social Conditions among the Negroes in Upper Canada before 1865". *Ontario Historical Society, Papers and Proceedings*, XXII (1925), pp. 144-59.

LeMoine, J.M. "Slavery at Quebec". *Canadian Antiquarian and Numismatic Journal*. April 1872, p. 158.

Lester, R.A. "Currency Issues to Overcome Depressions in Pennsylvania, 1723 and 1729". *Journal of Political Economy*, June 1938, pp. 324-75.

—."Currency Issues to Overcome Depressions in Delaware, New Jersey, New York and Maryland, 1715-37". *Journal of Political Economy*, April 1939, pp. 182-217.

Mackay, R.A. "The Political Ideas of William Lyon Mackenzie". *Canadian Journal of Economics and Political Science*, Feb. 1937, pp. 1-22.

Massicette, E.Z. "Les Forages de Sainte-Géneviève de Bastican". *Recherches historiques*, 41 (1935), pp. 564-67, 708-11.

— . "Les Métiers rares d'autrefois". *Recherches historiques*, 36 (1930), pp.609-13.

Moorehouse, Frances. "Canadian Migration in the Forties". *Canadian Historical Review*, Dec. 1928, pp. 309-29.

Murray, Elsie M. "An Upper Canada 'Bush Business' in the Fifties". *Ontario Historical Society, Papers and Proceedings*, XXXVI (1944), pp. 41-47.

Ostry, Bernard. "Conservatives, Liberals, and Labour in the 1870s". *Canadian Historical Review*, June 1960, pp. 93-127.

Paquet, L.A. "L'Esclavage au Canada". *Transactions of the Royal Society of Canada*, section I, 1913, pp. 139-49.

Parker, W.H. "A New Look at Unrest in Lower Canada in the 1830s". *Canadian Historical Review*, Sept. 1959, pp. 209-17.

— . "A Revolution in the Agricultural Geography of Lower Canada, 1833-38". *Revue canadienne de géographie*, 1957, pp. 189-94.

Pentland, H.C. "The Development of a Capitalistic Labour Market in Canada". *Canadian Journal of Economics and Political Science*, Nov. 1959, pp. 450-61.

— . "The Lachine Strike of 1843". *Canadian Historical Review*, Sept. 1948, pp. 255-77.

—."The Role of Capital in Canadian Economic Development before 1875". *Canadian Journal of Economics and Political Science*, Nov. 1950, pp. 457-74.

Rae, J.B. "Federal Land Grants in Aid of Canals". *Journal of Economic History*, Nov. 1944, pp. 167-77.

Rayback, J.C. "The American Workingman and the Antislavery Crusade". *Journal of Economic History*, Nov. 1943, pp.152-63.

Reid, W.S. "The Habitant's Standard of Living on the Seigneurie des Milles Isles, 1820-1850". *Canadian Historical Review*, Sept. 1947, pp. 266-78.

Reznick, S. "The Rise and Early Development of Industrial Consciousness in the United States, 1760-1830". *Journal of Economic and Business History*, Supplement to IV, no. 4 (Aug. 1932), pp. 784-811.

Rowe, R.C. "The St. Maurice Forges". *Canadian Geographical Journal*, July 1934, pp. 15-22.

Sexsmith, W.N. "The Buxton Settlement". *Kent Historical Society, Papers and Addresses*, IV (1919), pp. 40-44.

Shortt, Adam. "Railway Construction and National Prosperity: An Historic

Parallel". *Transactions of the Royal Society of Canada*, section III, 1914, pp. 295-308.

Smith, A.E. "Indentured Servants: New Light on Some of America's 'First' Families". *Journal of Economic History*, May 1942, pp. 40-52.

Tessier, A. "Une campagne antitrustarde il y a un siècle". *Les Cahiers des Dix*, 1937, pp. 198-203.

Wurtele, F.C. "Historical Record of the St. Maurice Forges, the Oldest Active Blast-Furnace on the Continent of America". *Transactions of the Royal Society of Canada*, section II, 1886, pp. 77-90.

Young, John H. "Comparative Economic Development: Canada and the United States". *American Economic Review, Papers and Proceedings*, May 1955, pp. 80-93.

Index